T0205846

LONDON MATHEMATICAL SOCIETY LECTURE NOTE SERIES

Managing Editor: Professor M. Reid, Mathematics Institute,
University of Warwick, Coventry CV4 7AL, United Kingdom

The titles below are available from booksellers, or from Cambridge University Press
at www.cambridge.org/mathematics

LONDON MATHEMATICAL SOCIETY LECTURE
NOTE SERIES: 385

Entropy of Hidden Markov Processes and Connections to Dynamical Systems

Papers from the Banff International Research Station Workshop

Edited by

BRIAN MARCUS
University of British Columbia, Vancouver

KARL PETERSEN
University of North Carolina, Chapel Hill

TSACHY WEISSMAN
Stanford University, California

CAMBRIDGE
UNIVERSITY PRESS

CAMBRIDGE
UNIVERSITY PRESS

University Printing House, Cambridge CB2 8BS, United Kingdom

Cambridge University Press is part of the University of Cambridge.

It furthers the University's mission by disseminating knowledge in the pursuit of education, learning and research at the highest international levels of excellence.

www.cambridge.org
Information on this title: www.cambridge.org/9780521111133

© Cambridge University Press 2011

First published 2011

A catalogue record for this publication is available from the British Library

Library of Congress Cataloguing in Publication data
Entropy of hidden Markov processes and connections to dynamical systems : papers from the Banff International Research Station workshop / edited by Brian Marcus, Karl Petersen, Tsachy Weissman.
p.   cm. – (London Mathematical Society lecture note series ; 385)
Includes bibliographical references and index.
ISBN 978-0-521-11113-3 (pbk.)
1. Markov processes–Congresses.   2. Entropy (Information theory)–Congresses.
3. Dynamics–Congresses.   I. Marcus, Brian, 1949–   II. Petersen, Karl Endel, 1943–
III. Weissman, Tsachy.   IV. Banff International Research Station for Mathematics Innovation & Discovery.   V. Title.   VI. Series.
QA274.7.E58 2011
519.2'33–dc22   2011005950

ISBN 978-0-521-11113-3 Paperback

Contents

Introduction

This volume is a collection of papers on hidden Markov processes (HMPs) involving connections with symbolic dynamics and statistical mechanics. The subject was the focus of a five-day workshop held at the Banff International Research Station (BIRS) in October 2007, which brought together thirty mathematicians, computer scientists, and electrical engineers from institutions throughout the world. Most of the papers in this volume are based either on work presented at the workshop or on problems posed at the workshop.

From one point of view, an HMP is a stochastic process obtained as the noisy observation process of a finite-state Markov chain; a simple example is a binary Markov chain observed in binary symmetric noise, i.e., each symbol (0 or 1) in a binary state sequence generated by a two-state Markov chain may be flipped with some small probability, independently from time instant to time instant. In another (essentially equivalent) viewpoint, an HMP is a process obtained from a finite-state Markov chain by partitioning its state set into groups and completely "hiding" the distinction among states within each group; more precisely, there is a deterministic function on the states of the Markov chain, and the HMP is the process obtained by observing the sequences of function values rather than sequences of states (and hence such a process is sometimes called a "function of a Markov chain").

HMPs are encountered in an enormous variety of applications involving phenomena observed in the presence of noise. These range from speech and optical character recognition, through target tracking, to biomolecular sequence analysis. HMPs are also important objects of study in their own right in many areas of pure and applied mathematics, including information theory, probability theory, and dynamical systems. An excellent survey of HMPs can be found in [3].

A central problem in the subject is computation of the entropy rate (sometimes known simply as entropy) of an HMP. The entropy rate of a process can

1

be regarded as the asymptotic exponential growth rate of the number of different sequences that can be generated by the process (after one discards certain sequences of abnormally small probability). The entropy rate is a measure of randomness of a process. It is one of the most fundamental concepts in information theory, because it also measures the incompressibility of a process. It is also closely related to important quantities in statistical mechanics [4].

There is a very simple closed-form formula for the entropy rate of a finite-state Markov chain, but there is no such simple formula for entropy rate of HMPs except in very special cases. However, more than fifty years ago Blackwell [2] discovered an expression for the entropy rate of an HMP as the integral of a very simple integrand with respect to a typically very complicated (and usually singular) measure on a simplex; his measure is the stationary measure for a continuous-valued Markov chain on the simplex. In some sense, Blackwell's formula demonstrates that computation of entropy rate of HMPs is intrinsically complicated. While his formula has been used to help estimate the entropy rate, it is primarily of theoretical interest. Shortly after publication of Blackwell's paper, Birch [1] discovered excellent general upper and lower bounds on the entropy rate. However, until recently, there had been very little progress, with only a few papers on the subject.

Closely related is the problem of computing the capacity of an information channel, especially a channel with memory. Roughly speaking, the capacity of a channel is defined as the maximum of a quantity known as mutual information rate between the input and corresponding output processes, over all possible input processes. For some channels, this amounts to maximizing the entropy rate of the output process, which would be an HMP if the input process were Markov.

Recently, the entropy rate problem and related problems have received a good deal of attention from people working in many different areas, primarily information theory, dynamical systems, statistical mechanics, and probability theory. In particular, there is considerable interest in symbolic dynamics on problems regarding the properties of images and pre-images of Markov and hidden Markov processes via factor maps between symbolic dynamical systems. These issues have also been studied in the wider context of Gibbs measures.

The papers in this volume address these and related themes.

Computation of entropy rate is the explicit focus of the papers by Ordentlich and Weissman, Peres and Quas, and Pollicott. Ordentlich and Weissman develop an alternative to Blackwell's continuous-valued Markov chain and use it to obtain improved bounds in various noise regimes over the binary symmetric channel; they also compare various approximation schemes via an analysis of complexity versus precision. Peres and Quas obtain explicit asymptotics for the entropy rate of HMPs obtained as the noisy observation processes of Markov chains observed in a certain noise regime ("rare transitions") over the binary symmetric channel, thereby solving an open problem posed in the workshop. Pollicott develops a new numerical technique for approximating entropy rate, using ideas from dynamical systems and statistical mechanics to obtain approximations which are provably superexponentially convergent in several cases.

The paper by Han, Marcus, and Peres develops a complex version of the Hilbert metric on the real simplex in order to obtain estimates of the domain of analyticity of entropy rate as a function of the underlying Markov chain transition probabilities.

Pfister focuses on computation of capacity for certain finite-state channels. He develops a formula for the derivative of entropy rate as a function of the underlying Markov chain and applies this to obtain estimates on mutual information rates for the channels.

Boyle and Petersen give an in-depth survey of results on hidden Markov processes relating to symbolic dynamics and connections with probability, automata theory, and thermodynamics. The survey contains many results and open problems regarding hidden Markov processes and factor maps. Chazottes and Ugalde show that under certain factor maps the image of every Gibbs measure, defined by a certain type of potential function Φ, is also a Gibbs measure, with a potential function of a type determined by regularity properties of Φ and the factor map. Pollicott and Kempton obtain similar results for Gibbs measures defined by a related class of potential functions. Verbitskiy explores the relationship between HMPs and the thermodynamic formalism. He surveys work on the problem of computing the decay rate of the conditional probability of the present given the past, computation of entropy rate, identification of a potential for a given measure known to be Gibbs, and relations to Markov random field models.

We thank the Banff International Research Station and its constituent institutions for running and supporting the workshop; the anonymous referees for their careful reading and reviewing of the papers; and the staff of Cambridge University Press for their expert handling of the publication.

References

[1] J. J. Birch. *Approximations for the entropy for functions of Markov chains.* Ann. Math. Statist. **33** (1962) 930–938

[2] D. Blackwell. *The entropy of functions of finite-state Markov chains.* In Trans. First Prague Conf. Information Theory, Statistical Decision Functions, Random Processes, 1957, pp. 13–20

[3] Y. Ephraim and N. Merhav. *Hidden Markov processes*. IEEE Trans. Inf. Theory **48** (2002) 1518–1569

[4] D. Ruelle. *Thermodynamic Formalism: The Mathematical Structures of Classical Equilibrium Statistical Mechanics*. Advanced Book Program, Addison-Wesley, Reading, MA, 1978

1

Hidden Markov processes in the context of symbolic dynamics

MIKE BOYLE

Department of Mathematics, University of Maryland, College Park,
MD 20742-4015, USA
E-mail address: mmb@math.umd.edu

KARL PETERSEN

Department of Mathematics, CB 3250, Phillips Hall, University of North Carolina,
Chapel Hill, NC 27599, USA
E-mail address: petersen@math.unc.edu

Abstract. In an effort to aid communication among different fields and perhaps facilitate progress on problems common to all of them, this article discusses hidden Markov processes from several viewpoints, especially that of symbolic dynamics, where they are known as sofic measures or continuous shift-commuting images of Markov measures. It provides background, describes known tools and methods, surveys some of the literature, and proposes several open problems.

1 Introduction

Symbolic dynamics is the study of shift (and other) transformations on spaces of infinite sequences or arrays of symbols and maps between such systems. A symbolic dynamical system, with a shift-invariant measure, corresponds to a stationary stochastic process. In the setting of information theory, such a system amounts to a collection of messages. Markov measures and hidden Markov measures, also called sofic measures, on symbolic dynamical systems have the desirable property of being determined by a finite set of data. But not all of their properties, for example the entropy, can be determined by finite algorithms. This article surveys some of the known and unknown properties of hidden Markov measures that are of special interest from the viewpoint of symbolic dynamics. To keep the article self contained, necessary background and related concepts are reviewed briefly. More can be found in [47, 56, 55, 71].

Entropy of Hidden Markov Processes and Connections to Dynamical Systems: Papers from the
Banff International Research Station Workshop, ed. B. Marcus, K. Petersen, and T. Weissman.
Published by Cambridge University Press. © Cambridge University Press 2011.

We discuss methods and tools that have been useful in the study of symbolic systems, measures supported on them, and maps between them. Throughout, we state several problems that we believe to be open and meaningful for further progress We review a swath of the complicated literature starting around 1960 that deals with the problem of recognizing hidden Markov measures, as closely related ideas were repeatedly rediscovered in varying settings and with varying degrees of generality or practicality. Our focus is on the probability papers that relate most closely to symbolic dynamics. We have left out much of the literature concerning probabilistic and linear automata and control, but we have tried to include the main ideas relevant to our problems. Some of the explanations that we give and connections that we draw are new, as are some results near the end of the article. In Section 5.2 we give bounds on the possible order (memory) if a given sofic measure is in fact a Markov measure, with the consequence that in some situations there is an algorithm for determining whether a hidden Markov measure is Markov. In Section 6.3 we show that every factor map is hidden Markovian, in the sense that every hidden Markov measure on an irreducible sofic subshift lifts to a fully supported hidden Markov measure.

2 Subshift background

2.1 Subshifts

Let $\mathcal{A}$ be a set, usually finite or sometimes countable, which we consider to be an alphabet of symbols.

$$\mathcal{A}^* = \bigcup_{k=0}^{\infty} \mathcal{A}^k \tag{1}$$

denotes the set of all finite blocks or words with entries from $\mathcal{A}$, including the empty word, ϵ; $\mathcal{A}^+$ denotes the set of all nonempty words in $\mathcal{A}^*$; $\mathbb{Z}$ denotes the integers, and $\mathbb{Z}_+$ denotes the nonnegative integers. Let $\Omega(\mathcal{A}) = \mathcal{A}^{\mathbb{Z}}$ and $\Omega^+(\mathcal{A}) = \mathcal{A}^{\mathbb{Z}_+}$ denote the sets of all two- or one-sided sequences with entries from $\mathcal{A}$. If $\mathcal{A} = \{0, 1, \ldots, d-1\}$ for some integer $d > 1$, we denote $\Omega(\mathcal{A})$ by Ω_d and $\Omega^+(\mathcal{A})$ by Ω_d^+. Each of these spaces is a metric space with respect to the metric defined by setting for $x \neq y$

$$k(x,y) = \min\{|j| : x_j \neq y_j\} \quad \text{and} \quad d(x,y) = e^{-k(x,y)}. \tag{2}$$

For $i \leq j$ and $x \in \Omega(\mathcal{A})$, we denote by $x[i,j]$ the block or word $x_i x_{i+1} \cdots x_j$. If $\omega = \omega_0 \cdots \omega_{n-1}$ is a block of length n, we define

$$\mathcal{C}_0(w) = \{y \in \Omega(\mathcal{A}) : y[0, n-1] = \omega\} \tag{3}$$

and, for $i \in \mathbb{Z}$,

$$C_i(\omega) = \{y \in \Omega(\mathcal{A}) : y[i, i+n-1] = \omega\}. \tag{4}$$

The cylinder sets $C_i(\omega), \omega \in \mathcal{A}^*, i \in \mathbb{Z}$, are open and closed and form a base for the topology of $\Omega(\mathcal{A})$.

In this article, a *topological dynamical system* is a continuous self map of a compact metrizable space. The *shift transformation* $\sigma : \Omega_d \to \Omega_d$ is defined by $(\sigma x)_i = x_{i+1}$ for all i. On Ω_d the maps σ and σ^{-1} are one-to-one, onto, and continuous. The pair (Ω_d, σ) forms a topological dynamical system which is called the *full d-shift*.

If X is a closed σ-invariant subset of Ω_d, then the topological dynamical system (X, σ) is called a *subshift*. In this article, with "σ-invariant" we include the requirement that the restriction of the shift be surjective. Sometimes we denote a subshift (X, σ) by only X, the shift map being understood implicitly. When dealing with several subshifts, their possibly different alphabets will be denoted by $\mathcal{A}(X), \mathcal{A}(Y)$, etc.

The *language* $\mathcal{L}(X)$ of the subshift X is the set of all finite words or blocks that occur as consecutive strings

$$x[i, i+k-1] = x_i x_{i+1} \cdots x_{i+k-1} \tag{5}$$

in the infinite sequences x which comprise X. Denote by $|w|$ the length of a string w. Then

$$\mathcal{L}(X) = \{w \in \mathcal{A}^* : \text{there are } n \in \mathbb{Z}, y \in X \text{ such that } w = y_n \cdots y_{n+|w|-1}\}. \tag{6}$$

Languages of (two-sided) subshifts are characterized by being *extractive* (or *factorial*) (which means that every subword of any word in the language is also in the language) and *insertive* (or *extendable*) (which means that every word in the language extends on both sides to a longer word in the language).

For each subshift (X, σ) of (Ω_d, σ) there is a set $\mathcal{F}(X)$ of finite "forbidden" words such that

$$X = \{x \in \Omega_d : \text{for each } i \leq j, x_i x_{i+1} \cdots x_j \notin \mathcal{F}(X)\}. \tag{7}$$

A *shift of finite type (SFT)* is a subshift (X, σ) of some $(\Omega(\mathcal{A}), \sigma)$ for which it is possible to choose the set $\mathcal{F}(X)$ of forbidden words defining X to be finite. (The choice of the set $\mathcal{F}(X)$ is not uniquely determined.) The SFT is *n-step* if it is possible to choose the set of words in $\mathcal{F}(X)$ to have length at most $n+1$. We will sometimes use "SFT" as an adjective describing a dynamical system.

One-step shifts of finite type may be defined by $0,1$ transition matrices. Let M be a $d \times d$ matrix with rows and columns indexed by $\mathcal{A} = \{0, 1, \ldots, d-1\}$ and entries from $\{0, 1\}$. Define

$$\Omega_M = \{\omega \in \mathcal{A}^{\mathbb{Z}} : \text{for all } n \in \mathbb{Z}, M(\omega_n, \omega_{n+1}) = 1\}. \tag{8}$$

These were called *topological Markov chains* by Parry [51]. A topological Markov chain Ω_M may be viewed as a *vertex shift*: its alphabet may be identified with the vertex set of a finite directed graph such that there is an edge from vertex i to vertex j if and only if $M(i,j) = 1$. (A square matrix with nonnegative integer entries can similarly be viewed as defining an *edge shift*, but we will not need edge shifts in this article.) A topological Markov chain with transition matrix M as above is called *irreducible* if for all $i, j \in \mathcal{A}$ there is k such that $M^k(i,j) > 0$. Irreducibility corresponds to the associated graph being strongly connected.

2.2 Sliding block codes

Let (X, σ) and (Y, σ) be subshifts on alphabets $\mathcal{A}, \mathcal{A}'$, respectively. For $k \in \mathbb{N}$, a *k-block code* is a map $\pi : X \to Y$ for which there are $m, n \geq 0$ with $k = m + n + 1$ and a function $\pi : \mathcal{A}^k \to \mathcal{A}'$ such that

$$(\pi x)_i = \pi(x_{i-m} \cdots x_i \cdots x_{i+n}). \tag{9}$$

We will say that π is a *block code* if it is a k-block code for some k.

Theorem 2.1. (Curtis–Hedlund–Lyndon theorem) [33] *For subshifts (X, σ) and (Y, σ), a map $\psi : X \to Y$ is continuous and commutes with the shift ($\psi\sigma = \sigma\psi$) if and only if it is a block code.*

If (X, T) and (Y, S) are topological dynamical systems, then a *factor map* is a continuous onto map $\pi : X \to Y$ such that $\pi T = S\pi$. (Y, S) is called a *factor* of (X, T), and (X, T) is called an *extension* of (Y, S). A one-to-one factor map is called an *isomorphism* or *topological conjugacy*.

Given a subshift (X, σ), $r \in \mathbb{Z}$, and $k \in \mathbb{Z}_+$, there is a block code $\pi = \pi_{r,k}$ onto the subshift which is the *k-block presentation* of (X, σ), by the rule

$$(\pi x)_i = x[i+r, i+r+1, \ldots, i+r+k-1] \quad \text{for all } x \in X. \tag{10}$$

Here π is a topological conjugacy between (X, σ) and its image $(X^{[k]}, \sigma)$ which is a subshift of the full shift on the alphabet $\mathcal{A}^k$.

Two factor maps ϕ, ψ are *topologically equivalent* if there exist topological conjugacies α, β such that $\alpha\phi\beta = \psi$. In particular, if ϕ is a block code with

$(\phi x)_0$ determined by $x[-m,n]$ and $k = m+n+1$ and ψ is the composition $(\pi_{m,k})^{-1}$ followed by ϕ, then ψ is a one-block code (i.e., $(\psi x)_0 = \psi(x_0)$) which is topologically equivalent to ϕ.

A *sofic* shift is a subshift which is the image of a shift of finite type under a factor map. A sofic shift Y is *irreducible* if it is the image of an irreducible shift of finite type under a factor map. (Equivalently, Y contains a point with a dense forward orbit. Equivalently, Y contains a point with a dense orbit, and the periodic points of Y are dense.)

2.3 Measures

Given a subshift (X,σ), we denote by $\mathcal{M}(X)$ the set of σ-invariant Borel probability measures on X. These are the measures for which the coordinate projections $\pi_n(x) = x_n$ for $x \in X, n \in \mathbb{Z}$ form a two-sided finite-state stationary stochastic process.

Let P be a $d \times d$ stochastic matrix and p a stochastic row vector such that $pP = p$. (If P is irreducible, then p is unique.) Define a $d \times d$ matrix M with entries from $\{0,1\}$ by $M(i,j) = 1$ if and only if $P(i,j) > 0$. Then P determines a one-step stationary (σ-invariant) Markov measure μ on the shift of finite type Ω_M by

$$\mu(\mathcal{C}_i(\omega[i,j])) = \mu\{y \in \Omega_M : y[i,j] = \omega_i \omega_{i+1} \cdots \omega_j\}$$
$$= p(\omega_i) P(\omega_i, \omega_{i+1}) \cdots P(\omega_{j-1}, \omega_j) \tag{11}$$

(by the Kolmogorov extension theorem [6, p. 3ff.]).

For $k \geq 1$, we say that a measure $\mu \in \mathcal{M}(X)$ is *k-step Markov* (or more simply *k-Markov*) if for all $i \geq 0$ and all $j \geq k-1$ and all x in X,

$$\mu(\mathcal{C}_0(x[0,i]) | \mathcal{C}_0(x[-j,-1])) = \mu(\mathcal{C}_0(x[0,i]) | \mathcal{C}_0(x[-k,-1])). \tag{12}$$

A measure is one-step Markov if and only if it is determined by a pair (p,P) as above. A measure is k-step Markov if and only if its image under the topological conjugacy taking (X,σ) to its k-block presentation is one-step Markov. We say that a measure is *Markov* if it is k-step Markov for some k. The set of k-step Markov measures is denoted by $\mathcal{M}_k$ (adding an optional argument to specify the system or transformation if necessary.) *From here on, "Markov" means "shift-invariant Markov with full support"*, that is, every nonempty cylinder subset of X has positive measure. With this convention, a Markov measure with defining matrix P is ergodic if and only if P is irreducible.

A probabilist might ask for motivation for bringing in the machinery of topological and dynamical systems when we want to study a stationary stochastic process. First, looking at $\mathcal{M}(X)$ allows us to consider and compare many measures in a common setting. By relating them to continuous functions ("thermodynamics" – see Section 3.2 below) we may find some distinguished measures, for example maximal ones in terms of some variational problem. Second, by topological conjugacy we might be able to simplify a situation conceptually; for example, many problems involving block codes reduce to problems involving just one-block codes. And, third, with topological and dynamical ideas we might see (and know to look for) some structure or common features, such as invariants of topological conjugacy, behind the complications of a particular example.

2.4 Hidden Markov (sofic) measures

If (X,σ) and (Y,σ) are subshifts and $\pi : X \to Y$ is a sliding block code (factor map), then each measure $\mu \in \mathcal{M}(X)$ determines a measure $\pi\mu \in \mathcal{M}(Y)$ by

$$(\pi\mu)(E) = \mu(\pi^{-1}E) \quad \text{for each measurable } E \subset Y. \tag{13}$$

(Some authors write $\pi_*\mu$ or $\mu\pi^{-1}$ for $\pi\mu$.)

If X is SFT, μ is a Markov measure on X, and $\pi : X \to Y$ is a sliding block code, then $\pi\mu$ on Y is called a *hidden Markov measure* or *sofic measure*. (Various other names, such as "submarkov" and "function of a Markov chain", have also been used for such a measure or the associated stochastic process.) Thus, $\pi\mu$ is a convex combination of images of ergodic Markov measures. *From here on, unless otherwise indicated, the domain of a Markov measure is assumed to be an irreducible SFT, and the Markov measure is assumed to have full support (and thus by irreducibility be ergodic). Likewise, unless otherwise indicated, a sofic measure is assumed to have full support and to be the image of an ergodic Markov measure.* Then the sofic measure is ergodic and it is defined on an irreducible sofic subshift. Hidden Markov measures provide a natural way to model systems governed by chance in which dependence on the past of probabilities of future events is limited (or at least decays, so that approximation by Markov measures may be reasonable) and complete knowledge of the state of the system may not be possible.

Hidden Markov processes are often defined as probabilistic functions of Markov chains (see for example [23]), but by enlarging the state space each such process can be represented as a deterministic function of a Markov chain, such as we consider here (see [3]).

The definition of hidden Markov measure raises several questions.

Problem 2.2. Let μ be a one-step Markov measure on (X,σ) and $\pi: X \to Y$ a one-block code. The image measure may not be Markov – see Example 2.8. What are necessary and sufficient conditions for $\pi\mu$ to be one-step Markov?

This problem has been solved, in fact several times. Similarly, given μ and π, it is possible to determine whether $\pi\mu$ is k-step Markov. Further, given π and a Markov measure μ, it is possible to specify k such that either $\pi\mu$ is k-step Markov or else is not Markov of any order. These results are discussed in Section 5.

Problem 2.3. Given a shift-invariant measure ν on (Y,σ), how can one tell whether or not ν is a hidden Markov measure? If it is, how can one construct Markov measures of which it is the image?

The answers to Problem 2.3 provided by various authors are discussed in Section 4. The next problem reverses the viewpoint.

Problem 2.4. Given a sliding block code $\pi: X \to Y$ and a Markov measure ν on (Y,σ), does there exist a Markov measure μ on X such that $\pi\mu = \nu$?

In Section 3, we take up Problem 2.4 (which apart from special cases remains open) and some theoretical background that motivates it.

Recall that a factor map $\pi: X \to Y$ between irreducible sofic shifts has a *degree*, which is the cardinality of the preimage of any doubly transitive point of Y [47]. (If the cardinality is infinite, it can only be the power of the continuum, and we simply write degree$(\pi) = \infty$.) If π has degree $n < \infty$, then an ergodic measure ν with full support on Y can lift to at most n ergodic measures on X. We say that the *degree of a hidden Markov measure* ν, also called its *sofic degree*, is the minimal degree of a factor map which sends some Markov measure to ν.

Problem 2.5. Given a hidden Markov measure ν on (Y,σ), how can one determine the degree of ν? If the degree is $n < \infty$, how can one construct Markov measures of which ν is the image under a degree n map?

We conclude this section with examples.

Example 2.6. An example was given in [49] of a code $\pi: X \to Y$ that is *non-Markovian*: some Markov measure on Y does not lift to any Markov measure on X, and hence (see Section 3.1) no Markov measure on Y has a Markov preimage on X. The following diagram presents a simpler example, due to Shin [67, 69], of such a map. Here π is a one-block code: $\pi(1) = 1$ and $\pi(j) = 2$ if $j \neq 1$.

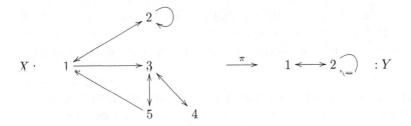

Example 2.7. Consider the shifts of finite type given by the graphs below, the one-block code π given by the rule $\pi(a)=a, \pi(b_1)=\pi(b_2)=b$, and the Markov measures μ, ν defined by the transition probabilities shown on the edges. We have $\pi\mu = \nu$, so the code is *Markovian* – some Markov measure maps to a Markov measure.

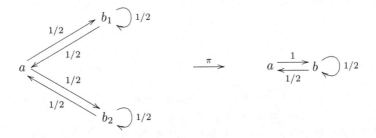

Example 2.8. This example uses the same shifts of finite type and one-block code as in Example 2.7, but we define a new one-step Markov measure on the upstairs shift of finite type X by assigning transition probabilities as shown.

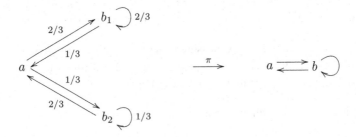

The entropy of the Markov measure μ (the definition is recalled in Section 3.2) is readily obtained from the familiar formula $-\sum p_i P_{ij} \log P_{ij}$, but there is no such simple rule for computing the entropy of ν. If ν were the finite-to-one image of some other Markov measure μ', maybe on some other shift of finite type, then we would have $h(\nu) = h(\mu')$ and the entropy of ν would be easily computed by applying the familiar formula to μ'. But for this example (due to Blackwell [8]) it can be shown [49] that ν is not the finite-to-one image of any Markov measure. Thus, Problem 2.5 is relevant to the much-studied problem of estimating the entropy of a hidden Markov measure (see [29, 30] and their references).

Example 2.9. In this example presented in [72], $X = Y = \Sigma_2 =$ full 2-shift, and the factor map is the two-block code

$$(\pi x)_0 = x_0 + x_1 \quad \mod 2. \tag{14}$$

Suppose that $0 < p < 1$ and μ_p is the Bernoulli (product) measure on X, with $\mu(\mathcal{C}_0(1)) = p$. Let ν_p denote the hidden Markov measure $\pi \mu_p = \pi \mu_{1-p}$. If $p \neq 1/2$, then ν_p is a hidden Markov measure strictly of degree two (it is not of degree one).

3 Factor maps and thermodynamical concepts

3.1 Markovian and non-Markovian maps

We have mentioned (Example 2.8) that the image under a factor map $\pi : X \to Y$ of a Markov measure need not be Markov, and (Example 2.6) that a Markov measure on Y need not have any Markov preimages. In this section we study maps that do not have the latter undesirable property. Recall our convention: a Markov measure is required to have full support.

Definition 3.1. [11] A factor map $\pi : \Omega_A \to \Omega_B$ between irreducible shifts of finite type (A and B are $0, 1$ transition matrices, see (8)) is *Markovian* if, for every Markov measure ν on Ω_B, there is a Markov measure on Ω_A such that $\pi \mu = \nu$.

Theorem 3.2. [11] *For a factor map $\pi : \Omega_A \to \Omega_B$ between irreducible shifts of finite type, if there exist any fully supported Markov measures μ and ν with $\pi \mu = \nu$, then π is Markovian.*

Note that if a factor map is Markovian, then so too is every factor map which is topologically equivalent to it, because a topological conjugacy takes Markov measures to Markov measures. We will see a large supply of Markovian maps (the "e-resolving factor maps") in Section 6.1.

These considerations lead to a reformulation of Problem 2.4:

Problem 3.3. Give a procedure to decide, given a factor map $\pi : \Omega_A \to \Omega_B$, whether π is Markovian

We sketch the proof of Theorem 3.2 for the one-step Markov case: if any one-step Markov measure on Ω_B lifts to a one-step Markov measure, then every one-step Markov measure on Ω_B lifts to a one-step Markov measure. For this, recall that if M is an irreducible matrix with spectral radius ρ, with positive right eigenvector r, then the *stochasticization* of M is the stochastic matrix

$$\text{stoch}(M) = \frac{1}{\rho} D^{-1} M D , \tag{15}$$

where D is the diagonal matrix with diagonal entries $D(i,i) = r(i)$.

Now suppose that $\pi : \Omega_A \to \Omega_B$ is a one-block factor map, with $\pi(i)$ denoted $\bar{i}$ for all i in the alphabet of Ω_A; that μ, ν are one-step Markov measures defined by stochastic matrices P, Q; and that $\pi\mu = \nu$. Suppose that $\nu' \in \mathcal{M}(\Omega_B)$ is defined by a stochastic matrix Q'. We will find a stochastic matrix P' defining μ' in $\mathcal{M}(\Omega_A)$ such that $\pi\mu' = \nu'$.

First define a matrix M of size matching P by $M(i,j) = 0$ if $P(i,j) = 0$ and otherwise

$$M(i,j) = Q'(\bar{i},\bar{j}) P(i,j)/Q(\bar{i},\bar{j}). \tag{16}$$

This matrix M will have spectral radius 1. Now set $P' = \text{stoch}(M)$. The proof that $\pi\mu' = \nu'$ is a straightforward computation that $\pi\mu' = \nu'$ on cylinders $C_0(y[0,n])$ for all $n \in \mathbb{N}$ and $y \in \Omega_B$. This construction is the germ of a more general thermodynamic result, the background for which we develop in the next section. We finish this section with an example.

Example 3.4. In this example one sees explicitly how being able to lift one Markov measure to a Markov measure allows one to lift other Markov measures to Markov measures.

Consider the one-block code π from $\Omega_3 = \{0, 1, 2\}^{\mathbb{Z}}$ to $\Omega_2 = \{0, 1\}^{\mathbb{Z}}$, via $0 \mapsto 0$ and $1, 2 \mapsto 1$. Let ν be the one-step Markov measure on Ω_2 given by the transition matrix

$$\begin{pmatrix} 1/2 & 1/2 \\ 1/2 & 1/2 \end{pmatrix} .$$

Given positive numbers $\alpha, \beta, \gamma < 1$, the stochastic matrix

$$\begin{pmatrix} 1/2 & \alpha(1/2) & (1-\alpha)(1/2) \\ 1/2 & \beta(1/2) & (1-\beta)(1/2) \\ 1/2 & \gamma(1/2) & (1-\gamma)(1/2) \end{pmatrix} \tag{17}$$

defines a one-step Markov measure on Ω_3 which π sends to ν.

Now, if ν' is any other one-step Markov measure on X_2, given by a stochastic matrix

$$\begin{pmatrix} p & q \\ r & s \end{pmatrix},$$

then ν' will lift to the one-step Markov measure defined by the stochastic matrix

$$\begin{pmatrix} p & \alpha q & (1-\alpha)q \\ r & \beta s & (1-\beta)s \\ r & \gamma s & (1-\gamma)s \end{pmatrix}. \tag{18}$$

3.2 Thermodynamics on subshifts

We recall the definitions of entropy and pressure and how the thermodynamical approach provides convenient machinery for dealing with Markov measures (and hence eventually, it is hoped, with hidden Markov measures).

Let (X, σ) be a subshift and $\mu \in \mathcal{M}(X)$ a shift-invariant Borel probability measure on X. The *topological entropy* of (X, σ) is

$$h(X) = \lim_{n \to \infty} \frac{1}{n} \log |\{x[0, n-1] : x \in X\}|. \tag{19}$$

The *measure-theoretic entropy* of the measure-preserving system (X, σ, μ) is

$$h(\mu) = h_\mu(X) = \lim_{n \to \infty} \frac{1}{n} \sum \{-\mu(\mathcal{C}_0(w)) \log \mu(\mathcal{C}_0(w)) : w \in \{x[0, n-1] : x \in X\}\}. \tag{20}$$

(For more background on these concepts, one could consult [56, 71].)

Pressure is a refinement of entropy which takes into account not only the map $\sigma : X \to X$ but also weights coming from a given "potential function" f on X. Given a continuous real-valued function $f \in C(X, \mathbb{R})$, we define the *pressure of f* (with respect to σ) to be

$$P(f, \sigma) = \lim_{n \to \infty} \frac{1}{n} \log \sum \{\exp[S_n(f, w)] : w \in \{x[0, n-1] : x \in X\}\}, \tag{21}$$

where

$$S_n(f,w) = \sum_{i=0}^{n-1} f(\sigma^i x) \quad \text{for some } x \in X \quad \text{such that } x[0, n-1] = w. \quad (22)$$

(In the limit, the choice of x does not matter.) Thus,

$$\text{if } f \equiv 0, \text{ then } P(f,\sigma) = h(X). \quad (23)$$

The pressure functional satisfies the important *variational principle*:

$$P(f,\sigma) = \sup \left\{ h(\mu) + \int f \, d\mu : \mu \in \mathcal{M}(X) \right\}. \quad (24)$$

An *equilibrium state* for f (with respect to σ) is a measure $\mu = \mu_f$ such that

$$P(f,\sigma) = h(\mu) + \int f \, d\mu. \quad (25)$$

Often (e.g., when the potential function f is Hölder continuous on an irreducible shift of finite type), there is a unique equilibrium state μ_f which is a *(Bowen) Gibbs measure* for f: i.e., $P(f,\sigma) = \log(\rho)$, and

$$\mu_f(\mathcal{C}_0(x[0, n-1])) \sim \rho^{-n} \exp S_n f(x). \quad (26)$$

Here "$\sim$" means that the ratio of the two sides is bounded above and away from zero, uniformly in x and n.

If $f \in C(\Omega_A, \mathbb{R})$ depends on only two coordinates, $f(x) = f(x_0 x_1)$ for all $x \in \Omega_A$, then f has a unique equilibrium state μ_f, and $\mu_f \in \mathcal{M}(\Omega_A)$. This measure μ_f is the one-step Markov measure defined by the stochastic matrix $P = \text{stoch}(Q)$, where

$$Q(i,j) = \begin{cases} 0 & \text{if } A(i,j) = 0, \\ \exp[f(ij)] & \text{otherwise.} \end{cases} \quad (27)$$

(For an exposition, see [52].)

The pressure of f is $\log \rho$, where ρ is the spectral radius of Q. Conversely, a Markov measure with stochastic transition matrix P is the equilibrium state of the potential function $f[ij] = \log P(i,j)$.

By passage to the k-block presentation, we can generalize to the case of k-step Markov measures: if $f(x) = f(x_0 x_1 \cdots x_k)$, then f has a unique equilibrium state μ, and μ is a k-step Markov measure.

Definition 3.5. We say that a function on a subshift X is *locally constant* if there is $m \in \mathbb{N}$ such that $f(x)$ depends only on $x[-m, m]$. $\mathrm{LC}(X, \mathbb{R})$ is the vector space of locally constant real-valued functions on X. $C_k(X, \mathbb{R})$ is the set of f in $\mathrm{LC}(X, \mathbb{R})$ such that $f(x)$ is determined by $x[0, k-1]$.

We can now express a viewpoint on Markov measures, due to Parry and Tuncel [70, 53], which follows from the previous results.

Theorem 3.6. [53] *Suppose that Ω_A is an irreducible shift of finite type; $k \geq 1$; and $f, g \in C_k(X, \mathbb{R})$. Then the following are equivalent.*

(1) $\mu_f = \mu_g$.
(2) *There are $h \in C(X, \mathbb{R})$ and $c \in \mathbb{R}$ such that $f = g + (h - h \circ \sigma) + c$.*
(3) *There are $h \in C_{k-1}(X, \mathbb{R})$ and $c \in \mathbb{R}$ such that $f = g + (h - h \circ \sigma) + c$.*

Proposition 3.7. [53] *Suppose that Ω_A is an irreducible shift of finite type. Let*

$$W = \{h - h \circ \sigma + c : h \in \mathrm{LC}(\Omega_A, \mathbb{R}), c \in \mathbb{R}\} . \tag{28}$$

Then the rule $[f] \mapsto \mu_f$ defines maps

$$C_k(\Omega_A, \mathbb{R})/W \;\to\; \mathcal{M}_k(\sigma_A),$$
$$\mathrm{LC}(\Omega_A, \mathbb{R})/W \;\to\; \bigcup_k \mathcal{M}_k(\sigma_A),$$

and these maps are bijections.

3.3 Compensation functions

Let $\pi : (X, T) \to (Y, S)$ be a factor map between topological dynamical systems. A *compensation function* for the factor map is a continuous function $\xi : X \to \mathbb{R}$ such that

$$P_Y(V) = P_X(V \circ \pi + \xi) \quad \text{for all } V \in \mathcal{C}(Y, \mathbb{R}). \tag{29}$$

Because $h(\pi \mu) \leq h(\mu)$ and $\int V \, d(\pi \mu) = \int V \circ \pi \, d\mu$, we always have

$$P_Y(V) = \sup \left\{ h(\nu) + \int_Y V \, d\nu : \nu \in \mathcal{M}(Y) \right\} \tag{30}$$

$$\leq \sup \left\{ h(\mu) + \int_X V \circ \pi \, d\mu : \mu \in \mathcal{M}(X) \right\} = P_X(V \circ \pi), \tag{31}$$

with possible strict inequality when π is infinite-to-one, in which case a strict inequality $h(\mu) > h(\pi \mu)$ can arise from (informally) the extra information/complexity arising from motion in fibers over points of Y. The pressure equality (29) tells us that the addition of a compensation function ξ to the functions $V \circ \pi$ takes into account (and exactly cancels out), for all potential functions V on Y at once, this measure of extra complexity. Compensation functions were introduced in [11] and studied systematically in [72]. A compensation function is a kind of oracle for how entropy can appear in a fiber. The Markovian case is the case in which the oracle has finite range, that is, there is a locally constant compensation function.

A compensation function for a factor map $\pi : X \to Y$ is *saturated* if it has the form $G \circ \pi$ for a continuous function G on Y.

Example 3.8. For the factor map in Examples 2.7 and 2.8, the formula

$$G(y) = \begin{cases} -\log 2 & \text{if } y = \cdot a \cdots, \\ 0 & \text{if } y = \cdot b \cdots \end{cases} \tag{32}$$

determines a saturated compensation function $G \circ \pi$ on Ω_A. The sum (or *cocycle*) $S_n G(y) = G(y) + G(\sigma y) + \cdots + G(\sigma^{n-1} y)$ measures the growth of the number of preimages of initial blocks of y:

$$|\pi^{-1}(y_0 \cdots y_{n-1})| = 2^{\#\{i : y_i = a, 0 \leq i < n\} \pm 1} \sim 2^{\#\{i : y_i = a, 0 \leq i < n\}} = e^{-S_n G(y)}. \tag{33}$$

Example 3.9. In the situation described at the end of Section 3.1, in which a one-step Markov measure maps to a one-step Markov measure under a one-block map, an associated compensation function is

$$\xi(x) = \log P(i,j) - \log Q(\bar{i}, \bar{j}) \quad \text{when } x_0 x_1 = ij. \tag{34}$$

Theorem 3.10. [11, 72] *Suppose that* $\pi : \Omega_A \to \Omega_B$ *is a factor map between irreducible shifts of finite type, with* $f \in LC(\Omega_A)$ *and* $g \in LC(\Omega_B)$, *and* $\pi \mu_f = \mu_g$. *Then there is a constant* c *such that* $f - g \circ \pi + c$ *is a compensation function. Conversely, if* ξ *is a locally constant compensation function, then* $\mu_{\xi + g \circ \pi}$ *is Markov and* $\pi \mu_{\xi + g \circ \pi} = \mu_g$.

In Theorem 3.10, the locally constant compensation function ξ relates potential functions on Ω_B to their lifts by composition on Ω_A in the same way that the corresponding equilibrium states are related:

$$\begin{aligned} LC(\Omega_B) &\hookrightarrow LC(\Omega_A) \quad \text{via } g \to (g \circ \pi) + \xi, \\ \mathcal{M}(\Omega_B) &\hookrightarrow \mathcal{M}(\Omega_A) \quad \text{via } \mu_g \to \mu_{(g \circ \pi) + \xi}. \end{aligned} \tag{35}$$

Theorem 3.10 holds if we replace the class of locally constant functions with the class of Hölder (exponentially decaying) functions, or with functions in the larger and more complicated "Walters class" (defined in [72, Section 4]). More generally, the arguments in [72, Theorem 4.1] go through to prove the following.

Theorem 3.11. *Suppose that $\pi : \Omega_A \to \Omega_B$ is a factor map between irreducible shifts of finite type. Let $\mathcal{V}_A, \mathcal{V}_B$ be real vector spaces of functions in $C(\Omega_A, \mathbb{R}), C(\Omega_B, \mathbb{R})$ respectively such that the following hold.*

(1) *$\mathcal{V}_A$ and $\mathcal{V}_B$ contain the locally constant functions.*
(2) *If f is in $\mathcal{V}_A$ or $\mathcal{V}_B$, then f has a unique equilibrium state μ_f, and μ_f is a Gibbs measure.*
(3) *If $f \in \mathcal{V}_B$, then $f \circ \pi \in \mathcal{V}_A$.*

Suppose that $f \in \mathcal{V}_A$ and $g \in \mathcal{V}_B$, and $\pi \mu_f = \mu_g$. Then there is a constant C such that $f - g \circ \pi + C$ is a compensation function. Conversely, if ξ in $\mathcal{V}_A$ is a compensation function, then for all $g \in \mathcal{V}_B$ we have $\pi \mu_{\xi + g \circ \pi} = \mu_g$.

Moreover, if $G \in \mathcal{V}_B$, then $G \circ \pi$ is a compensation function if and only if there is $c \geq 1$ such that

$$\frac{1}{c} \leq e^{S_n G(y)} |\pi^{-1}(y_0 \cdots y_{n-1})| \leq c \quad \text{for all } y, n. \tag{36}$$

Problem 3.12. Determine whether there exists a factor map $\pi : X \to Y$ between mixing SFTs and a potential function $F \in \mathcal{C}(X)$ which is *not* a compensation function but has a unique equilibrium state μ_F whose image $\pi \mu_F$ is the measure of maximal entropy on Y. If there were such an example, it would show that the assumptions on function classes in Theorem 3.11 cannot simply be dropped.

We finish this section with some more general statements about compensation functions for factor maps between shifts of finite type.

Proposition 3.13. [72] *Suppose that $\pi : \Omega_A \to \Omega_B$ is a factor map between irreducible shifts of finite type. Then the following hold.*

(1) *There exists a compensation function.*
(2) *If ξ is a compensation function, $g \in C(\Omega_B, \mathbb{R})$, and μ is an equilibrium state of $\xi + g \circ \pi$, then $\pi \mu$ is an equilibrium state of g.*
(3) *The map π takes the measure of maximal entropy (see Section 3.5) of Ω_A to that of Ω_B if and only if there is a constant compensation function.*

Yayama [74] has begun the study of compensation functions which are bounded Borel functions.

3.4 Relative pressure

When studying factor maps, relativized versions of entropy and pressure are relevant concepts. Given a factor map $\pi : \Omega_A \to \Omega_B$ between shifts of finite type, for each $n = 1, 2, \ldots$ and $y \in Y$, let $D_n(y)$ be a set consisting of exactly one point from each nonempty set $[x_0 \cdots x_{n-1}] \cap \pi^{-1}(y)$. Let $V \in C(\Omega_A, \mathbb{R})$ be a potential function on Ω_A. For each $y \in \Omega_B$, the *relative pressure of V at y with respect to π* is defined to be

$$P(\pi, V)(y) = \limsup_{n \to \infty} \frac{1}{n} \log \left[\sum_{x \in D_n(y)} \exp \left(\sum_{i=0}^{n-1} V(\sigma^i x) \right) \right]. \tag{37}$$

The *relative topological entropy function* is defined for all $y \in Y$ by

$$P(\pi, 0)(y) = \limsup_{n \to \infty} \frac{1}{n} \log |D_n(y)|, \tag{38}$$

the relative pressure of the potential function $V \equiv 0$.

For the relative pressure function, a *relative variational principle* was proved by Ledrappier and Walters ([46], see also [21]): for all ν in $\mathcal{M}(\Omega_B)$ and all V in $C(\Omega_A)$,

$$\int P(\pi, V) d\nu = \sup \left\{ h(\mu) + \int V d\mu : \pi \mu = \nu \right\} - h(\nu). \tag{39}$$

In particular, for a fixed $\nu \in \mathcal{M}(\Omega_B)$, the maximum measure-theoretic entropy of a measure on Ω_A that maps under π to ν is given by

$$h(\nu) + \sup \{ h_\mu(X | Y) : \pi \mu = \nu \} = h(\nu) + \sup \{ h(\mu) - h(\nu) : \pi \mu = \nu \} \tag{40}$$

$$= h(\nu) + \int_Y P(\pi, 0) d\nu .$$

In [58] a finite-range, combinatorial approach was developed for the relative pressure and entropy, in which instead of examining entire infinite sequences x in each fiber over a given point $y \in \Omega_B$, it is enough to deal just with preimages of finite blocks (which may or may not be extendable to full sequences in the fiber). For each $n = 1, 2, \ldots$ and $y \in Y$, let $E_n(y)$ be a set consisting of exactly one point from each nonempty cylinder $x[0, n-1] \subset \pi^{-1} y[0, n-1]$. Then, for each $V \in C(\Omega_A)$,

$$P(\pi, V)(y) = \limsup_{n \to \infty} \frac{1}{n} \log \left[\sum_{x \in E_n(y)} \exp \left(\sum_{i=0}^{n-1} V(\sigma^i x) \right) \right] \tag{41}$$

almost everywhere with respect to every ergodic invariant measure on Y. Thus, we obtain the value of $P(\pi, V)(y)$ almost everywhere with respect to every ergodic invariant measure on Y if we delete from the definition of $D_n(y)$ the requirement that $x \in \pi^{-1}(y)$.

In particular, the relative topological entropy is given by

$$P(\pi, 0)(y) = \limsup_{n \to \infty} \frac{1}{n} \log |\pi^{-1}y[0, n-1]| \tag{42}$$

almost everywhere with respect to every ergodic invariant measure on Y.

And, if μ is relatively maximal over ν, in the sense that it achieves the supremum in (40), then

$$h_\mu(X | Y) = \int_Y \lim_{n \to \infty} \frac{1}{n} \log |\pi^{-1}y[0, n-1]| \, d\nu(y). \tag{43}$$

3.5 Measures of maximal and relatively maximal entropy

Shannon [66] constructed the measures of maximal entropy on irreducible shifts of finite type. Parry [51] independently and from the dynamical viewpoint rediscovered the construction and proved uniqueness. For an irreducible shift of finite type, the unique measure of maximal entropy is a one-step Markov measure whose transition probability matrix is the stochasticization, as in (15), of the $0, 1$ matrix that defines the subshift. When studying factor maps $\pi : \Omega_A \to \Omega_B$ it is natural to look for *measures of maximal relative entropy*, which we also call *relatively maximal measures*: for fixed ν on Ω_B, look for the $\mu \in \pi^{-1}\nu$ which have maximal entropy in that fiber. Such measures always exist by compactness and upper semicontinuity, but, in contrast to the Shannon–Parry case (when Ω_B consists of a single point), they need not be unique. For example, in Example 2.9, the two-to-one map π respects entropy, and for $p \neq 1/2$ there are exactly two ergodic measures (the Bernoulli measures μ_p and μ_{1-p}) which π sends to ν_p. Moreover, there exists some $V_p \in C(Y)$ which has ν_p as a unique equilibrium state [38, 59], and $V_p \circ \pi$ has exactly two ergodic equilibrium states, μ_p and μ_{1-p}.

Here is a useful characterization of relatively maximal measures due to Shin.

Theorem 3.14. [68] *Suppose that $\pi : X \to Y$ is a factor map of shifts of finite type, $\nu \in \mathcal{M}(Y)$ is ergodic, and $\pi \mu = \nu$. Then μ is relatively maximal over ν if and only if there is $V \in C(Y, \mathbb{R})$ such that μ is an equilibrium state of $V \circ \pi$.*

If there is a *locally constant* saturated compensation function $G \circ \pi$, then every Markov measure on Y has a unique relatively maximal lift, which is

Markov, because then the relatively maximal measures over an equilibrium state of $V \in \mathcal{C}(Y, \mathbb{R})$ are the equilibrium states of $V \circ \pi + G \circ \pi$ [72]. Further, the measure of maximal entropy $\max_X$ is the unique equilibrium state of the potential function 0 on X; and the relatively maximal measures over $\max_Y$ are the equilibrium states of $G \circ \pi$.

It was proved in [57] that for each ergodic ν on Y, there are only a finite number of relatively maximal measures over ν. In fact, for a one-block factor map π between one-step shifts of finite type X, Y, the number of ergodic invariant measures of maximal entropy in the fiber $\pi^{-1}\{\nu\}$ is at most

$$N_\nu(\pi) = \min\{|\pi^{-1}\{b\}| : b \in \mathcal{A}(Y), \nu[b] > 0\}. \tag{44}$$

This follows from the theorem in [57] that for each ergodic ν on Y, any two distinct ergodic measures on X of maximal entropy in the fiber $\pi^{-1}\{\nu\}$ are *relatively orthogonal*. This concept is defined as follows.

For $\mu_1, \ldots, \mu_n \in \mathcal{M}(X)$ with $\pi\mu_i = \nu$ for all i, their *relatively independent joining* $\hat{\mu}$ over ν is defined by: if $A_1, \ldots, A_n$ are measurable subsets of X and $\mathcal{F}$ is the σ-algebra of Y, then

$$\hat{\mu}(A_1 \times \cdots \times A_n) = \int_Y \prod_{i=1}^n \mathbb{E}_{\mu_i}(\mathbf{1}_{A_i} | \pi^{-1}\mathcal{F}) \circ \pi^{-1} d\nu, \tag{45}$$

in which $\mathbb{E}$ denotes conditional expectation. Two ergodic measures μ_1, μ_2 with $\pi\mu_1 = \pi\mu_2 = \nu$ are *relatively orthogonal* (over ν), $\mu_1 \perp_\nu \mu_2$, if

$$(\mu_1 \otimes_\nu \mu_2)\{(u, v) \in X \times X : u_0 = v_0\} = 0. \tag{46}$$

This means that with respect to the relatively independent joining or coupling, there is zero probability of coincidence of symbols in the two coordinates.

That the second theorem (distinct ergodic relatively maximal measures in the same fiber are relatively orthogonal) implies the first (no more than $N_\nu(\pi)$ relatively maximal measures over ν) follows from the pigeonhole principle. If we have $n > N_\nu(\pi)$ ergodic measures $\mu_1, \ldots, \mu_n$ on X, each projecting to ν and each of maximal entropy in the fiber $\pi^{-1}\{\nu\}$, we form the relatively independent joining $\hat{\mu}$ on X^n of the measures μ_i as above. Write p_i for the projection $X^n \to X$ onto the ith coordinate. For $\hat{\mu}$-almost every $\hat{x}$ in X^n, $\pi(p_i(\hat{x}))$ is independent of i; abusing notation for simplicity, denote it by $\pi(\hat{x})$. Let b be a symbol in the alphabet of Y such that b has $N_\nu(\pi)$ preimages $a_1, \ldots, a_{N_\nu(\pi)}$ under the block map π. Since $n > N_\nu(\pi)$, for every $\hat{x} \in \pi^{-1}[b]$ there are $i \neq j$ with $(p_i\hat{x})_0 = (p_j\hat{x})_0$. At least one of the sets $S_{i,j} = \{\hat{x} \in X^n : (p_i\hat{x})_0 = (p_j\hat{x})_0\}$ must have positive $\hat{\mu}$-measure, and then also

$$(\mu_i \otimes_v \mu_j)\{(u,v) \in X \times X : \pi u = \pi v, u_0 = v_0\} > 0, \tag{47}$$

contradicting relative orthogonality. (Briefly, if you have more measures than preimage symbols, two of those measures have to coincide on one of the symbols: with respect to each measure, that symbol almost surely appears infinitely many times in the same place.)

The second theorem is proved by "interleaving" measures to increase entropy. If there are two relatively maximal measures over v which are not relatively orthogonal, then the measures can be "mixed" to give a measure with greater entropy. We concatenate words from the two processes, using the fact that the two measures are supported on sequences that agree infinitely often. Since X is a one-step SFT, we can switch over whenever a coincidence occurs. That the switching increases entropy is seen by using the strict concavity of the function $-t \log t$ and lots of calculations with conditional expectations.

Example 3.15. Here is an example (also discussed in [57, Example 1]) showing that to find relatively maximal measures over a Markov measure it is not enough to consider only sofic measures which map to it. We describe a factor map π which is both left and right e-resolving (see Section 6.1) and such that there is a unique relatively maximal measure μ above any fully supported Markov measure v, but the measure μ is not Markov, and it is not even sofic.

We use vertex shifts of finite type. The alphabet for the domain subshift is $\{a_1, a_2, b\}$ (in that order for indexing purposes), and the factor map (onto the 2-shift (Ω_2, σ)) is the one-block code π which erases subscripts. The transition diagram and matrix A for the domain shift of finite type (Ω_A, σ) are

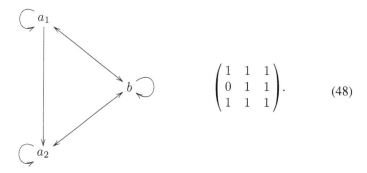

$$\begin{pmatrix} 1 & 1 & 1 \\ 0 & 1 & 1 \\ 1 & 1 & 1 \end{pmatrix}. \tag{48}$$

Above the word $ba^n b$ in Ω_2 there are $n+1$ words in Ω_A: above a^n we see k a_1's followed by $n-k$ a_2's, where $0 \le k \le n$. Let us for simplicity consider the maximal measure v on (Ω_2, T); so, $v(\mathcal{C}_0(ba^n b)) = 2^{-n-2}$. Now the maximal entropy lift μ of v will assign equal measure $2^{-(n+2)}/(n+1)$ to each of the preimage blocks of $ba^n b$. If μ is sofic, then (as in Section 4.1.4) there are vectors u, v and a square matrix Q such that $\mu(\mathcal{C}_0(b(a_1)^n b)) = uQ^n v$ for all $n > 0$. Then the function $n \mapsto uQ^n v$ is some finite sum of terms of the form $rn^j(\lambda^n)$, where $j \in \mathbb{Z}_+$ and r, λ are constants. The function $n \mapsto 2^{-(n+2)}/(n+1)$ is not a function of this type.

Problem 3.16. Is it true that for every factor map $\pi : \Omega_A \to \Omega_B$ every (fully supported) Markov measure v on Ω_B has a unique relatively maximal measure that maps to it, and this is also a measure with full support?

Remark 3.17. After the original version of this article was posted on the Math ArXiv and submitted for review, we received the preprint [75] of Yoo containing the following result: "Given a factor map from an irreducible SFT X to a sofic shift Y and an invariant measure v on Y with full support, every measure on X of maximal relative entropy over v is fully supported." This solves half of Problem 3.16.

3.6 Finite-to-one codes

Suppose that $\pi : \Omega_A \to \Omega_B$ is a finite-to-one factor map of irreducible shifts of finite type. There are some special features of this case which we collect here for mention. Without loss of generality, after recoding we assume that π is a one-block code. Given a Markov measure μ and a periodic point x, we define the *weight-per-symbol* of x (with respect to μ) to be

$$\mathrm{wps}_\mu(x) := \lim_{n \to \infty} \frac{1}{n} \log \mu\{y : x_i = y_i, 0 \le i < n\}. \tag{49}$$

Proposition 3.18. *Suppose that $\pi : \Omega_A \to \Omega_B$ is a finite-to-one factor map of irreducible shifts of finite type. Then the following hold.*

(1) *The measure of maximal entropy on Ω_B lifts to the measure of maximal entropy on Ω_A.*
(2) *Every Markov measure on Ω_B lifts to a unique Markov measure of equal order on Ω_A.*
(3) *If μ, v are Markov measures on Ω_A, Ω_B, respectively, then the following are equivalent:*
   (a) $\pi\mu = v$,
   (b) for every periodic point x in Ω_A, $\mathrm{wps}_\mu(x) = \mathrm{wps}_v(\pi x)$.

Proofs can be found in, for example, [41]. For infinite-to-one codes, we do not know an analogue of Proposition 3.18(3).

3.7 The semigroup measures of Kitchens and Tuncel

There is a hierarchy of sofic measures according to their sofic degree. Among the degree-one sofic measures, there is a distinguished and very well behaved subclass, properly containing the Markov measures. These are the *semigroup measures* introduced and studied by Kitchens and Tuncel in their memoir [42]. Roughly speaking, semigroup measures are to Markov measures as sofic subshifts are to SFTs.

A sofic subshift can be presented by a semigroup [73, 42]. Associated to this are nonnegative transition matrices R_0, L_0. A semigroup measure (for the semigroup presentation) is defined by a state probability vector and a pair of stochastic matrices R, L with 0/+ pattern matching R_0, L_0 and satisfying certain consistency conditions. These matrices can be multiplied to compute measures of cylinders. A measure is a semigroup measure if there exist a semigroup and apparatus as above which can present it. We will not review this constructive part of the theory, but we mention some alternate characterizations of these measures.

For a sofic measure μ on X and a periodic point x in X, the weight-per-symbol of x with respect to μ is still well defined by (49). Let us say a factor map π *respects μ-weights* if whenever x, y are periodic points with the same image we have $\mathrm{wps}_\mu(x) = \mathrm{wps}_\mu(y)$. Given a word $U = U[-n \cdots 0]$ and a measure μ, let μ_U denote the conditional measure on the future, i.e., if UW is an allowed word then $\mu_U(W) = \mu(UW)/\mu(U)$.

Theorem 3.19. [42] *Let v be a shift-invariant measure on an irreducible sofic subshift Y. Then the following are equivalent:*

(1) *v is a semigroup measure.*
(2) *v is the image of a Markov measure μ under a finite-to-one factor map which respects μ-weights.*
(3) *v is the image of a Markov measure μ under a degree-one resolving factor map which respects μ-weights.*
(4) *The collection of conditional measures μ_U, as U ranges over all Y-words, is finite.*

There is also a thermodynamic characterization of these measures as unique equilibrium states of bounded Borel functions which are locally constant on doubly transitive points, very analogous to the characterization of Markov measures as unique equilibrium states of continuous locally constant functions. The semigroup measures satisfy other nice properties as well.

Theorem 3.20. [42] *Suppose that* $\pi : X \to Y$ *is a finite-to-one factor map of irreducible sofic subshifts and* μ *and* ν *are semigroup measures on X and Y respectively. Then*

(1) ν *lifts by* π *to a unique semigroup measure on X, and this is the unique ergodic measure on X which maps to* ν*;*
(2) $\pi\mu$ *is a semigroup measure if and only if* π *respects* μ*-weights;*
(3) *there is an irreducible sofic subshift X' of X such that* π *maps X' finite-to-one onto X [49], and therefore* ν *lifts to a semigroup measure on X'.*

In contrast to the last statement, it can happen for an infinite-to-one factor map between irreducible SFTs that there is a Markov measure on the range which cannot lift to a Markov measure on any subshift of the domain [49].

We finish here with an example. There are others in [42].

Example 3.21. This is an example of a finite-to-one, one-to-one almost everywhere one-block code $\pi : \Omega_A \to \Omega_B$ between mixing vertex shifts of finite type, with a one-step Markov measure μ on Ω_A, such that the following hold:

(1) For all periodic points x, y in Ω_A, $\pi x = \pi y$ implies that $\mathrm{wps}_\mu(x) = \mathrm{wps}_\mu(y)$.
(2) $\pi\mu$ is not Markov on Ω_B.

Here the alphabet of Ω_A is $\{1, 2, 3\}$; the alphabet of Ω_B is $\{1, 2\}$;

$$A = \begin{pmatrix} 0 & 1 & 0 \\ 1 & 0 & 1 \\ 1 & 1 & 0 \end{pmatrix} \quad \text{and} \quad B = \begin{pmatrix} 0 & 1 \\ 1 & 1 \end{pmatrix};$$

and π is the one-block code sending 1 to 1 and sending 2 and 3 to 2. The map π collapses the points in the orbit of $(23)^*$ to a fixed point and collapses no other periodic points. (Given a block B, we let B^* denote a periodic point obtained by infinite concatenation of the block B.)

Let f be the function on Ω_A such that $f(x) = \log 2$ if $x_0 x_1 = 23$, $f(x) = \log(1/2)$ if $x_0 x_1 = 32$, and $f(x) = 0$ otherwise. Let μ be the one-step Markov measure which is the unique equilibrium state for f, defined by the stochasticization P of the matrix

$$M = \begin{pmatrix} 0 & 1 & 0 \\ 1 & 0 & 2 \\ 1 & 1/2 & 0 \end{pmatrix}.$$

Let λ denote the spectral radius of M. Suppose that $\nu = \pi\mu$ is Markov, of any order. Then $\text{wps}_\nu(2^*) = \text{wps}_\mu((23)^*) = -\log\lambda$. Also, there must be a constant c such that, for all large n,

$$\text{wps}_\nu((12^n)^*) = \frac{1}{n+1}(c + (n+1)\text{wps}_\nu(2^*)) = \frac{c}{n+1} - \log\lambda . \qquad (50)$$

So, for all large n,

$$\frac{c}{2n+1} - \log\lambda = \text{wps}_\nu((12^{2n})^*) = \text{wps}_\mu((1(23)^n)^*) = \frac{1}{2n+1}\log(2\lambda^{-(2n+1)}) \qquad (51)$$

and

$$\frac{c}{2n+2} - \log\lambda = \text{wps}_\nu((12^{2n+1})^*) = \text{wps}_\mu((1(23)^n2)^*) = \frac{1}{2n+2}\log(\lambda^{-(2n+2)}).$$

Thus, $c = \log 2$ and $c = 0$, a contradiction. Therefore, $\pi\mu$ is not Markov.

4 Identification of hidden Markov measures

Given a finite-state stationary process, how can we tell whether it is a hidden Markov process? If it is, how can we construct some Markov process of which it is a factor by means of a sliding block code? When is the image of a Markov measure under a factor map again a Markov measure? These questions are of practical importance, since scientific measurements often capture only partial information about systems under study, and in order to construct useful models the significant hidden variables must be identified and included. Beginning in the 1960s some criteria were developed for recognizing a hidden Markov process: loosely speaking, an abstract algebraic object constructed from knowing the measures of cylinder sets should be in some sense finitely generated. Theorem 4.20 below gives equivalent conditions, in terms of formal languages and series (the series is "rational"), linear algebra (the measure is "linearly representable"), and abstract algebra (some module is finitely generated), that a shift-invariant probability measure be the image under a one-block map of a shift-invariant one-step Markov measure. In the following, we briefly explain this result, including the terminology involved.

Kleene [44] characterized rational languages as the linearly representable ones, and this was generalized to formal series by Schützenberger [65]. In the study of stochastic processes, functions of Markov chains were analyzed by Gilbert [26], Furstenberg [25], Dharmadhikari [15, 16, 17, 18, 19, 20], Heller [34, 35], and others. For the connection between rational series and continuous

images of Markov chains, we follow Berstel and Reutenauer [4] and Hansel and
Perrin [31], with an addition to explain how to handle zero entries. Subsequent
sections describe the approaches of Furstenberg and Heller and related work.

Various problems around these ideas were (and continue to be) explored
and solved. In particular, it is natural to ask when is the image of a Markov
measure μ under a continuous factor map π a Gibbs measure (see (26)), or
when is the image of a Gibbs measure again a Gibbs measure? Chazottes and
Ugalde [13] showed that if μ is k-step Markov on a full shift Ω_d and π maps
Ω_d onto another full shift Ω_D, then the image $\pi\mu$ is a Gibbs measure which
is the unique equilibrium state of a Hölder continuous potential which can
be explicitly described in terms of a limit of matrix products and computed
at periodic points. They also gave sufficient conditions in the more general
case when the factor map is between SFTs. The case when μ is Gibbs but not
necessarily Markov is considered in [14]. For higher-dimensional versions, see
for example [45, 48, 28].

Among the extensive literature that we do not cite elsewhere, we can mention
in addition [32, 24, 5, 64].

4.1 Formal series and formal languages

4.1.1 Basic definitions

As in Section 2.1, continue to let $\mathcal{A}$ be a finite alphabet, $\mathcal{A}^*$ the set of all finite
words on $\mathcal{A}$, and $\mathcal{A}^+$ the set of all finite nonempty words on $\mathcal{A}$. Let ϵ denote
the empty word. A *language* on $\mathcal{A}$ is any subset $\mathcal{L} \subset \mathcal{A}^*$.

Recall that a *monoid* is a set S with a binary operation $S \times S \to S$ which is
associative and has a neutral element (identity). This means we can think of
$\mathcal{A}^*$ as the multiplicative free monoid generated by $\mathcal{A}$, where the operation is
concatenation and the neutral element is ϵ.

A *formal series* (nonnegative real-valued, based on $\mathcal{A}$) is a function $s : \mathcal{A}^* \to$
$\mathbb{R}_+$. For all $w \in \mathcal{A}^*$, $s(w) = (s, w) \in \mathbb{R}_+$, which can be thought of as the coefficient
of w in the series s. We will think of this s as $\sum_{w \in \mathcal{A}^*} s(w)w$, and this will be
justified later. If $v \in \mathcal{A}^*$ and s is the series such that $s(v) = 1$ and $s(w) = 0$
otherwise, then we sometimes use simply v to denote s.

Associated with any language $\mathcal{L}$ on $\mathcal{A}$ is its *characteristic series* $F_{\mathcal{L}} : \mathcal{A}^* \to$
$\mathbb{R}_+$ which assigns 1 to each word in $\mathcal{L}$ and 0 to each word in $\mathcal{A}^* \setminus \mathcal{L}$. Associated
to any Borel measure μ on $\mathcal{A}^{\mathbb{Z}_+}$ is its *corresponding series* F_μ defined by

$$F_\mu(w) = \mu(\mathcal{C}_0(w)) = \mu\{x \in \mathcal{A}^{\mathbb{Z}_+} : x[0, |w| - 1] = w\}. \tag{52}$$

It is sometimes useful to consider formal series with values in any *semiring* K, which is just a ring without subtraction. That is, K is a set with operations $+$ and $\cdot$ such that $(K,+)$ is a commutative monoid with identity element 0; $(K,\cdot)$ is a monoid with identity element 1; the product distributes over the sum; and for $k \in K$, $0k = k0 = 0$.

We denote the set of all K-valued formal series based on $\mathcal{A}$ by $K\langle\langle\mathcal{A}\rangle\rangle$ or $\mathcal{F}_K(\mathcal{A})$. We further abbreviate $\mathbb{R}_+\langle\langle\mathcal{A}\rangle\rangle = \mathcal{F}(\mathcal{A})$.

Then $\mathcal{F}(\mathcal{A})$ is a semiring in a natural way: for $f_1, f_2 \in \mathcal{F}(\mathcal{A})$, define

(1) $(f_1 + f_2)(w) = f_1(w) + f_2(w)$,
(2) $(f_1 f_2)(w) = \sum f_1(u) f_2(v)$, where the sum is over all $u, v \in \mathcal{A}^*$ such that $uv = w$, a finite sum.

The neutral element for multiplication in $\mathcal{F}(\mathcal{A})$ is

$$s_1(w) = \begin{cases} 1 & \text{if } w = \epsilon, \\ 0 & \text{otherwise.} \end{cases} \tag{53}$$

As discussed above, we will usually write simply ϵ for s_1. There is a natural injection $\mathbb{R}_+ \hookrightarrow \mathcal{F}(\mathcal{A})$ defined by $t \mapsto t\epsilon$ for all $t \in \mathbb{R}_+$.

Note the following points.

- $\mathbb{R}_+$ acts on $\mathcal{F}(\mathcal{A})$ on both sides:
  $(ts)(w) = ts(w)$, $(st)(w) = s(w)t$, for all $w \in \mathcal{A}^*$, for all $t \in \mathbb{R}_+$.
- There is a natural injection $\mathcal{A}^* \hookrightarrow \mathcal{F}(\mathcal{A})$ as a multiplicative submonoid: for $w \in \mathcal{A}^*$ and $v \in \mathcal{A}^*$, define

$$w(v) = \delta_{wv} = \begin{cases} 1 & \text{if } w = v, \\ 0 & \text{otherwise.} \end{cases}$$

This is a one-term series.

Definition 4.1. The *support* of a formal series $s \in \mathcal{F}(\mathcal{A})$ is

$$\text{supp}(s) = \{w \in \mathcal{A}^* : s(w) \neq 0\}.$$

Note that $\text{supp}(s)$ is a language. A language corresponds to a series with coefficients 0 and 1, namely its characteristic series.

Definition 4.2. A *polynomial* is an element of $\mathcal{F}(\mathcal{A})$ whose support is a finite subset of $\mathcal{A}^*$. Denote the K-valued polynomials based on $\mathcal{A}$ by $\wp_K(\mathcal{A}) = K\langle\mathcal{A}\rangle$. The *degree* of a polynomial p is $\deg(p) = \max\{|w| : p(w) \neq 0\}$ and is $-\infty$ if $p \equiv 0$.

Definition 4.3. A family $\{f_\lambda : \lambda \in \Lambda\} \subset \mathcal{F}(\mathcal{A})$ of series is called *locally finite* if for all $w \in \mathcal{A}^*$ there are only finitely many $\lambda \in \Lambda$ for which $f_\lambda(w) \neq 0$. A series $f \in \mathcal{F}(\mathcal{A})$ is called *proper* if $f(\epsilon) = 0$.

Proposition 4.4. *If $f \in \mathcal{F}(\mathcal{A})$ is proper, then $\{f^n : n = 0, 1, 2, \dots\}$ is locally finite.*

Proof. If $n > |w|$, then $f^n(w) = 0$, because

$$f^n(w) = \sum_{\substack{u_1 \cdots u_n = w \\ u_i \in \mathcal{A}^*, i=1,\dots,n}} f(u_1) \cdots f(u_n)$$

and at least one u_i is ϵ. $\square$

Definition 4.5. If $f \in \mathcal{F}(\mathcal{A})$ is proper, define

$$f^* = \sum_{n=0}^{\infty} f^n \text{ and } f^+ = \sum_{n=1}^{\infty} f^n \text{ (a pointwise finite sum)},$$

with $f^0 = 1 = 1 \cdot \epsilon = \epsilon$.

4.1.2 Rational series and languages

Definition 4.6. The *rational operations* in $\mathcal{F}(\mathcal{A})$ are sum (+), product (·), multiplication by real numbers (tw), and $* : f \to f^*$. The family of *rational series* consists of those $f \in \mathcal{F}(\mathcal{A})$ that can be obtained by starting with a finite set of polynomials in $\mathcal{F}(\mathcal{A})$ and applying a finite number of rational operations.

Definition 4.7. A language $\mathcal{L} \subset \mathcal{A}^*$ is *rational* if and only if its characteristic series

$$F(w) = \begin{cases} 1 & \text{if } w \in \mathcal{L}, \\ 0 & \text{if } w \notin \mathcal{L} \end{cases} \tag{54}$$

is rational.

Recall that regular languages correspond to *regular expressions*: the set of regular expressions includes $\mathcal{A}$, ϵ, $\emptyset$ and is closed under $+$, $\cdot$, *. A language recognizable by a finite-state automaton, or consisting of words obtained by reading off sequences of edge labels on a finite labeled directed graph, is regular.

Proposition 4.8. *A language $\mathcal{L}$ is rational if and only if it is regular. Thus, a nonempty insertive and extractive language is rational if and only if it is the language of a sofic subshift.*

4.1.3 Distance and topology in $\mathcal{F}(\mathcal{A})$

If $f_1, f_2 \in \mathcal{F}(\mathcal{A})$, define

$$D(f_1, f_2) = \inf\{n \geq 0 : \text{there is } w \in \mathcal{A}^n \text{ such that } f_1(w) \neq f_2(w)\} \quad (55)$$

and

$$d(f_1, f_2) = \frac{1}{2^{D(f_1, f_2)}}. \quad (56)$$

Note that $d(f_1, f_2)$ defines an *ultrametric* on $\mathcal{F}(\mathcal{A})$:

$$d(f, h) \leq \max\{d(f, g), d(g, h)\} \leq d(f, g) + d(g, h). \quad (57)$$

With respect to the metric d, $f_k \to f$ if and only if for each $w \in \mathcal{A}^*$, $f_k(w) \to f(w)$ in the discrete topology on $\mathbb{R}$, i.e., $f_k(w)$ eventually equals $f(w)$.

Proposition 4.9. $\mathcal{F}(\mathcal{A})$ *is* complete *with respect to the metric d and is a* topological semiring *with respect to the metric d (that is, $+$ and $\cdot$ are continuous as functions of two variables).*

Definition 4.10. A family $\{F_\lambda : \lambda \in \Lambda\}$ of formal series is called *summable* if there is a series $F \in \mathcal{F}(\mathcal{A})$ such that for every $\delta > 0$ there is a finite set $\Lambda_\delta \subset \Lambda$ such that for each finite set $I \subset \Lambda$ with $\Lambda_\delta \subset I$, $d(\sum_{i \in I} F_i, F) < \delta$. Then F is called the *sum* of the series and we write $F = \sum_{\lambda \in \Lambda} F_\lambda$.

Proposition 4.11. *If $\{F_\lambda : \lambda \in \Lambda\}$ is locally finite, then it is summable, and conversely.*

Thus, any $F \in \mathcal{F}(\mathcal{A})$ can be written as $F = \sum_{w \in \mathcal{A}^*} F(w)w$, where the formal series is a convergent infinite series of polynomials in the metric of $\mathcal{F}(\mathcal{A})$. Recall that

$$(F(w)w)(v) = \begin{cases} F(w) & \text{if } w = v, \\ 0 & \text{if } w \neq v, \end{cases}$$

where $F(w)w \in \mathcal{F}(\mathcal{A})$ and $w \in \mathcal{A}^*$, so that $\{F(w)w : w \in \mathcal{A}^*\}$ is a locally finite, and hence summable, subfamily of $\mathcal{F}(\mathcal{A})$.

We note here that the set $\wp(\mathcal{A})$ of all polynomials is dense in $\mathcal{F}(\mathcal{A})$.

4.1.4 Recognizable (linearly representable) series

Definition 4.12. $F \in \mathcal{F}(\mathcal{A})$ is *linearly representable* if there exists an $n \geq 1$ (the *dimension* of the representation) such that there are a $1 \times n$ nonnegative row vector $x \in \mathbb{R}_+^n$, an $n \times 1$ nonnegative column vector $y \in \mathbb{R}_+^n$, and a morphism of multiplicative monoids $\phi : \mathcal{A}^* \to \mathbb{R}_+^{n \times n}$ (the multiplicative monoids of nonnegative $n \times n$ matrices) such that for all $w \in \mathcal{A}^*$, $F(w) = x\phi(w)y$ (matrix multiplication). A *linearly representable measure* is one whose associated series is linearly representable. The triple (x, ϕ, y) is called a *linear representation* of the series (or measure).

Example 4.13. Consider a Bernoulli measure $\mathcal{B}(p_0, p_1, \ldots, p_{d-1})$ on $\Omega_+(\mathcal{A}) = \mathcal{A}^{\mathbb{Z}_+}$, where $\mathcal{A} = \{a_0, a_1, \ldots, a_{d-1}\}$ and $p = (p_0, p_1, \ldots, p_{d-1})$ is a probability vector. Let $f = \sum_{i=0}^{d-1} p_i a_i \in \mathcal{F}(\mathcal{A})$. Then

$$f(w) = \begin{cases} p_i & \text{if } w = a_i, \\ 0 & \text{if } w \neq a_i. \end{cases}$$

Define $F_p = f^* = \sum_{n \geq 0} f^n$. Note that f is proper since we have $f(\epsilon) = 0$. Consider the particular word $w = a_2 a_0$. Then $f^0(w) = f(w) = 0$ and, for $n \geq 3$, we have $f^n(w) = 0$ because any factorization $w = u_1 u_2 u_3$ includes ϵ and $f(\epsilon) = 0$. Thus, $F_p(w) = f^*(w) = f^2(w) = \sum_{uv=w} f(u)f(v) = f(a_2)f(a_0) = p_2 p_0$. Continuing in this way, we see that for $w_i \in \mathcal{A}$, $F_p(w_1 w_2 \cdots w_n) = p_{w_1} p_{w_2} \cdots p_{w_n}$.

Example 4.14. Consider a Markov measure μ on $\Omega_+(\mathcal{A})$ defined by a $d \times d$ stochastic matrix P and a d-dimensional probability row vector $p = (p_0, p_1, \ldots, p_{d-1})$. Define $F_{p,P} \in \mathcal{F}(\mathcal{A})$ by $F_{p,P}(w_1 \cdots w_n) = \mu(C_0(w_1 \cdots w_n))$ for all $w_1, \ldots, w_n \in \mathcal{A}$. Put $y = (1, \ldots, 1)^{\mathrm{T}} \in \mathbb{R}_+^d$, $x = p \in \mathbb{R}_+^d$, and let ϕ be generated by $\phi(a_j), j = 0, 1, \ldots, d-1$, where

$$\phi(a_j) = \begin{pmatrix} 0 & \cdots & P_{0j} & 0 & \cdots & 0 \\ 0 & \cdots & P_{1j} & 0 & \cdots & 0 \\ \vdots & \ddots & \vdots & \vdots & \ddots & \vdots \\ 0 & \cdots & P_{d-1,j} & 0 & \cdots & 0 \end{pmatrix} \quad \text{for each} \quad a_j \in \mathcal{A}. \tag{58}$$

Then the triple (x, ϕ, y) represents the given Markov measure μ. In this Markov case each matrix $\phi(a_j)$ has at most one nonzero column and thus has rank at most 1.

Example 4.15. Now we show how to obtain a linear representation of a sofic measure that is the image under a one-block map π of a one-step Markov measure. Let μ be a one-step Markov measure determined by a $d \times d$ stochastic matrix P and fixed vector p as in Example 4.14. Let $\pi : X \to Y$ be a one-block map from the SFT X to a subshift Y. For each a in the alphabet $B = \mathcal{A}(Y)$, let

P_a be the $d \times d$ matrix such that

$$P_a(i',j') = \begin{cases} P(i',j') & \text{if } \pi(j') = a, \\ 0 & \text{otherwise.} \end{cases} \tag{59}$$

Thus, P_a just zeroes out all the columns of P except the ones corresponding to indices in the π-preimage of the symbol a in the alphabet of Y. Again let $y = (1, \ldots, 1)^\mathrm{T}$. For each $a \in B$, define $\phi(a) = P_a$. That the ν-measure of each cylinder in Y is the sum of the μ-measures of its preimages under π says that the triple (x, ϕ, y) represents $\nu = \pi \mu$.

In working with linearly representable measures, it is useful to know that the nature of the vectors and matrices involved in the representation can be assumed to have a particular restricted form. Below, we say that a matrix P is a direct sum of irreducible stochastic matrices if the index set for the rows and columns of P is the disjoint union of sets for which the associated principal submatrices of P are irreducible stochastic matrices. (Equivalently, there are irreducible stochastic matrices $P_1, \ldots, P_k$ and a permutation matrix Q such that QPQ^{-1} is the block diagonal matrix whose successive diagonal blocks are $P_1, \ldots, P_k$.)

Proposition 4.16. *A formal series $F \in \mathcal{F}(\mathcal{A})$ corresponds to a linearly representable shift-invariant probability measure μ on $\Omega_+(\mathcal{A})$ if and only if F has a linear representation (x, ϕ, y) with $P = \sum_{a \in \mathcal{A}} \phi(a)$ a stochastic matrix, y a column vector of all ones, and $xP = x$. Moreover, in this case the vector x can be chosen to be positive, with the matrix P a direct sum of irreducible stochastic matrices.*

Proof. It is straightforward to check that any (x, ϕ, y) of the specified form linearly represents a shift-invariant measure. Conversely, given a linear representation (x, ϕ, y) as in Definition 4.12 of a shift-invariant probability measure μ, define $P = \sum_{a \in \mathcal{A}} \phi(a)$ and note that, by induction, for all $w \in \mathcal{A}^*$, $\mu(C_0(w)) = x\phi(w)P^k y = xP^k \phi(w)y$ for all natural numbers k.

Next, one shows that it is possible to reduce to a linear representation (x, ϕ, y) of μ such that each entry of x and y is nonzero and, with P defined as $P = \sum_{a \in \mathcal{A}} \phi(a)$, $xP = x$ and $Py = y$. This requires some care. If indices corresponding to zero entries in x or y, or to zero rows or columns in P, are jettisoned nonchalantly, the resulting new ϕ may no longer be a morphism.

Definition 4.17. A triple (x', ϕ', y') is obtained from (x, ϕ, y) by *deleting a set I of indices* if the following holds: the indices for (x, ϕ, y) are the disjoint union of the set I and the indices for (x', ϕ', y'); and, for every symbol a and all indices

i,j not in I, we have $x'_i = x_i, y'_i = y_i$, and $\phi'(a)(i,j) = \phi(a)(i,j)$. Then we let ϕ' also denote the morphism determined by the map on generators $a \mapsto \phi'(a)$.

First, suppose that j is an index such that column j of P (and therefore column j of every $\phi(a) := M_a$) is zero. By shift invariance of the measure, (xP, ϕ, y) is still a representation, so we may assume without loss of generality that $x_j = 0$. Let (x', ϕ', y) be obtained from (x, ϕ, y) by deleting the index j. We claim that (x', ϕ', y) still gives a linear representation of μ. This is because for any word $a_1 \cdots a_m$, the difference $[x\phi(a_1) \cdots \phi(a_m)y] - [x'\phi'(a_1) \cdots \phi'(a_m)y']$ is a sum of terms of the form

$$x(i_0)M_{a_1}(i_0, i_1)M_{a_2}(i_1, i_2) \cdots M_{a_m}(i_{m-1}, i_m)y(i_m), \qquad (60)$$

in which at least one index i_t equals j. If $i_0 = j$, then $x(i_0) = 0$; if $i_t = j$ with $t > 0$, then $M_{a_t}(i_{t-1}, i_t) = 0$. In either case, the product is zero.

By the analogous argument involving y rather than x, we may pass to a new representation by deleting the index of any zero row of P. We repeat until we arrive at a representation in which no row or column of P is zero.

An *irreducible component* of P is a maximal principal submatrix C which is an irreducible matrix. C is an *initial component* if for every index j of a column through C, $P(i,j) > 0$ implies that (i,j) indexes an entry of C. C is a *terminal component* if for every index i of a row through C, $P(i,j) > 0$ implies that (i,j) indexes an entry of C.

Now suppose that $\mathcal{I}$ is the index set of an initial irreducible component of P, and $x(i) = 0$ for every i in $\mathcal{I}$. Define (x', ϕ', y) by deleting the index set $\mathcal{I}$. By an argument very similar to the argument for deleting the index of a zero column, the triple (x', ϕ', y') still gives a linear representation of μ. Similarly, if $\mathcal{J}$ is the index set of a terminal irreducible component of P, and $y(j) = 0$ for every j in $\mathcal{J}$, we may pass to a new representation by deleting the index set $\mathcal{J}$.

Iterating these moves, we arrive at a representation for which P has no zero row and no zero column; every initial component has an index i with $x(i) > 0$; and every terminal component has an index j with $y(j) > 0$. We now claim that for this representation the set of matrices $\{P^n\}$ is bounded. Suppose not. Then there is a pair of indices i,j for which the entries $P^n(i,j)$ are unbounded. There are some initial component index i_0, and some $k \geq 0$, such that $x(i_0) > 0$ and $P^k(i_0, i) > 0$. Likewise, there are a terminal component index j_0 and an $m \geq 0$ such that $y(j_0) > 0$ and $P^m(j, j_0) > 0$. Appealing to shift invariance of μ, for all $n > 0$ we have

$$1 = xP^{n+k+m}y \geq x(i_0)P^k(i_0, i)P^n(i,j)P^m(j, j_0)y(j_0), \qquad (61)$$

which is a contradiction to the unboundedness of the entries $P^n(i,j)$. This proves that the family of matrices P_n is bounded.

Next let Q_n be the Cesàro sum, $(1/n)(P + \cdots + P^n)$. Let Q be a limit of a subsequence of the bounded sequence $\{Q_n\}$. Then $PQ = Q = QP$; xQ and Qy are fixed vectors of P; and (xQ, ϕ, Qy) is a linear representation of μ. It could be that xQ vanishes on all indices through some initial component, or that Qy vanishes on all indices through some terminal component. In this case we simply cycle through our reductions until finally arriving at a linear representation (x, ϕ, y) of μ such that: $xP = x$; $Py = y$; the set of matrices $\{P^n\}$ is bounded; P has no zero row or column; x does not vanish on all indices of any initial component; and y does not vanish on all indices of any terminal component.

If C is an initial component of P, then the restriction of x to the indices of C is a nontrivial fixed vector of C. Thus, this restriction is positive, and the spectral radius of C is at least 1. The spectral radius of C must then be exactly 1, because the set $\{P^n\}$ is bounded.

We are almost done. Suppose that P is not the direct sum of irreducible matrices. Then there must be an initial component with index set $\mathcal{I}$ and a terminal component with index set $\mathcal{J} \neq \mathcal{I}$, with some $i \in \mathcal{I}$, $j \in \mathcal{J}$, and m minimal in $\mathbb{N}$ such that $P^m(i,j) > 0$. Because $\mathcal{I}$ indexes an initial component, for any $k \in \mathbb{N}$ we have that $(xP^k)_i$ is the sum of the terms $x_{i_0} P(i_0, i_1) \cdots P(i_{k-1}, i)$ such that $i_t \in \mathcal{I}$, $0 \leq t \leq k-1$. Because $\mathcal{J}$ indexes a terminal component, for any $k \in \mathbb{N}$ we have that $(P^k y)_j$ is the sum of the terms $P(j, i_1) \cdots P(i_{k-1}, i_k) y(i_k)$ such that $i_t \in \mathcal{J}$, $1 \leq t \leq k$. Because $\mathcal{I} \neq \mathcal{J}$, by the minimality of m we have for all $n \in \mathbb{N}$ that

$$xy = xP^{m+n}y \geq \sum_{k=0}^{n} (xP^k)_i P^m(i,j)(P^{n-k}y)_j = (n+1)x_i P^m(i,j)y_j, \qquad (62)$$

a contradiction.

Consequently, P is now a direct sum of irreducible matrices, each of which has spectral radius 1. The eigenvectors x, y are now positive. Let D be the diagonal matrix with $D(i,i) = y(i)$. Define $(x', \phi', y) = (xD, D^{-1}\phi D, D^{-1}y)$. Then (x', ϕ', y) is the linear representation satisfying all the conditions of the proposition. $\qquad \square$

Example 4.18. The conclusion of the proposition does not follow without the hypothesis of stationarity: there need *not* be any linear representation with positive vectors x, y, and there need not be any linear representation in which the nonnegative vectors x, y are fixed vectors of P. For example, consider the nonstationary Markov measure μ on two states a, b with initial vector $p = (1, 0)$

and transition matrix

$$T = \begin{pmatrix} 1/2 & 1/2 \\ 0 & 1 \end{pmatrix} = \begin{pmatrix} 1/2 & 1/2 \\ 0 & 0 \end{pmatrix} + \begin{pmatrix} 0 & 0 \\ 0 & 1 \end{pmatrix} = N_a + N_b. \qquad (63)$$

If q is the column vector $(1,1)^T$, then p, N_a, N_b, q generate a linear representation of μ, e.g., $1 = \mu(C_0(a)) = pN_aq$ and $(1/2)^k = \mu(C_0(a^k b^m)) = p(N_a)^k (N_b)^m q$ when $k, m > 0$.

Now suppose that there is a linear representation of μ generated by positive vectors x, y and nonnegative matrices M_a, M_b. Then

$$\begin{aligned} 1 &= \mu(C_0(a)) = xM_a y, \\ 0 &= \mu(C_0(b)) = xM_b y. \end{aligned} \qquad (64)$$

From the second of these equations, $M_b = 0$, since $x > 0$ and $y > 0$. But this contradicts $0 < \mu(C_0(ab)) = xM_a M_b y$.

Next, suppose that there is a linear representation for which x, y could be chosen as eigenvectors of $P = M_a + M_b$ (necessarily with eigenvalue 1, since $xP^n y = 1$ for all $n > 0$). Then

$$\frac{1}{2} = \mu(C_0(ab)) = xM_a M_b y \le xPM_b y = xM_b y = \mu(C_0(b)) = 0, \qquad (65)$$

which is a contradiction.

4.2 Equivalent characterizations of hidden Markov measures

4.2.1 Sofic measures – formal series approach

The semiring $\mathcal{F}(\mathcal{A})$ of formal series on the alphabet $\mathcal{A}$ is an $\mathbb{R}_+$-*module* in a natural way. On this module we have a *(linear) action of* $\mathcal{A}^*$ defined as follows: for $F \in \mathcal{F}(\mathcal{A})$ and $w \in \mathcal{A}^*$, define $(w, F) \to w^{-1}F$ by

$$(w^{-1}F)(v) = F(wv) \text{ for all } v \in \mathcal{A}^*.$$

Thus,

$$w^{-1}F = \sum_{v \in \mathcal{A}^*} F(wv)v.$$

If $F = u \in \mathcal{A}^*$, then

$$(w^{-1}F)(v) = u(wv) = \begin{cases} 1 & \text{if } wv = u, \\ 0 & \text{if } wv \ne u. \end{cases}$$

Thus, $w^{-1}u \neq 0$ if and only if $u = wv$ for some $v \in \mathcal{A}^*$, and then $w^{-1}u = v$ (in the sense that they are the same function on $\mathcal{A}^*$): $w^{-1}v$ erases w from v if v has w as a prefix, otherwise $w^{-1}v$ gives 0. Note also that this is a *monoid action*:

$$(vw)^{-1}F = w^{-1}(v^{-1}F). \tag{66}$$

Definition 4.19. A submodule M of $\mathcal{F}(\mathcal{A})$ is called *stable* if $w^{-1}F \in M$ for all $F \in M$, i.e., $w^{-1}M \subset M$ for all $w \in \mathcal{A}^*$.

Theorem 4.20. *Let $\mathcal{A}$ be a finite alphabet. For a formal series $F \in \mathcal{F}_{\mathbb{R}_+}(\mathcal{A})$ that corresponds to a shift-invariant probability measure v in $\Omega_+(\mathcal{A})$, the following are equivalent:*

(1) *F is linearly representable.*
(2) *F is a member of a stable finitely generated submodule of $\mathcal{F}_{\mathbb{R}_+}(\mathcal{A})$.*
(3) *F is rational.*
(4) *The measure v is the image under a one-block map of a shift-invariant one-step Markov probability measure μ.*

In the latter case, the measure v is ergodic if and only if it is possible to choose μ ergodic.

In the next few sections we sketch the proof of this theorem.

4.2.2 Proof that a series is linearly representable if and only if it is a member of a stable finitely generated submodule of $\mathcal{F}(\mathcal{A})$

Suppose that F is linearly representable by (x, ϕ, y). For each $i = 1, 2, \dots, n$ (where n is the dimension of the representation) and each $w \in \mathcal{A}^*$, define

$$F_i(w) = [\phi(w)y]_i.$$

Let $M = \langle F_1, \dots, F_n \rangle$ be the span of the F_i with coefficients in $\mathbb{R}_+$, which is a submodule of $\mathcal{F}(\mathcal{A})$. Since

$$F(w) = x\phi(w)y = \sum_{i=1}^{n} x_i [\phi(w)y]_i = \sum_{i=1}^{n} x_i F_i(w),$$

we have $F = \sum_{i=1}^{n} x_i F_i$, which means that $F \in M$.

We next show that M is stable. Let $w \in \mathcal{A}^*$. Then, for $u \in \mathcal{A}^*$,

$$(w^{-1}F_i)(u) = F_i(wu) = [\phi(wu)y]_i = [\phi(w)\phi(u)y]_i$$

$$= \sum_{j=1}^{n} \phi(w)_{ij}[\phi(u)y]_j = \sum_{j=1}^{n} \phi(w)_{ij}F_j(u).$$

Since $\phi(w)_{ij} \in \mathbb{R}_+$, we have $\sum_{j=1}^{n} \phi(w)_{ij} F_j(u) \in M$, so

$$w^{-1} F_i = \sum_{j=1}^{n} x_i \phi(w)_{ij} F_j \in \langle F_1, \ldots, F_n \rangle = M.$$

Conversely, let M be a stable finitely generated left submodule, and assume that $F \in \langle F_1, \ldots, F_n \rangle = M$. Then there are $x_1, \ldots, x_n \in \mathbb{R}_+$ such that $F = \sum_{i=1}^{n} x_i F_i$. Since M is stable, for each $a \in \mathcal{A}$ and each $i = 1, 2, \ldots, n$, we have $a^{-1} F_i \in \langle F_1, \ldots, F_n \rangle$. So, there exist $c_{ij} \in \mathbb{R}_+, j = 1, 2, \ldots, n$, such that $a^{-1} F_i = \sum_{j=1}^{n} c_{ij} F_j$. Define $\phi(a)_{ij} = c_{ij}$ for $i, j = 1, 2, \ldots, n$. Note by linearity that for any nonnegative row vector $(t_1, \ldots, t_n)$ we have

$$a^{-1} \left(\sum_{i=1}^{n} t_i F_i \right) = \sum_{j=1}^{n} \left((t_1, \ldots, t_n) \phi(a) \right)_j F_j. \tag{67}$$

Extend ϕ to a monoid morphism $\phi : \mathcal{A}^* \to \mathbb{R}_+^{n \times n}$ by defining $\phi(a_1 \cdots a_n) = \phi(a_1) \cdots \phi(a_n)$. Because the action of $\mathcal{A}^*$ on $\mathcal{F}(\mathcal{A})$ satisfies the monoidal condition (66), we have from (67) that for any $w = a_1 a_2 \cdots a_n \in \mathcal{A}^*$,

$$w^{-1} \left(\sum_{i=1}^{n} t_i F_i \right) = (a_1 \cdots a_n)^{-1} \left(\sum_{i=1}^{n} t_i F_i \right) = \left(a_n^{-1} \cdots \left(a_1^{-1} \sum_{i=1}^{n} t_i F_i \right) \cdots \right)$$

$$= \sum_{j} \left((t_1, \ldots, t_n) \phi(a_1) \cdots \phi(a_n) \right)_j F_j = \sum_{j} \left((t_1, \ldots, t_n) \phi(w) \right)_j F_j.$$

Define the column vector y by $y_j = F_j(1)$ for $j = 1, 2, \ldots, n$ and let x be the row vector $(x_1, \ldots, x_n)$. Then

$$F(w) = w^{-1} F(1) = \left(\sum_{j} \left(x\phi(w) \right)_j F_j \right)(1) = \sum_{j} \left(x\phi(w) \right)_j F_j(1) = x\phi(w) y,$$

$$\tag{68}$$

showing that (x, ϕ, y) is a linear representation for F.

4.2.3 Proof that a formal series is linearly representable if and only if it is rational

This equivalence is from [44, 65]. Recall that a series is rational if and only if it is in the closure of the polynomials under the rational operations $+$ (union), $\cdot$ (concatenation), $*$, and multiplication by elements of $\mathbb{R}_+$.

First we prove by a series of steps that every rational series F is linearly representable.

Proposition 4.21. *Every polynomial is linearly representable.*

Proof. If $w \in \mathcal{A}$ and $|w|$ is greater than the degree of the polynomial F, then $w^{-1} \equiv 0$. Let $S = \{w^{-1}F : w \in \mathcal{A}^*\}$. Then S is finite and stable; hence, S spans a finitely generated stable submodule M to which F belongs. (Take $\epsilon^{-1}F = F$.) By Section 4.2.2, F is linearly representable. □

The next observation follows immediately from the definition of stability. The proof of the following lemma is included for practice.

Proposition 4.22. *If F_1 and F_2 are in stable finitely generated submodules of $\mathcal{F}(\mathcal{A})$ and $t \in \mathbb{R}_+$, then $(F_1 + F_2)$ and (tF_1) are in stable finitely generated submodules of $\mathcal{F}(\mathcal{A})$.*

Lemma 4.23. *For $F, G \in \mathcal{F}(\mathcal{A})$ and $a \in \mathcal{A}$, $a^{-1}(FG) = (a^{-1}F)G + F(\epsilon)a^{-1}G$.*

Proof. For any $w \in \mathcal{A}^*$,

$$\begin{aligned}
(a^{-1}(FG))(w) = (FG)(aw) &= \sum_{uv=aw} F(u)G(v) \\
&= F(\epsilon)G(aw) + \sum_{u'v'=w} F(au')G(v') \\
&= F(\epsilon)G(aw) + \sum_{u'v'=w} (a^{-1}F)(u')G(v') \\
&= F(\epsilon)(a^{-1}G)(w) + ((a^{-1}F)(G))(w).
\end{aligned} \tag{69}$$

□

Proposition 4.24. *Suppose that for $i = 1, 2$, $F_i \in M_i$, where each M_i is a stable, finitely generated submodule. Let $M = M_1F_2 + M_2$. Then M is finitely generated and stable and contains F_1F_2.*

Proof. The facts that $F_1F_2 \in M$ and M is finitely generated are immediate. The proof that M is stable is a consequence of the lemma. For, if $f_1F_2 + f_2$ is an element of M and $a \in \mathcal{A}$, then

$$a^{-1}(f_1F_2 + f_2) = (a^{-1}f_1)F_2 + f_1(\epsilon)(a^{-1}F_2) + a^{-1}f_2. \tag{70}$$

Note that $a^{-1}f_1 \in M_1$ and $a^{-1}f_2$, $a^{-1}F_2 \in M_2$. Thus, $f_1(\epsilon)(a^{-1}F_2) + f_2 \in M_2$, so we conclude that M is stable. □

Lemma 4.25. *If F is proper (that is, $F_1(\epsilon) = 0$) and $a \in \mathcal{A}$, then $a^{-1}(F^*) = (a^{-1}F)F^*$.*

Proof. Recall that $F_1^* = \sum_{n \geq 0} F_1^n$. Thus, $a^{-1}(F^*) = a^{-1}(1 + FF^*) = a^{-1}(\epsilon + FF^*) = a^{-1}\epsilon + (a^{-1}F)F^* + F(\epsilon)a^{-1}(F^*)$.

Because $(a^{-1}\epsilon)(w) = \epsilon(aw) = 0$ for all $w \in \mathcal{A}^*$ and $F(\epsilon) = 0$, we get $a^{-1}F^* = (a^{-1}F)F^*$. $\qquad\square$

Proposition 4.26. *Suppose that M_1 is finitely generated and stable, and that $F_1 \in M_1$ is proper. Then F_1^* is in a finitely generated stable submodule.*

Proof. Define $M = \mathbb{R}_+ + M_1 F_1^*$. We have

$$F_1^* = 1 + \sum_{n \geq 1} F_1^n = (1 + F_1 F_1^*) \in M.$$

Also, M is finitely generated (by 1 and the $f_i F_1^*$ if the f_i generate M_1).

To show that M is stable, suppose that $t \in \mathbb{R}_+$ and $a \in \mathcal{A}$. Then for any $u \in \mathcal{A}^*$ we have $(a^{-1}t)(u) = t(au) = 0$, so $a^{-1}t = 0 \in \mathbb{R}_+$. And, for any $f_1 \in M_1$ and $a \in \mathcal{A}$, $a^{-1}(f_1 F_1^*) = (a^{-1}f_1)F_1^* + f_1(\epsilon)a^{-1}(F_1^*)$. Since M_1 is stable, $a^{-1}f_1 \in M_1$ and the first term is in $M_1 F_1^*$. By the lemma, the second term is $f_1(\epsilon)(a^{-1}F_1)F_1^*$, which is again in $M_1 F_1^*$. $\qquad\square$

These observations show that if F is rational, then F lies in a finitely generated stable submodule, so by Section 4.2.2 F is linearly representable.

Now we turn our attention to proving the statement in the title of this section in the other direction. So, assume that $F \in \mathcal{F}(\mathcal{A})$ is linearly representable. Then $F(w) = x\phi(w)y$ for all $w \in \mathcal{A}$ for some (x, ϕ, y). Consider the semiring of formal series $\mathcal{F}_K(\mathcal{A}) = K^{\mathcal{A}^*}$, where K is the semiring $\mathbb{R}_+^{n \times n}$ of $n \times n$ nonnegative real matrices and n is the dimension of the representation. Let $D = \sum_{a \in \mathcal{A}} \phi(a)a \in \mathcal{F}_K(\mathcal{A})$. The series D is proper, so we can form

$$D^* = \sum_{h \geq 0} D^h = \sum_{h \geq 0} \left(\sum_{a \in \mathcal{A}} \phi(a)a \right)^h = \sum_{h \geq 0} \left(\sum_{w \in \mathcal{A}^h} \phi(w)w \right) = \sum_{w \in \mathcal{A}} \phi(w)w. \quad (71)$$

This series D^* is a rational element of $\mathcal{F}_K(\mathcal{A})$, since we started with a polynomial and formed its *. By Lemma 4.27 below, each entry $(D^*)_{ij}$ is rational in $\mathcal{F}_{\mathbb{R}_+}(\mathcal{A})$.

With D and D^* now defined, we have

$$F(w) = x\phi(w)y = \sum_{i,j} x_i \phi(w)_{ij} y_j = \sum_{i,j} x_i D^*(w)_{ij} y_j, \quad (72)$$

and each $D^*(w)_{ij}$ is a rational series applied to w. Thus, $F(w)$ is a finite linear combination of rational series D_{ij}^* applied to w and hence is rational.

Lemma 4.27. *Suppose that D is an $n \times n$ matrix whose entries are proper rational formal series (e.g., polynomials). Then the entries of D^* are also rational.*

Proof. We use induction on n. The case $n = 1$ is trivial. Suppose that the lemma holds for $n - 1$, and D is $n \times n$ with block form $D = \begin{pmatrix} a & u \\ v & Y \end{pmatrix}$, with a a rational series. The entries of D can be thought of as labels on a directed graph; a path in the graph has a label which is the product of the labels of its edges; and then $D^*(i,j)$ represents the sum of the labels of all paths from i to j (interpret the term "1" in $D(i,i)$ as the label of a path of length zero). With this view, one can see that $D^* = \begin{pmatrix} b & w \\ x & Z \end{pmatrix}$, where

(1) $b = (a + uY^*v)^*$,
(2) $Z = (Y + va^*u)^*$,
(3) $w = buY^*$,
(4) $x = Y^*vb$.

Now Y^* and Z have rational entries by the induction hypothesis, and consequently all entries of D^* are rational. □

4.2.4 Linearly representable series correspond to sofic measures

The (topological) support of a measure is the smallest closed set of full measure. Recall our convention (Section 2.4) that Markov and sofic measures are ergodic with full support.

Theorem 4.28. [25, 31, 34] *A shift-invariant probability measure ν on $\Omega_+(\mathcal{A})$ corresponds to a linearly representable (equivalently, rational) formal series $F = F_\nu \in \mathcal{F}_{\mathbb{R}_+}(\mathcal{A})$ if and only if it is a convex combination of measures which (restricted to their supports) are sofic measures. Moreover, if (x, ϕ, y) is a representation of F_ν such that x and y are positive and the matrix $\sum_{i \in B} \phi(i)$ is irreducible, then ν is a sofic measure.*

Proof. Suppose that ν is the image under a one-block map (determined by a map $\pi : \mathcal{A} \to B$ between the alphabets) of a one-step Markov measure μ. Then ν is linearly representable by the construction in Example 4.15.

Alternatively, if F_μ is represented by (x, ϕ, y), then for each $w \in \mathcal{A}^*$ we have

$$F_\mu(w) = \sum_{i,j} x_i \phi(w)_{ij} y_j = \sum_{i,j} x_i \left(\left[\sum_{a \in \mathcal{A}} \phi(a)a \right]^* (w) \right)_{ij} y_j. \tag{73}$$

For $u \in \mathcal{B}^*$, define

$$F_\nu(u) = \sum_{i,j} x_i \left(\left[\sum_{h \in \mathcal{B}} \left(\sum_{a \in \mathcal{A}, \phi(a)=h} \phi(a) \right) b \right]^*(u) \right)_{ij} y_j \qquad (74)$$

to see that F_ν is a linear combination of rational series and to see its linear representation.

Conversely, suppose that ν corresponds to a rational (and hence linearly representable) formal series $F = F_\nu \in \mathcal{F}_{\mathbb{R}_+}(\mathcal{B})$ with dimension n. Let (x, ϕ, y) represent F. To indicate an ordering of the alphabet $\mathcal{B}$, we use notation $\mathcal{B} = \{1, 2, \ldots, k\}$ and $\phi(i) = P_i$. First assume that the $n \times n$ matrix P is irreducible and the vectors x and y are positive. We will construct a Markov measure μ and a one-block map π such that $\nu = \pi \mu$.

Applying the standard stochasticization trick as in the last paragraph of the proof of Proposition 4.16, we may assume that the irreducible matrix P is stochastic, every entry of y is 1, and x is stochastic. Define matrices with block forms

$$M = \begin{pmatrix} P_1 & P_2 & \cdots & P_k \\ P_1 & P_2 & \cdots & P_k \\ \vdots & \vdots & \ddots & \vdots \\ P_1 & P_2 & \cdots & P_k \end{pmatrix}, \qquad R = \begin{pmatrix} I \\ I \\ \vdots \\ I \end{pmatrix},$$

$$C = \begin{pmatrix} P_1 & P_2 & \cdots & P_k \end{pmatrix}, \qquad M_i = \begin{pmatrix} 0 & \cdots & P_i & \cdots & 0 \\ 0 & \cdots & P_i & \cdots & 0 \\ \vdots & \ddots & \vdots & \ddots & \vdots \\ 0 & \cdots & P_i & \cdots & 0 \end{pmatrix},$$

where each P_i is $n \times n$; R is $nk \times k$; I is the $n \times n$ identity matrix; C and the M_i are $nk \times nk$; and M_i is zero except in the ith block column, where it is RP_i. The matrix M is stochastic, but it can have zero columns. (We thank Uijin Jung for pointing this out.) Let M' be the largest principal submatrix of M with no zero column or row.

We have a strong shift equivalence $M = RC$, $P = CR$, and it then follows from the irreducibility of P that M' is irreducible. Therefore, there is a unique left stochastic fixed vector X for M. Let Y be the $nk \times 1$ column vector with every entry 1. We have $MR = RP$, and consequently $XR = x$. Also, $M_i R = RP_i$ for each i. So, for any word $i_1 \cdots i_j$, we have

$$x P_{i_1} \cdots P_{i_j} y = X R P_{i_1} \cdots P_{i_j} y$$

$$= X M_{i_1} \cdots M_{i_j} R Y = X M_{i_1} \cdots M_{i_j} Y.$$

This shows that (X, Φ, Y) is also a representation of F_ν, where $\Phi(i) = M_i$. Let $X', \Phi'(i) = M_i', Y'$ be the restrictions of $X, \Phi(i), Y$ to the vectors/matrices on the indices of M'. Then (X', Φ', Y') is also a representation of F_ν. Let A' be the $0, 1$ matrix of size matching M' whose zero entries are the same as for M'. Then (X', M', Y') defines an ergodic Markov measure μ on $\Omega_{A'}$ and there is a one-block code π such that $\pi\mu = \nu$. Explicitly, π is the restriction of the code which sends $\{1, 2, \ldots, n\}$ to 1; $\{n+1, n+2, \ldots, 2n\}$ to 2; and so on. Thus, ν is a sofic measure.

Now, for the representation (x, ϕ, y) of F_ν, we drop the assumption that the matrix P is irreducible. However, by Proposition 4.16, without loss of generality we may assume that P is the direct sum of irreducible stochastic matrices $P^{(j)}$; x is a positive stochastic left fixed vector of P; and y is the column vector with every entry 1. Restricted to the indices through $P^{(j)}$, x is a fixed vector of $P^{(j)}$ and therefore is a multiple $c_j x^{(j)}$ of the stochastic left fixed vector $x^{(j)}$ of $P^{(j)}$. Note that $\sum_j c_j = 1$. If $y^{(j)}$ denotes the column vector with every entry 1 such that $P^{(j)} y^{(j)} = y^{(j)}$, then

$$(x, \phi, y) = \sum_j c_j (x_j, P^{(j)}, y^{(j)}) .$$

If follows from the irreducible case that μ is a convex combination of sofic measures. $\qquad\square$

4.3 Sofic measures – Furstenberg's approach

Below we are extracting from [25, Sections 18 and 19] only what we need to describe Furstenberg's approach to the identification of sofic measures and compare it to the others. This leaves out a lot. We follow Furstenberg's notation, apart from change of symbols, except that we refer to shift-invariant measures as well as finite-state stationary processses.

Furstenberg begins with the following definition.

Definition 4.29. [25, Definition 18.1] A *stochastic semigroup of order* r is a semigroup S having an identity e (i.e., a monoid), together with a set of r elements $\mathcal{A} = \{a_1, a_2, \ldots, a_r\}$ generating S, and a real-valued function F defined on S satisfying

(1) $F(e) = 1$,
(2) $F(s) \geq 0$ for each $s \in S$ and $F(a_i) > 0$, $i = 1, 2, \ldots, r$,
(3) $\sum_{i=1}^r F(a_i s) = \sum_{i=1}^r F(sa_i) = F(s)$ for each $s \in S$.

Given a subshift X on an alphabet $\{a_1, a_2, \ldots, a_r\}$ with shift-invariant Borel probability μ and $\mu(a_i) > 0$ for every i, let S be the free semigroup of all formal products of the a_i, with the empty product taken as the identity e. Define F on S by $F(e) = 1$ and $F(a_{i_1} a_{i_2} \cdots a_{i_k}) = \mu(C_0(a_{i_1} a_{i_2} \cdots a_{i_k}))$. Clearly, the triple $(\{a_1, a_2, \ldots, a_r\}, S, F)$ is a stochastic semigroup, which we denote $S(X)$.

Conversely, any stochastic semigroup $(\{a_1, a_2, \ldots, a_r\}, S, F)$ determines a unique shift-invariant Borel probability μ for which $F(a_{i_1} a_{i_2} \cdots a_{i_k}) = \mu(C_0((a_{i_1} a_{i_2} \cdots a_{i_k})))$ for all $a_{i_1} a_{i_2} \cdots a_{i_k}$. We denote by $X(S)$ this finite-state stationary process (equivalently the full shift on r symbols with invariant measure μ). Two stochastic semigroups are called *equivalent* if they define the same finite-state stationary process modulo a bijection of their alphabets. A *cone* in a linear space is a subset closed under addition and multiplication by positive real numbers [25, Section 15.1].

Definition 4.30. [25, Definition 19.1] Let D be a linear space, D^* its dual, and let C be a cone in D such that for all x, y in D, if $x + \lambda y \in C$ for all real λ, then $y = 0$. Let $\theta \in C$ and $\theta^* \in D^*$, and suppose that θ^* is nonnegative on C. A *linear stochastic semigroup* S on (C, θ, θ^*) is a stochastic semigroup $(\{a_1, \ldots, a_r\}, S, F)$ whose elements are linear transformations *from C to C* satisfying:

(1) $\sum a_i \theta = \theta$;
(2) $\sum a_i^* \theta^* = \theta^*$ (where L^* denotes the transformation of D^* adjoint to a transformation L of D);
(3) $F(s) = (\theta^*, s\theta)$ for $s \in S$, where $(\cdot, \cdot)$ denotes the dual pairing of D^* and D;
(4) $(\theta^*, a_i \theta) > 0$, $i = 1, 2, \ldots, r$.

$(S, D, C, \theta, \theta^*)$ was called *finite dimensional* by Furstenberg if there is $m \in \mathbb{N}$ such that $D = \mathbb{R}^m$, C is the cone of vectors in $\mathbb{R}^m$ with all entries nonnegative, and each element of S is an $m \times m$ matrix with nonnegative entries.

A semigroup S of transformations satisfying conditions (1) to (4) does define a stochastic semigroup if $(\theta^*, \theta) = 1$.

Theorem 4.31. [25, Theorem 19.1] *Every stochastic semigroup S is equivalent to some linear stochastic semigroup.*

Proof. Let $A_0(S)$ be the real semigroup algebra of S, i.e., the real vector space with basis S and multiplication determined by the semigroup multiplication in S and the distributive property

$$\left(\sum \alpha_s s \right) \left(\sum \beta_t t \right) = \sum \alpha_s \beta_t st. \tag{75}$$

(Each sum above has finitely many terms.)

If S is the free monoid generated by r symbols, then $A_0(S)$ is isomorphic to the set $\wp_{\mathbb{R}}(\mathcal{A})$ of real-valued polynomials, i.e., finitely supported formal series $\mathcal{A}^* \to \mathbb{R}$ (see Definition 4.2).

Extend F from S to a linear functional on $A_0(S)$, i.e., $F(\sum \alpha_s s) = \sum \alpha_s F(s)$. Define $I = \{u \in A_0(S) : F(u) = 0\}$, an ideal in $A_0(S)$, and the algebra $A = A(S) = A_0(S)/I$. Define the element $\tau = a_1 + a_2 + \cdots + a_r$ in $A(S)$ (here a_i abbreviates $a_i + I$) and set $D = A/A(e - \tau)$.

The elements of A and in particular those of S operate on D by left multiplication. Let a_i' denote the operator induced by left multiplication by $a_i \in S$. Take V to be the image in D of the set of elements of A that can be represented as positive linear combinations of elements in S. Denote by $\bar{u}$ the image in D of an element u in A. Set $\theta = \bar{e}$ and let θ^* be the functional induced on D by F on A (F vanishes on $A(e - \tau)$).

Then the four conditions in the definition of linear stochastic semigroup are satisfied. This linear stochastic semigroup given by

$$(\{a_1', \ldots, a_r'\}, D, V, \theta, \theta^*) \tag{76}$$

is equivalent to the given S because $F(s') = (\theta^*, s'\theta) = F(s)$. (We will see later that this construction is closely related to Heller's "stochastic module" construction.) $\qquad\square$

Given a shift-invariant sofic measure on the set of two-sided sequences on the alphabet $\{1, \ldots, r\}$ which assigns positive measure to each symbol, it is possible to associate an explicit finite-dimensional linear stochastic semigroup to μ in the same way that we attached a linear representation in Example 4.15. Here μ is the image under some one-block code π of a Markov measure defined from some $m \times m$ stochastic matrix P. For $1 \leq i \leq r$, let P_i be the $m \times m$ matrix such that $P_i(i', j') = P(i', j')$ if $\pi(j') = i$ and otherwise $P_i(i', j') = 0$. Let θ^* be a stochastic (probability) left fixed vector for P and let θ be the column vector with every entry 1. Let C be the cone of all nonnegative vectors in $D = \mathbb{R}^m$. If we identify P_i with the symbol i, then these data give a finite-dimensional linear stochastic semigroup equivalent to $S(X)$. Along with this observation, Furstenberg established the converse.

Theorem 4.32. [25, Theorem 19.2] *A linear stochastic semigroup S is finite dimensional if and only if the stochastic process that it determines is a one-block factor of a one-step stationary finite-state Markov process.*

In the statement of Theorem 4.32, "Markov" does not presume ergodic. The construction for the theorem is essentially the one given in Theorem 4.28, with a simplification. Because of the definition of linear stochastic semigroup

(Definition 4.30), Furstenberg can begin with θ^*, θ actual fixed vectors of $P :=$ $\sum_i P_i$. The triple (P, θ^*, θ) corresponds to (P, x, y) in Proposition 4.16, where x, y need not be fixed vectors. Thus, Furstenberg can reduce more quickly to the form where θ^* and θ are *positive* fixed vectors of P. Note that "finite dimensional" in Theorem 4.32 means more than having the cone C of the linear stochastic semigroup generating a finite-dimensional space D: here C is a cone in $\mathbb{R}^m$ with exactly m (in particular, finitely many) extreme rays.

4.4 Sofic measures – Heller's approach

Repeating some problems already stated, but with some refinements, here are the natural questions about sofic measures which we are currently discussing, in subshift language.

Problem 4.33. Let $\pi : \Omega_A \to Y$ be a one-block map from a shift of finite type to a (sofic) subshift and let μ be a (fully supported) one-step Markov measure on Ω_A. When is $\pi \mu$ Markov? Can one determine what the *order* (a k such that the measure is k-step Markov) of the image measure might be?

Problem 4.34. Given a shift-invariant probability measure ν on a subshift Y, when are there a shift of finite type Ω_A, a factor map $\pi : \Omega_A \to Y$, and a one-step shift-invariant fully supported Markov measure μ on Ω_A such that $\pi \mu = \nu$?

Problem 4.35. If ν is a sofic measure, how can one explicitly construct Markov measures of which ν is a factor? Are there procedures for constructing Markov measures that map to ν which have a minimal number of states or minimal entropy?

Problem 4.33 was discussed in [12], for the reversible case. Later complete solutions depend on Heller's solution of Problem 4.34, so we discuss that first. Effective answers to the first part of Problem 4.35 are given by Furstenberg and in the proof of Theorem 4.28.

Problem 4.34 goes back at least to a 1959 paper of Gilbert [26]. Following Gilbert and Dharmadhikari [15, 16, 17, 18], Heller [35, 34] created his stochastic module theory and within this gave a characterization of sofic measures. We describe this next.

4.4.1 Stochastic module

We describe the stochastic module machinery setup of Heller [34] (with some differences in notation). Let $S = \{1, 2, ..., s\}$ be a finite state space for a stochastic process. Let A_S be the associative real algebra with free generating set S. An A_S-*module* is a real vector space V on which A_S acts by linear transformations, such that for each $i \in S$ there is a linear transformation $M_i : V \to V$ such that a word $u_1 \cdots u_k$ sends $v \in V$ to $M_{u_1}(M_{u_2}(\cdots M_{u_k}(v))\cdots)$. We denote an A_S-module as $(\{M_i\}, V)$ or for brevity just $\{M_i\}$, where the M_i are the associated generating linear transformations $V \to V$ as above.

Definition 4.36. A *stochastic S-module* for a stochastic process with state space S is a triple $(l, \{M_i\}, r)$, where $(\{M_i\}, V)$ is an A_S-module, $r \in V$, $l \in V^*$, and for every word $u = u_1 \cdots u_t$ on S its probability $\mathrm{Prob}(u) = \mathrm{Prob}(\mathcal{C}_0(u))$ is given by

$$\mathrm{Prob}(u) = l M_{u_1} M_{u_2} \cdots M_{u_t} r. \tag{77}$$

Given an A_S-module M, an $l \in V^*$, and $r \in V$, a few axioms are required to guarantee that they define a stochastic process with state space S. Define $\sigma = \sum \{a_i : a_i \in S\}$ and denote by $\mathcal{C}_S$ the cone of polynomials in A_S with nonnegative coefficients. Then the axioms are that

(1) $lr = 1$;
(2) $l(\mathcal{C}_S r) \subset [0, \infty)$;
(3) for all $f \in A_S$, $l(f(\sigma - 1)r) = 0$.

Example 4.37. *A stochastic module for a sofic measure.* As we saw in Section 4.3, this setup of a stochastic module arises naturally when a one-block map π is applied to a one-step Markov measure μ with state space S given by an $s \times s$ stochastic transition matrix P and row probability vector l. For each $i \in S$, let M_i be the matrix whose jth column equals column j of P if $\pi(j) = i$ and whose other columns are zero. The probability of an S-word $u = u_1 \cdots u_t$ is $l M_{u_1} M_{u_2} \cdots M_{u_t} r$, where r is the vector of all ones. With $V = \mathbb{R}^s$, presented as column vectors, $(l, \{M_i\}, r)$ is a stochastic module for the process given by $\pi \mu$.

4.4.2 The reduced stochastic module

A stochastic module $(l, (\{M_i\}, V), r)$ is *reduced* if (i) V is the smallest invariant (under the operators M_i) vector space containing r and (ii) l annihilates no nonzero invariant subspace of V. Given a stochastic module $(l, \{M_i\}, r)$ for a stochastic process, with its operators M_i operating on the real vector space V, a smallest stochastic module $(l', \{M_i'\}, r')$ describing the stochastic process may be defined as follows. Let R_1 be the cyclic submodule of V generated by the action on r; let L_1 be the cyclic submodule of V^* generated by the (dual) action on l; let V' be R_1 modulo the subspace annihilated by L_1; for each $i \in S$, let M_i' be the (well-defined) transformation of V' induced by M_i; let r', l' be the elements of V' and $(V')^\perp$ determined by r, l. Now (l', M', r') is *the reduced stochastic module* of the process. V' is the subspace generated by the action of the M_i' on r', and no nontrivial submodule of V' is annihilated by l'. The reduced stochastic module is still a stochastic module for the original stochastic process. We say "the" reduced stochastic module because any stochastic modules describing the same stochastic process have isomorphic reduced stochastic modules.

4.4.3 Heller's answer to Problem 4.34

We give some preliminary notation. A process is "induced from a Markov chain" if its states are lumpings of states of a finite-state Markov process, that is, there is a one-block code which sends the associated Markov measure to the measure associated to the stochastic process. Let $(A_S)_+$ be the subset of A_S consisting of linear combinations of words with all coefficients nonnegative. A *cone* in a real vector space V is a union of rays from the origin. A convex cone $\mathcal{C}$ is *strongly convex* if it contains no line through the origin. It is *polyhedral* if it is the convex hull of finitely many rays.

Theorem 4.38. *Let $(l, (\{M_i\}, V), r)$ be a reduced stochastic module. The associated stochastic process is induced from a Markov chain if and only if there is a cone $\mathcal{C}$ contained in the vector space V such that the following hold:*

(1) $r \in \mathcal{C}$,
(2) $l\mathcal{C} \subset [0, \infty)$,
(3) $(A_S)_+ \mathcal{C} \subset \mathcal{C}$,
(4) $\mathcal{C}$ is strongly convex and polyhedral.

Heller stated this result in [34, Theorem 1]. The proof there contained a minor error which was corrected in [35]. Heller defined a process to be *finitary* if its associated reduced stochastic module is finite dimensional. (We will call the corresponding measure finitary.) A consequence of Theorem 4.38 is the (obvious) fact that the reduced stochastic module of a sofic measure must be finitary. Heller gave an example [34] of a finitary process which is not a one-block factor of a one-step Markov measure, and therefore is not a factor of any Markov measure. (However, a subshift with a weakly mixing finitary measure is measure-theoretically isomorphic to a Bernoulli shift [7].)

4.5 Linear automata and the reduced stochastic module for a finitary measure

The 1960s and 1970s saw the development of the theory of probabilistic automata and linear automata. We have not thoroughly reviewed this literature, and we may be missing from it significant points of contact with and independent invention of the ideas under review. However, we mention at least one. A finite-dimensional stochastic module is a special case of a linear space automaton, as developed in [37] by Inagaki, Fukumura, and Matuura, following earlier work on probabilistic automata (e.g., [54, 61]). They associated to each linear space automaton its canonical (up to isomorphism) equivalent irreducible linear space automaton. When the linear space automaton is a stochastic module, its irreducible linear space automaton corresponds exactly to Heller's canonical (up to isomorphism) reduced stochastic module. Following [37] and Nasu's paper [50], we will give some concrete results on the reduced stochastic module.

We continue Example 4.37 and produce a concrete version of the reduced stochastic module in the case that a measure on a subshift is presented by a stochastic module which is finite dimensional as a real vector space (for example, in the case of a sofic measure). Correspondingly, in this section we will reverse Heller's roles for row and column vectors and regard the stochastic module as generated by row vectors.

So, let $(u, \{M_i\}, v)$ be a finite-dimensional stochastic module on finite alphabet $\mathcal{A}$. We take the presentation so that there is a positive integer n such that the M_i are $n \times n$ matrices; u and v are n-dimensional row and column vectors; and the map $a \mapsto M_a$ induces a monoid homomorphism ϕ from $\mathcal{A}^*$, sending a word $w = a_1 \cdots a_j$ to the matrix $\phi(w) = M_{a_1} \cdots M_{a_j}$.

Let $\mathcal{U}$ be the vector space generated by vectors of the form $u\phi(w)$, $w \in \mathcal{A}^*$. Similarly, define $\mathcal{V}$ as the vector space generated by vectors of the form $\phi(w)v$, $w \in \mathcal{A}^*$. Let $k = \dim(\mathcal{U})$. If $k < n$, then construct a smaller module (presenting the same measure) as follows. Let L be a $k \times n$ matrix whose rows form a basis of $\mathcal{U}$. For each symbol a there exists a $k \times k$ matrix $\widehat{M}_a$ such that $LM_a = \widehat{M}_a L$. Define $\widehat{u}$ to be the k-dimensional row vector such that $\widehat{u}L = u$ and set $\widehat{v} = Lv$. Let $a \to \widehat{M}_a$ induce a monoid homomorphism $\widehat{\phi}$ from $\mathcal{A}^*$, sending a word $w = a_1 \cdots a_j$ to $\widehat{\phi}(w) = \widehat{M}_{a_1} \cdots \widehat{M}_{a_j}$. The subspace $\widehat{\mathcal{U}}$ of $\mathbb{R}^k$ generated by vectors of the form $\widehat{u}\widehat{\phi}(w)$ is equal to $\mathbb{R}^k$ because $\widehat{\mathcal{U}}L = \mathcal{U}$ and $\dim(\mathcal{U}) = k$. It is easily checked that $\widehat{u}\widehat{\phi}(w)\widehat{v} = u\phi(w)v$ for every w in $\mathcal{A}^*$. Let $\widehat{\mathcal{V}}$ be the subspace of $\mathbb{R}^k$ generated by column vectors $\widehat{\phi}(w)\widehat{v}$. We have for each a that $LM_a v = \widehat{M}_a Lv = \widehat{M}_a \widehat{v}$, so L maps $\mathcal{V}$ onto $\widehat{\mathcal{V}}$. Also, L maps the space of n-dimensional column vectors onto $\mathbb{R}^k$. It follows that if $\dim(\mathcal{V}) = n$, then $\dim(\widehat{\mathcal{V}}) = k$.

If $\dim(\widehat{\mathcal{V}}) < k$, then repeat the reduction move, but applying it to v (column vectors) rather than to u. This will give a stochastic module $(\overline{u}, \{\overline{M}_a\}, \overline{v})$, say with $m \times m$ matrices $\overline{M}_a$ and invariant subspaces $\overline{\mathcal{U}}, \overline{\mathcal{V}}$ generated by the action on $\overline{u}, \overline{v}$. By construction, we have $\dim(\overline{\mathcal{V}}) = m$. And, because $\widehat{\mathcal{U}}$ had full dimension, we have $\dim(\overline{\mathcal{U}}) = m$ also. Regarding $\mathcal{V}$ as a space of functionals on $\mathcal{U}$, and letting $\ker(\mathcal{V})$ denote the subspace of $\mathcal{U}$ annihilated by all elements of $\mathcal{V}$, we see that $u \mapsto \overline{u}$ is a presentation of the map $\pi : \mathcal{U} \to \mathcal{U}/\ker(\mathcal{V})$. Thus, $(\overline{u}, \{\overline{M}_a\}, \overline{v})$ is a presentation of the reduced stochastic module. Also, for all a, $\pi M_a = \overline{M}_a \pi$, and therefore the surjection π (acting from the right) also satisfies

$$\left(\sum_a M_a \right) \pi = \pi \left(\sum_a \overline{M}_a \right). \tag{78}$$

If $(\tilde{u}, \{\tilde{M}_a\}, \tilde{v})$ is another such presentation of the reduced stochastic module, then it must have the same (minimal) dimension m, and there will be an invertible matrix G (giving the isomorphism of the two presentations) such that for all a,

$$\left(\tilde{u}, \{\tilde{M}_a\}, \tilde{v} \right) = \left(\overline{u}G, \{G^{-1}\overline{M}_a G\}, G^{-1}\overline{v} \right). \tag{79}$$

To find G, simply take m words w such that the vectors $\overline{u}\overline{\phi}(w)$ are a basis for $\overline{\mathcal{U}}$, and let G be the matrix such that for each of these w,

$$\overline{u}\overline{\phi}G = \tilde{u}\tilde{\phi}. \tag{80}$$

The rows of the matrix L above (a basis for the space $\mathcal{U}$) may be obtained by examining vectors $u\phi(w)$ in some order, with the length of w nondecreasing, and including as a row any vector not in the span of previous vectors. Let $\mathcal{U}_m$ denote the space spanned by vectors $u\phi(w)$ with w of length at most m. If for some m we have $\mathcal{U}_m = \mathcal{U}_{m+1}$, then $\mathcal{U}_m = \mathcal{U}$. In particular, if n is the dimension of the original stochastic module, then the matrix L can be found by considering words of length at most $n-1$.

One can check that if two equivalent stochastic modules have dimensions n_1 and n_2, then they are equivalent (define the same measure) if and only if they assign the same measure to words of length $n_1 + n_2 - 1$. (This is a special case of [37, Theorem 5.2].) If the reduced stochastic module of a measure has dimension at most n, then one can also construct the reduced stochastic module from the measures of words of length at most $2n - 1$ (one construction is given in [37, Theorem 6.2]). However, without additional information about the measure, this forces the examination of a number of words which for a fixed alphabet can grow exponentially as a function of n, as indicated by the following example.

Example 4.39. Let X be the full shift on the three symbols $0, 1, 2$. Given $k \in \mathbb{N}$, define a stochastic matrix P indexed by X-words of length $k+1$ by $P(10^k, 0^k 1) = 1/6 = P(20^k, 0^k 2)$; $P(10^k, 0^k 2) = 1/2 = P(20^k, 0^k 1)$; $P(a_0 \cdots a_k, a_1 \cdots a_{k+1}) = 1/3$ otherwise; and all other entries of P are zero. This matrix defines a $(k+1)$-step Markov measure μ on X which agrees with the

Bernoulli $(1/3, 1/3, 1/3)$ measure on all words of length at most $k+2$ except the four words $10^k 1, 10^k 2, 20^k 1, 10^k 2$. The reduced stochastic module has dimension at most $2k+1$, because for any word U the conditional probability function on X-words defined by $\rho_U : W \mapsto \mu(UW|U)$ will be a constant multiple of ρ_V for one of the words $V = 0^{k+1}, 10^j, 20^j$, with $0 \leq j \leq k$. The number of X-words of length $k+2$ is 3^{k+2}.

4.6 Topological factors of finitary measures, and Nasu's core matrix

The content of this section is essentially taken from Nasu's paper [50], as we explain in more detail below. Given a square matrix M, in this section we let M^* denote any square matrix similar to one giving the action of M on the maximal invariant subspace on which the action of M is nonsingular.

Adapting terminology from [50], we define the *core matrix* of a finite-dimensional stochastic module give by matrices, $(l, \{M_i\}, r)$, to be $\sum_i M_i$. A core matrix for a finitary measure μ is any matrix which is the core matrix of a reduced stochastic module for μ. This matrix is well defined only up to similarity, but for simplicity of language we refer to *the core matrix of* μ, denoted Core(μ). Similarly, we define *the eventual core matrix of* μ to be Core$(\mu)^*$, denoted Core$^*(\mu)$. For example, if Core(μ) is $\begin{pmatrix} \frac{1}{2} & 0 & 0 & 0 \\ 0 & 1 & 0 & 0 \\ 0 & 0 & 0 & 1 \\ 0 & 0 & 0 & 0 \end{pmatrix}$, then Core$^*(\mu)$ is $\begin{pmatrix} \frac{1}{2} & 0 \\ 0 & 1 \end{pmatrix}$.

Considering square matrices M and N as linear endomorphisms, we say that N is a quotient of M if there is a linear surjection π such that, writing action from the right, $M\pi = \pi N$. (Equivalently, by duality, the action of N is isomorphic to the action of M on some invariant subspace.) In this case, the characteristic polynomial of M divides that of N (but, e.g., $\begin{pmatrix} 2 & 0 \\ 0 & 2 \end{pmatrix}$ is a principal submatrix of but not a quotient of $\begin{pmatrix} 2 & 1 & 0 \\ 0 & 2 & 1 \\ 0 & 0 & 2 \end{pmatrix}$).

Theorem 4.40. *Suppose that ϕ is a continuous factor map from a subshift X onto a subshift Y, $\mu \in \mathcal{M}(X)$, and $\phi\mu = \nu \in \mathcal{M}(Y)$. Suppose that μ is finitary. Then ν is finitary and Core$^*(\nu)$ is a quotient of Core$^*(\mu)$. In particular, if ϕ is a topological conjugacy, then Core$^*(\nu) = $ Core$^*(\mu)$.*

The key to the topological invariance in Theorem 4.40 is the following lemma (a measure version of [50, Lemma 5.2]).

Lemma 4.41. *Suppose that μ is a finitary measure on a subshift X and $n \in \mathbb{N}$. Let $X^{[n]}$ be the n-block presentation of X; let $\psi : X^{[n]} \to X$ be the one-block factor map defined on symbols by $[a_1 \cdots a_n] \mapsto a_1$; let $\mu^{[n]} \in \mathcal{M}(X^{[n]})$ be the measure such that $\psi \mu^{[n]} = \mu$. Then $\mu^{[n]}$ is finitary and $\mathrm{Core}^*(\mu^{[n]})$ is a quotient of $\mathrm{Core}^*(\mu)$.*

Proof of Lemma 4.41. For $n > 1$, the n-block presentation of X is (after a renaming of the alphabet) equal to the two-block presentation of $X^{[n-1]}$. So, by induction, it suffices to prove the lemma for $n = 2$.

Let $(l, \{P_i\}, r)$ be a reduced stochastic module for μ, where the P_i are $k \times k$ and $\mathcal{A}(X) = \{1, 2, \ldots, m\}$. For each symbol ij of $\mathcal{A}(X^{[2]})$, define an $mk \times mk$ matrix P'_{ij} as an $m \times m$ system of $k \times k$ blocks, in which the (i,j)th block is P_i and the other entries are zero. Define $l' = (l, \ldots, l)$ (m copies of l) and define $r' = \begin{pmatrix} P_1 r \\ \vdots \\ P_m r \end{pmatrix}$.

Then $(l', \{P'_{ij}\}, r')$ is a stochastic module for $\mu^{[2]}$, which is therefore finitary. Also, we have an elementary strong shift equivalence of the core matrices P and P',

$$P' = \begin{pmatrix} P_1 \\ \vdots \\ P_m \end{pmatrix} (I \quad \cdots \quad I), \qquad P = (I \quad \cdots \quad I) \begin{pmatrix} P_1 \\ \vdots \\ P_m \end{pmatrix},$$

and therefore $P^* = (P')^*$. Because $\mathrm{Core}(\mu^{[2]})$ is a quotient of P', it follows that $\mathrm{Core}^*(\mu^{[2]})$ is a quotient of $(P')^* = P^* = \mathrm{Core}^*(\mu)$. $\qquad\square$

If $\phi : X \to Y$ is a factor map of irreducible sofic shifts of equal entropy, then ϕ must send the unique measure of maximal entropy of X, μ_X, to that for Y. These are sofic measures, and consequently Theorem 4.40 gives computable obstructions to the existence of such a factor map between given X and Y. In his work, Nasu associated to a given X a certain linear (not stochastic) automaton. If we denote it $(l, \{M_i\}, r)$, and let $\log(\lambda)$ denote the topological entropy of X, then $(l, \{(1/\lambda)M_i\}, r)$ would be a stochastic module for μ_X. In the end, Nasu's core matrix is $\lambda\mathrm{Core}(\mu_X)$. Nasu remarked in [50] that his arguments could as well be carried out with respect to measures to obtain his results, and that is what we have done here.

Eigenvalue relations between core matrices (not so named) of equivalent linear automata already appear in [37, Section 7]. Also, Kitchens [40] earlier used the (Markov) measure of maximal entropy for an irreducible shift of finite type in a similar way to show that the existence of a factor map of equal-entropy irreducible SFTs, $\Omega_A \to \Omega_B$, implies (in our terminology) that B^* is a quotient of A^*. This is a special case of Nasu's constraint.

5 When is a sofic measure Markov?

5.1 When is the image of a one-step Markov measure under a one-block map one-step Markov?

We return to considering Problem 4.33. In this subsection, suppose that μ is a one-step Markov measure, that is, a one-step fully supported shift-invariant Markov measure on an irreducible shift of finite type Ω_A. Suppose that π is a one-block code with domain Ω_A. How does one characterize the case when the measure $\pi \mu$ is again one-step Markov?

To our knowledge, this problem was introduced, in the language of Markov processes, by Burke and Rosenblatt in 1958 [12], who solved it in the reversible case [12, Theorem 1]. Kemeny and Snell [39, Theorems 6.4.8 and 6.3.2] gave another exposition and introduced the "lumpability" terminology. Kemeny and Snell defined a (not necessarily stationary) finite-state Markov process X to be *lumpable* with respect to a partition of its states if for every initial distribution for X the corresponding quotient process is Markov. They defined X to be *weakly lumpable* with respect to the partition if there exists an initial distribution for X for which the quotient process Y is Markov. In all of this, by Markov they mean one-step Markov. Various problems around these ideas were (and continue to be) explored and solved. For now we restrict our attention to the question of the title of this subsection and describe two answers.

5.1.1 Stochastic module answer

Theorem 5.1. *Let (l, M, r) be a presentation of the reduced stochastic module of a sofic measure ν on Y, in which M_i denotes the matrix by which a symbol i of $\mathcal{A}(Y)$ acts on the module. Suppose that $k \in \mathbb{N}$. Then the sofic measure ν is k-step Markov if and only if every product $M_{i(1)} \cdots M_{i(k)}$ of length k has rank at most 1.*

The case $k = 1$ of Theorem 5.1 was proved by Heller [34, Proposition 3.2] An equivalent characterization was given a good deal later, evidently without awareness of Heller's work, by Bosch [9], who worked from the papers of Gilbert [26] and Dharmadhikari [15]. The case of general k in Theorem 5.1 was proved by Holland [36, Theorem 4], following Heller.

5.1.2 Linear algebra answer

One can approach the problem of deciding whether a sofic measure is Markov with straight linear algebra. There is a large literature using such ideas in the context of automata, control theory, and the "lumpability" strand of literature emanating from Kemeny and Snell (see e.g. [27] and its references). Propositions 5.2 and 5.3 and Theorem 5.4 below are taken from Gurvits and Ledoux [27]. As with previous references, we are considering only a fragment of this one.

Let N be the size of the alphabet of the irreducible shift of finite type Ω_A. Let π be a one-block code mapping Ω_A onto a subshift Y. Let P be an $N \times N$ irreducible stochastic matrix defining a one-step Markov measure μ on Ω_A. Let p be the positive stochastic row fixed vector of P. Let U be the matrix such that $U(i,j) = 1$ if π maps the state i to the state j, and $U(i,j) = 0$ otherwise. Given $i \in \mathcal{A}(\Omega_A)$, let $\bar{i}$ be its image symbol in Y. Given $j \in \mathcal{A}(Y)$, let P_j be the matrix of size P which equals P in columns i such that $\bar{i} = j$, and is zero in other entries. Likewise, define p_j. Given a Y-word $w = j_1 \cdots j_k$, we let $P_w = P_{j_1} \cdots P_{j_k}$.

Alert: we are using parenthetical notation for matrix and vector entries and subscripts for lists. If $\pi\mu$ is a one-step Markov measure on Y, then it is defined using a stochastic row vector q and a stochastic matrix Q. The vector q can only be pU, and the entries of Q are determined by $Q(j,k) = (p_j P_k U)(k)/q(j)$. Let ν denote the Markov measure defined using q, Q. Define q_j, Q_j by replacing entries of q, Q with zero in columns not indexed by j. For a word $w = j_0 \cdots j_k$ on symbols from $\mathcal{A}(Y)$, we have $(\pi\mu)(\mathcal{C}_0(w)) = \nu(\mathcal{C}_0(w))$ if and only if

$$p_{j_0} P_{j_1} \cdots P_{j_k} U = p_{j_0} U Q_{j_1} \cdots Q_{j_k} \tag{81}$$

(since $q_{j_0} = p_{j_0} U$). Thus, $\pi\mu = \nu$ if and only if (81) holds for all Y-words w. This remark is already more or less in Kemeny and Snell [39, Theorem 6.4.1].

For the additional argument which produces a finite procedure, we define certain vector spaces (an idea already in [22, 41, 62, 63, 27] and elsewhere).

Let $\mathcal{V}_k$ denote the real vector space generated by the row vectors $p_{j_0} P_{j_1} \cdots P_{j_k}$ such that $j_0 j_1 \cdots j_t$ is a Y-word and $0 \le t \le k$. So, $\mathcal{V}_0$ is the vector space generated by the vectors p_{j_0}, and $\mathcal{V}_{k+1}$ is the subspace generated by $\mathcal{V}_k \cup \{vP_j : v \in \mathcal{V}_k, j \in \mathcal{A}(Y)\}$. In fact, for $k \ge 0$, we claim that

$$\mathcal{V}_k = \langle \{p_{j_0} P_{j_1} \cdots P_{j_k} : j_0 \cdots j_k \in \mathcal{A}(Y)^{k+1}\} \rangle, \tag{82}$$

$$\mathcal{V}_{k+1} = \langle \{vP_j : v \in \mathcal{V}_k, j \in \mathcal{A}(Y)\} \rangle, \tag{83}$$

where $\langle \ \rangle$ is used to denote span. Clearly, (83) follows from (82), which is a consequence of stationarity, as follows. Because $\sum_j p_j = p = pP = \sum_j pP_j$, and for $i \neq j$ the vectors p_i and pP_j cannot both be nonzero in any coordinate, we have $p_j = pP_j$. So, given t and $j_1 \cdots j_t$, we have

$$p_{j_1} P_{j_2} \cdots P_{j_t} = pP_{j_1} P_{j_2} \cdots P_{j_t}$$
$$= \sum_{j_0} p_{j_0} P_{j_1} P_{j_2} \cdots P_{j_t},$$

from which (83) easily follows. Let $\mathcal{V} = \langle \bigcup_{k \geq 0} \mathcal{V}_k \rangle$.

Proposition 5.2. *Suppose that P is an $N \times N$ irreducible stochastic matrix and ϕ is a one-block code. Let the vector spaces $\mathcal{V}_k$ be defined as above, and let n be the smallest positive integer such that $\mathcal{V}_n = \mathcal{V}_{n+1}$. Then $n \leq N - |\mathcal{A}(Y)|$, $\mathcal{V}_n = \mathcal{V}$, and the following are equivalent:*

(1) $\phi\mu$ is a one-step Markov measure on the image of ϕ.
(2) $p_{j_0} P_{j_1} \cdots P_{j_n} U = p_{j_0} U Q_{j_1} \cdots Q_{j_n}$, for all $j_0 \cdots j_n \in \mathcal{A}(Y)^{n+1}$.

Proof. For $k \geq 1$, we have $\mathcal{V}_k \subset \mathcal{V}_{k+1}$, and also

$$\mathcal{V}_k = \mathcal{V}_{k+1} \quad \text{implies that} \quad \mathcal{V}_k = \mathcal{V}_t = \mathcal{V} \quad \text{for all } t \geq k . \tag{84}$$

Because $\dim(\mathcal{V}_0) = |\mathcal{A}(Y)|$, it follows that $n \leq N - |\mathcal{A}(Y)|$.

Because (1) is equivalent to (81) holding for all Y-words $j_0 j_1 \cdots j_k$, $k \geq 0$, we have that (1) implies (2).

Now suppose that (2) holds. For $K \geq 1$, the linear condition (81) holds for all Y-words of length k less than or equal to K if and only if $vUQ_j = vP_jU$ for all j in $\mathcal{A}(Y)$ and all v in $\mathcal{V}_K$. (U is the matrix defined above.) Because $\mathcal{V}_K = \mathcal{V}_n$ for $K \geq n$, we conclude from (2) and (82) that (81) holds for all Y-words $j(0)j(1) \cdots j(k)$, $k \geq 0$, and therefore (1) holds. $\qquad \square$

Next we consider an irrreducible $N \times N$ matrix P defining a one-step Markov measure μ on Ω_A and a one-block code ϕ from Ω_A onto a subshift Y. Given a positive integer $k \geq 1$, we are interested in understanding when $\phi\mu$ is k-step Markov. We use notation $U, p, p_j, P_j, \mathcal{V}_t$ and $\mathcal{V}_n = \mathcal{V}$ as above. Define a stochastic row vector q indexed by Y-words of length k, with $q(j_0 \cdots j_{k-1}) = (p_{j_0} P_{j_1} \cdots P_{j_{k-1}} U)(j_{k-1})$. Let Q be the square matrix indexed by Y-words of length k whose nonzero entries are defined by

$$Q(j_0 \cdots j_{k-1}, j_1 \cdots j_k) = \frac{(p_{j_0} P_{j_1} \cdots P_{j_k} U)(j_k)}{q(j_0 \cdots j_{k-1})} .$$

Then Q is an irreducible stochastic matrix and q is a positive stochastic vector such that $qQ = q$. Let ν be the k-step Markov measure defined on Y by (q, Q).

The measures v and $\phi\mu$ agree on cylinders $C_0(j_0 \cdots j_k)$ and therefore on all cylinders $C_0(j_0 \cdots j_t)$ with $0 \le t \le k$. Clearly, if $\phi\mu$ is k-step Markov, then $\phi\mu$ must equal v.

Proposition 5.3. [27] *Suppose that P is an $N \times N$ irreducible stochastic matrix defining a one-step Markov measure μ on Ω_A and $\phi : \Omega_A \to Y$ is a one-block code. Let k be a fixed positive integer. With the notation above, the following are equivalent:*

(1) *$\phi\mu$ is a k-step Markov measure (i.e., $\phi\mu = v$).*
(2) *For every Y-word $w = w_0 \cdots w_{k-1}$ of length k and every $v \in \mathcal{V}$,*

$$v P_w (PU - 1Q^w) = 0, \tag{85}$$

where $P_w = P_{w_0} \cdots P_{w_{k-1}}$; 1 is the size-N column vector with every entry 1; and Q^w is the stochastic row vector defined by

$$Q^w(j) = Q(w_0 \cdots w_{k-1}, w_1 \cdots w_{k-1}j), \quad j \in \mathcal{A}(Y). \tag{86}$$

Proof. We continue to denote by $z(j)$ the entry in the jth coordinate of a row vector z. By construction of v, we have for $t = 0$ that

$$(\pi\mu)C_0(j_0 \cdots j_{t+k}) = v C_0(j_0 \cdots j_{t+k}) \quad \text{for all } j_0 \cdots j_{t+k} \in \mathcal{A}^{t+k+1}. \tag{87}$$

Now suppose that t is a nonnegative integer and (87) holds for t. Given $j_0 \cdots j_{t+k}$, let w be its terminal word of length k. Then, for $j \in \mathcal{A}(Y)$,

$$
\begin{aligned}
&(\pi\mu)C_0(j_0 \cdots j_{t+k}j) - v C_0(j_0 \cdots j_{t+k}j) \\
&= \left(p_{j_0} P_{j_1} \cdots P_{j_{t+k}} P_j U \right)(j) - \left(v C_0(j_0 \cdots j_{t+k}) Q^w \right)(j) \\
&= \left(p_{j_0} P_{j_1} \cdots P_{j_{t+k}} P_j U \right)(j) - \left((p_{j_0} P_{j_1} \cdots P_{j_{t+k}} 1) Q^w \right)(j) \\
&= \left(p_{j_0} P_{j_1} \cdots P_{j_{t+k}} [P_j U - 1Q^w] \right)(j) \\
&= \left(p_{j_0} P_{j_1} \cdots P_{j_t} P_w [PU - 1Q^w] \right)(j),
\end{aligned}
$$

where the term $P_{j_1} \cdots P_{j_t}$ is included only if $t > 0$, and the last equality holds because the jth columns of PU and $P_j U$ are equal. Thus, given (87) for t, by (82) we have (87) for $t+1$ if and only if $v P_w [PU - 1Q^w] = 0$ for all $v \in \mathcal{V}_t$ and all w of length k. It follows from induction that (87) holds for all $t \ge 0$ (i.e., $\pi\mu = v$) if and only if (85) holds for all $v \in \mathcal{V}$. $\qquad \square$

Because $\mathcal{V}$ can be computed, Proposition 5.3 gives an algorithm, given k, for determining whether the image of a one-step Markov measure is a k-step Markov measure. The next result gives a criterion which does not require computation of the matrix Q.

Theorem 5.4. [27] *Let notation be as in Proposition 5.3. Then $\phi\mu$ is a k-step Markov measure on Y if and only if for every Y-word w of length k,*

$$\big((\mathcal{V}P_w)\cap\ker(U)\big)P \subset \ker(U) . \tag{88}$$

Proof. Let $w = w_0 \cdots w_{k-1}$ be a Y-word of length k. Using the computations of the proof of Proposition 5.3, we obtain for $j \in \mathcal{A}(Y)$ that

$$
\begin{aligned}
0 &= \pi\mu\mathcal{C}_0(w_0\cdots w_{k-1}j) - v\mathcal{C}_0(w_0\cdots w_{k-1}j) \\
&= \big(p_{w_0}P_{w_1}\cdots P_{w_{k-1}}[PU-1Q^w]\big)(j) \\
&= \big(pP_{w_0}P_{w_1}\cdots P_{w_{k-1}}[PU-1Q^w]\big)(j) \\
&= \big(pP_w[PU-1Q^w]\big)(j) .
\end{aligned}
$$

Consequently, the vector $v = p$ satisfies (85). Moreover,

$$(pP_wU)(w_{k-1}) = (p_{w_0}P_{w_1}\cdots P_{w_{k-1}}U)(w_{k-1}) = \pi\mu\mathcal{C}_0(w) > 0,$$

and therefore $pP_w \notin \ker(U)$. Because $vP_w = 0$ if and only if $vP_w1 = 0$, the space $\mathcal{V}P_w$ is spanned by pP_w and $(\mathcal{V}P_w)\cap\ker(U)$. Thus, (85) holds for all $v \in \mathcal{V}$ if and only if (85) holds for all $v \in \mathcal{V}$ such that $vP_w \in \ker(U)$, which is equivalent to (88). $\qquad\square$

Gurvits and Ledoux [27, Section 2.2.2] explained how Theorem 5.4 can be used to produce an algorithm, polynomial in the number N of states, for deciding whether $\pi\mu$ is a one-step Markov measure.

5.2 Orders of Markov measures under codes

This section includes items relevant to the second part of Problem 4.33.

Definition 5.5. Given positive integers m, n, k with $1 \leq k \leq n$, recursively define integers $N(k, m, n)$ by setting

$$N(n, m, n) = 1, \tag{89}$$

$$N(k, m, n) = (1 + m^{N(k+1, m, n)})N(k+1, m, n) \quad \text{if } 1 \leq k < n . \tag{90}$$

Proposition 5.6. *Suppose that* $\pi : \Omega_A \to Y$ *is a one-block code and* μ *is a one-step Markov measure on* Ω_A. *Let* n *be the dimension of the reduced stochastic module of* $\pi\mu$ *and let* $m = |\mathcal{A}(Y)|$. *Suppose that* $n \geq 2$. *(In the case* $n = 1$, $\pi\mu$ *is Bernoulli.) Let* $K - N(2, m, n)$. *If* $\pi\mu$ *is not* K-step Markov, then it is not k-step Markov for any k.

Before proving Proposition 5.6, we state our main interest in it.

Corollary 5.7. *Suppose that* μ *is a one-step Markov measure on an irreducible SFT* Ω_A *determined by a stochastic matrix* P, *and that there are algorithms for doing arithmetic in the field generated by the entries of* P. *Suppose that* ϕ *is a block code on* Ω_A. *Then there is an algorithm for deciding whether the measure* $\phi\mu$ *is Markov.*

Proof. The corollary is an easy consequence of Propositions 5.2 and 5.6. □

The proof of Proposition 5.6 uses two lemmas.

Lemma 5.8. *Suppose that* $P_1, \ldots, P_t$ *are* $n \times n$ *matrices such that* $\mathrm{rank}(P_1 \cdots P_t P_1) = \mathrm{rank}(P_1) = r$. *Then, for all positive integers* m, $\mathrm{rank}(P_1 \cdots P_t)^m P_1 = r$.

Proof. It follows from the rank equality that $(P_1 \cdots P_k)$ defines an isomorphism from the image of P_1 (a vector space of column vectors) to itself. □

Lemma 5.9. *Suppose that* k, m, n *are positive integers and* $1 \leq k \leq n$. *Suppose that* $\mathcal{Q}$ *is a collection of* m *matrices of size* $n \times n$, *and there exists a product of* $N(k, m, n)$ *matrices from* $\mathcal{Q}$ *with rank at least* k. *Then there are arbitrarily long products of matrices from* $\mathcal{Q}$ *with rank at least* k.

Proof. We prove the proposition by induction on k, for k decreasing from n. The case $k = n$ is clear. Suppose now that $1 \leq k < n$ and the lemma holds for $k + 1$. Suppose that a matrix M is a product $Q_{i(1)} \cdots Q_{i(N(k,m,n))}$ of $N(k, m, n)$ matrices from $\mathcal{Q}$ and has rank at least k. We must show there are arbitrarily long products from $\mathcal{Q}$ with rank at least k.

The given product is a concatenation of products of length $N(k+1, m, n)$, and we define corresponding matrices

$$P_j = Q_{1+(j-1)(N(k+1,m,n))} \cdots Q_{j(N(k+1,m,n))}, \quad 1 \leq j \leq 1 + m^{N(k+1,m,n)} . \quad (91)$$

If any P_j has rank at least $k + 1$, then by the induction hypothesis there are arbitrarily long products with rank at least $k + 1$, and we are done. So, suppose that every P_j has rank at most k. Because $\mathrm{rank}(P_j) \geq \mathrm{rank}(M) \geq k$, it follows that M, and every P_j, and every subproduct of consecutive P_j, has rank k.

There are only $m^{N(k+1,m,n)}$ words of length $N(k+1,m,n)$ on m symbols, so two of the matrices P_j must be equal. The conclusion now follows from Lemma 5.8. □

Proof of Proposition 5.6. As described in Example 4.37 and Section 4.5, there are algorithms for producing the reduced stochastic module for $\pi\mu$ as a set of matrices M_a (one for each symbol from $\mathcal{A}(Y)$) and a pair of vectors u, v such that for any Y-word $a_1 \cdots a_t$, $(\pi\mu)\mathcal{C}_0(a_1 \cdots a_t) = uM_{a_1} \cdots M_{a_t} v$. By Theorem 5.1, $\pi\mu$ is k-step Markov if and only if every product $M_{a_1} \cdots M_{a_k}$ has rank at most 1. Let $K = N(2,m,n)$. If $\pi\mu$ is not K-step Markov, then some matrix $\prod_{i=1}^{K} M_{a(i)}$ has rank at least 2, and by Lemma 5.9 there are then arbitrarily long products of M_a with rank at least 2. By Theorem 5.1, this shows that $\pi\mu$ is not k-step Markov for any k. □

Remark 5.10. Given m and n, the numbers $N(k,m,n)$ grow very rapidly as k decreases. Consequently, the bound K in Proposition 5.6 (and consequently the algorithm of Corollary 5.7) is not practical. However, in an analogous case (Problem 5.13 below) we do not even know the existence of an algorithm.

Problem 5.11. Find a reasonable bound K for Proposition 5.6.

Example 5.12. This is an example to show that the cardinality of the domain alphabet cannot be used as the bound K in Proposition 5.6. Given $n > 1$ in $\mathbb{N}$, let A be the adjacency matrix of the directed graph $\mathcal{G}$ which is the union of two cycles, $a_1 b_1 b_2 \cdots b_{n+4} a_1$ and $a_2 b_3 b_4 \cdots b_{n+3} a_2$. The vertex set $\{a_1, a_2, b_1, \ldots, b_{n+4}\}$ is the alphabet $\mathcal{A}$ of Ω_A. Let ϕ be the one-block code defined by erasing subscripts, and let Y be the subshift which is the image of ϕ, with alphabet $\{a, b\}$. Let μ be any one-step Markov measure on Ω_A. In $\mathcal{G}$, there are exactly four first return paths from $\{a_1, a_2\}$ to $\{a_1, a_2\}$: $a_1 b_1 \cdots b_{n+4} a_1$, $a_1 b_1 \cdots b_{n+3} a_2$, $a_2 b_3 \cdots b_{n+4} a_1$, and $a_2 b_3 \cdots b_{n+3} a_2$. Thus, in a point of Y, successive occurrences of the symbol a must correspondingly be separated by m b's, with $m \in \{n+4, n+3, n+2, n+1\}$. Each Y-word $ab^m a$ has a unique preimage word, so $\phi : \Omega_A \to Y$ is a topological conjugacy. Thus, $\phi\mu$ is k-step Markov for some k. We have

$$\phi(b_1 \cdots b_{n+3} a_2 b_3 \cdots b_{n+3} a_2) = (b^{n+3} ab^{n+1})a \quad \text{and}$$
$$\phi(a_1 b_1 \cdots b_{n+4} a_1 b_1 \cdots b_{n+1}) = ab(b^{n+3} ab^{n+1}) \,.$$

So, $(b^{n+3}ab^{n+1})a$ and $ab(b^{n+3}ab^{n+1})$ are Y-words, but $ab(b^{n+3}ab^{n+1})a$ is not a Y-word. Consequently, we have conditional probabilities

$$\phi\mu[y_0 = a \mid y_{-(2n+5)} \cdots y_{-1} = (b^{n+3}ab^{n+1})] > 0 \,,$$

$$\phi\mu[y_0 = a \mid y_{-(2n+7)} \cdots y_{-1} = ab(b^{n+3}ab^{n+1})] = 0 \,,$$

which show that $\phi\mu$ cannot be $(2n+5)$-Markov. In contrast, $|\mathcal{A}| = n+6 < 2n+5$.

With regard to Problem 3.3 of determining whether a given factor map is Markovian, the analogue of Proposition 5.6 is the following open problem.

Problem 5.13. Find (or prove there does not exist) an algorithm for attaching to any one-block code ϕ from an irreducible shift of finite type a number N with the following property: if a one-step Markov measure μ on the range of ϕ has no preimage measure which is N-step Markov, then μ has no preimage measure which is Markov.

Remark 5.14. (The persistence of memory) Suppose that $\phi : \Omega_A \to \Omega_B$ is a one-block code from one irreducible one-step SFT onto another. We collect some facts on how the memory of a Markov measure and a Markov image must or can be related.

(1) The image of a one-step Markov measure can be Markov but not one-step Markov. (For example, the standard map from the k-block presentation to the one-block presentation takes the one-step Markov measures onto the k-step Markov measures.)

(2) If ϕ is finite-to-one and ν is k-step Markov on Ω_B, then there is a unique Markov measure μ on Ω_A such that $\phi\mu = \nu$, and μ is also k-step Markov (Proposition 3.18).

(3) If any one-step Markov measure on Ω_B lifts to a k-step Markov measure on Ω_A, then, for every n, every n-step Markov measure on Ω_B lifts to an $(n+k)$-step Markov measure on Ω_A. (This follows from the explicit construction sketched in the proof of Theorem 3.2 in Section 3.1 and passage as needed to a higher block presentation.)

(4) If ϕ is infinite-to-one then it can happen [11, Section 2] ("peculiar memory example") that every one-step Markov measure on Ω_B lifts to a two-step Markov measure on Ω_A but not to a one-step Markov measure, while every one-step Markov on Ω_A maps to a two-step Markov measure on Ω_B.

6 Resolving maps and Markovian maps

In this section, Ω_A denotes an irreducible one-step shift of finite type defined by an irreducible matrix A.

6.1 Resolving maps

In this section, $\pi : \Omega_A \to Y$ is a one-block code onto a subshift Y, with Y not necessarily a shift of finite type, unless specified. U denotes the $0, 1, |\mathcal{A}(\Omega_A)| \times |\mathcal{A}(Y)|$ matrix such that $U(i,j) = 1$ if and only if $\pi(i) = j$. Denote a symbol $(\pi x)_0$ by $\overline{x_0}$.

Definition 6.1. The factor map π as above is *right resolving* if for all symbols $i, \overline{i}, k$ such that $\overline{i}k$ occurs in Y, there is at most one j such that ij occurs in Ω_A and $\overline{j} = k$. In other words, for any diagram

$$
\begin{array}{c}
i \\
\downarrow \\
\overline{i} \longrightarrow k
\end{array}
\tag{92}
$$

there is at most one j such that

$$
\begin{array}{ccc}
i & \longrightarrow & j \\
\downarrow & & \downarrow \\
\overline{i} & \longrightarrow & k
\end{array}
\tag{93}
$$

Definition 6.2. A factor map π as above is *right e-resolving* if it satisfies the definition above, with "at most one" replaced by "at least one".

Reverse the roles of i and j above to define *left resolving* and *left e-resolving*. A map π is *resolving (e-resolving)* if it is left or right resolving (e-resolving).

Proposition 6.3.

(1) *If π is resolving, then $h(\Omega_A) = h(Y)$.*
(2) *If $Y = \Omega_B$ and $h(\Omega_A) = h(\Omega_B)$, then π is e-resolving if and only if π is resolving.*
(3) *If π is e-resolving, then Y is a one-step shift of finite type, Ω_B.*
(4) *If π is e-resolving and $k \in \mathbb{N}$, then every k-step Markov measure on $Y = \Omega_B$ lifts to a k-step Markov measure on Ω_A.*

Proof. (1) This holds because a resolving map must be finite-to-one [47, 43].

(2) We argue as in [47, 43]. Suppose that π is right resolving. This means precisely that $AU \leq UB$. If $AU \neq UB$, then it would be possible to increase some entry of A by one and have a resolving map onto Ω_R from some irreducible SFT Ω_C properly containing Ω_A. But now $h(\Omega_C) > h(\Omega_A)$, while $h(\Omega_C) = h(\Omega_B) = h(\Omega_A)$ because the resolving maps respect entropy. This is a contradiction. The other direction holds by a similar argument.

(3) This is an easy exercise [11].

(4) We consider $k = 1$ (the general case follows by passage to the higher block presentation). Suppose that π is right e-resolving. This means that $AU \geq UB$. Suppose that Q is a stochastic matrix defining a one-step Markov measure μ on Ω_B. For each positive entry $B(k, \ell)$ of B and i such that $\pi(i) = k$, let $\mathcal{J}(i, k, l)$ be the set of indices j such that $A(i,j) > 0$ and $\pi(j) = \ell$. Now simply choose P to be any nonnegative matrix of size and zero/positive pattern matching A such that for each i, k, l, $\sum_{j \in \mathcal{J}(i,k,l)} P(i,j) = Q(k, \ell)$. Then $PU = UQ$, and this guarantees that $\pi \mu = \nu$. The condition on the +/0 pattern guarantees that μ has full support on Ω_A. (The code π in Example 3.4 is right e-resolving, and (18) gives an example of this construction.) $\square$

The resolving maps, and the maps which are topologically equivalent to them (the *closing* maps), form the only class of finite-to-one maps between nonconjugate irreducible shifts of finite type which we know how to construct in significant generality [1, 2, 47, 43, 10]. The e-resolving maps, and the maps topologically equivalent to them (the *continuing* maps), are similarly the Markovian maps we know how to construct in significant generality [11]. If Ω_A, Ω_B are mixing shifts of finite type with $h(\Omega_A) > h(\Omega_B)$ and there exists any factor map from Ω_A to Ω_B (as there will given a trivially necessary condition), then there will exist infinitely many continuing (hence Markovian) factor maps from Ω_A to Ω_B. However, the most obvious hope, that the factor map send the maximal entropy measure of Ω_A to that of Ω_B, can rarely be realized. Given Ω_A, there are only finitely many possible values of topological entropy for Ω_B for which such a map can exist [11].

6.2 All factor maps lift one-to-one almost everywhere to Markovian maps

Here "all factor maps" means "all factor maps between irreducible sofic subshifts". Factor maps between irreducible SFTs need not be Markovian, but they are in the following strong sense close to being Markovian, even if the subshifts X and Y are only sofic.

Theorem 6.4. [10] *Suppose that* $\pi : X \to Y$ *is a factor map of irreducible sofic subshifts. Then there are irreducible SFTs* Ω_A, Ω_B *and a commuting diagram of factor maps*

$$
\begin{array}{ccc}
\Omega_A & \xrightarrow{\ \gamma\ } & \Omega_B \\
{\scriptstyle \alpha}\downarrow & & \downarrow{\scriptstyle \beta} \\
X & \xrightarrow[\ \pi\ ]{} & Y
\end{array}
\tag{94}
$$

such that α, β *are degree one right resolving and* γ *is e-resolving. In particular,* γ *is Markovian. If* Y *is SFT, then the composition* $\beta\gamma$ *is also Markovian.*

The Markovian claims in Theorem 6.4 hold because finite-to-one maps are Markovian (Proposition 3.18), e-resolving maps are Markovian (Proposition 6.3), and a composition of Markovian maps is Markovian. In the case when π is degree one between irreducible SFTs, the "Putnam diagram" (94) is a special case of Putnam's work in [60], which was the stimulus for [10].

6.3 Every factor map between SFTs is hidden Markovian

A factor map $\pi : \Omega_A \to \Omega_B$ is Markovian if some (and therefore every) Markov measure on Ω_B lifts to a Markov measure on Ω_A. There exist factor maps between irreducible SFTs which are not Markovian. In this section we will show in contrast that all factor maps between irreducible SFTs (and more generally between irreducible sofic subshifts) are *hidden Markovian*: every sofic (i.e., hidden Markov) measure lifts to a sofic measure. The terms Markov measure and sofic measure continue to include the requirement of full topological support.

Theorem 6.5. *Let* $\pi : X \to Y$ *be a factor map between irreducible sofic subshifts and suppose that* ν *is a sofic measure on* Y. *Then* ν *lifts to a sofic measure* μ *on* X. *Moreover,* μ *can be chosen to satisfy* degree$(\mu) \leq$ degree(ν).

Proof. We consider two cases.

Case I: ν is a Markov measure on Y. Consider the Putnam diagram (94) associated to π in Theorem 6.4. The measure ν lifts to a Markov measure μ^* on Ω_A. Set $\mu = \alpha\mu^*$. Then $\pi\mu = \nu$, and degree$(\mu) = 1 \leq$ degree(ν).

Case II: ν is a degree-n sofic measure on Y. (Possibly $n = \infty$.) Then there are an irreducible SFT Ω_C with a Markov measure μ' and a degree-n factor map $g : \Omega_C \to Y$ which sends μ' to ν. By Lemma 6.8 below, there exist another irreducible SFT Ω_F and factor maps $\tilde{g}$ and $\tilde{\pi}$ with degree$(\tilde{g}) \leq$ degree(g) such that the following diagram commutes:

$$\begin{array}{ccc} \Omega_F & \xrightarrow{\ \tilde{\pi}\ } & \Omega_C \\ \tilde{g}\downarrow & & \downarrow g \\ X & \xrightarrow{\ \pi\ } & Y \end{array} \qquad (95)$$

Apply Case I to $\tilde{\pi}$ to get a degree one sofic measure ν^* on Ω_F which $\tilde{\pi}$ sends to μ'. Then $\tilde{g}(\nu^*)$ is a sofic measure of degree at most n which π sends to ν. □

To complete the proof of Theorem 6.5 by proving Lemma 6.8, we must recall some background on magic words. Suppose that $X = \Omega_A$ is SFT and $\pi : \Omega_A \to Y$ is a one-block factor map. Any X-word v is mapped to a Y-word πv of equal length. Given a Y-word $w = w[1,n]$ and an integer i in $[1,n]$, set $d(w,i) = |\{w'_i : \pi w' = w\}|$. As in [10], the *resolving degree* $\delta(\pi)$ of π is defined as the minimum of $d(w,i)$ over all allowed w, i, and w is a *magic word* for π if for some i, $d(w,i) = \delta(\pi)$. (For finite-to-one maps, these are the standard magic words of symbolic dynamics [47, 43]; some of their properties are still useful in the infinite-to-one case. The junior author confesses an error: [10, Theorem 7.1] is wrong. The resolving degree is not in general invariant under topological conjugacy, in contrast to the finite-to-one case.)

If a magic word has length 1, then it is a *magic symbol*. As remarked in [10, Lemma 2.4], the argument of [43, Proposition 4.3.2] still works in the infinite-to-one case to show that π is topologically equivalent to a one-block code from a one-step irreducible SFT for which there is a magic symbol. (Factor maps π, ϕ are topologically equivalent if there exist topological conjugacies α, β such that $\alpha\phi\beta = \pi$.)

Proposition 6.6. *Suppose that X is SFT; $\pi : X \to Y$ is a one-block factor map; a is a magic symbol for π; aQa is a Y-word; and $a'Q'a''$ is an X-word such that $\pi(a'Q'a'') = aQa$. Then the image of the cylinder $C_0[a'Q'a'']$ equals the cylinder $C_0[aQa]$.*

Proof. Suppose that $PaQaR$ is a Y-word, with preimage X-words $P^j a^j Q^j (a_*)^j R^j$, say $1 \le j \le J$, with the one-block code acting by erasing $*$ and superscripts. Because a is a magic symbol, there must exist some j such that $a_j = a'$, and there must exist some k such that $(a_*)^k = a''$. Because X is a one-step SFT, $P^j a'Q'a''R^k$ is an X-word, and it maps to $PaQaR$. This shows that the image of $C_0[a'Q'a']$ is dense in $C_0[aQa]$ and, therefore, by compactness, equal to it. □

Corollary 6.7. *Suppose that* $\pi : X \to Y$ *is a factor map from an irreducible SFT X to a sofic subshift Y. Then there is a residual set of points in Y which lift to doubly transitive points in X.*

Proof. Without loss of generality, we assume that π is a one-block factor map, X is a one-step SFT, and there is a magic symbol a for π. Let $v_n = a'P_n a'$, $n \in \mathbb{N}$, be a set of X-words such that every X-word occurs as a subset of some P_n and a' is a symbol sent to a. The set E_n of points in X which see the words $v_1, v_2, \ldots, v_n$ both in the future and in the past is a dense open subset of X. It follows from Proposition 6.6 that each πE_n is open. For every n, E_n contains E_{n+1}, so $\pi(\bigcap_n E_n) = \bigcap_n \pi E_n$. Thus, the set $\bigcap_n E_n$ of doubly transitive points in X maps to a residual subset of Y. $\qquad\square$

We do not know whether in Corollary 6.7 every doubly transitive point of Y must lift to a doubly transitive point of X.

Lemma 6.8. *Suppose that* $\alpha : X \to Z$ *and* $\beta : Y \to Z$ *are factor maps of irreducible sofic subshifts. Then there is an irreducible SFT W with factor maps* $\tilde{\alpha}$ *and* $\tilde{\beta}$ *such that* degree$(\tilde{\beta}) \le$ degree(β) *and the following diagram commutes.*

$$
\begin{array}{ccc}
W & \xrightarrow{\ \tilde{\alpha}\ } & Y \\[4pt]
{\scriptstyle \tilde{\beta}}\Big\downarrow & & \Big\downarrow{\scriptstyle \beta} \\[4pt]
X & \xrightarrow[\ \alpha\ ]{} & Z
\end{array}
\qquad (96)
$$

Proof. First, suppose that X and Y are SFT. The intersection of any two residual sets in Z is nonempty, so by Corollary 6.7 we may find x and y, doubly transitive in X and Y, respectively, such that $\alpha x = \beta y$. Let Ω_F be the irreducible component of the fiber product $\{(u, v) \in X \times Y : \alpha x = \beta y\}$ built from α and β to which the point (x, y) is forward asymptotic, and let $\tilde{\beta}, \tilde{\alpha}$ be restrictions to Ω_F of the coordinate projections. These restrictions must be surjective. Note that degree$(\tilde{\beta}) \le$ degree(β).

If X and Y are not necessarily SFT, then there are degree one factor maps from irreducible SFTs, $\rho_1 : \Omega_A \to X$ and $\rho_2 : \Omega_B \to Y$, and we can apply the first case to find $\widetilde{\alpha \rho_1}$ and $\widetilde{\beta \rho_2}$ in the diagram with respect to the pair $\alpha \rho_1, \beta \rho_2$. Now for $\tilde{\alpha}$ and $\tilde{\beta}$ we use the maps $\rho_1 \widetilde{\alpha \rho_1}$ and $\rho_2 \widetilde{\beta \rho_2}$. $\qquad\square$

Acknowledgments. This article arose from the October 2007 workshop "Entropy of Hidden Markov Processes and Connections to Dynamical Systems" at the Banff International Research Station, and we thank BIRS, PIMS, and MSRI for hospitality and support. We thank Jean-René Chazottes, Masakazu Nasu, Sujin Shin, Peter Walters, and Yuki Yayama for very helpful comments.

We are especially grateful to Uijin Jung and the two referees for extremely thorough comments and corrections. Both authors thank the Departamento de Ingeniería Matemática, Center for Mathematical Modeling, of the University of Chile and the CMM-Basal Project, and the second author also the Université Pierre et Marie Curie (University of Paris 6) and Queen Mary University of London, for hospitality and support during the preparation of this article. Much of Section 4 is drawn from lectures given by the second author in a graduate course at the University of North Carolina, and we thank the students who wrote up the notes: Rika Hagihara, Jessica Hubbs, Nathan Pennington, and Yuki Yayama.

References

[1] J. Ashley. *Resolving factor maps for shifts of finite type with equal entropy.* Ergodic Theory Dyn. Syst. **11**(2) (1991) 219–240. MR 1116638 (92d:58056)

[2] J. Ashley. *An extension theorem for closing maps of shifts of finite type.* Trans. Amer. Math. Soc. **336**(1) (1993) 389–420. MR 1105064 (93e:58048)

[3] L. E. Baum and T. Petrie. *Statistical inference for probabilistic functions of finite state Markov chains.* Ann. Math. Statist. **37** (1966) 1554–1563. MR 0202264 (34 #2137)

[4] J. Berstel and C. Reutenauer. *Rational Series and Their Languages.* Springer, Berlin, 1988

[5] Z. I. Bezhaeva and V. I. Oseledets. *Erdős measures, sofic measures, and Markov chains.* Zap. Nauchn. Sem. S.-Peterburg. Otdel. Mat. Inst. Steklov. (POMI) **326** (2005) 28–47, 279–280. MR 2183214 (2006h:60119)

[6] P. Billingsley. *Ergodic Theory and Information.* Wiley, New York, 1965. MR 0192027 (33 #254)

[7] M. Binkowska and B. Kaminski. *Classification of ergodic finitary shifts.* Ann. Sci. Univ. Clermont-Ferrand II Prob. Appl. **2** (1984) 25–37

[8] D. Blackwell. *The entropy of functions of finite state Markov chains.* In Trans. First Prague Conf. Information Theory, Statistical Decision Functions, Random Processes, 1957, pp. 13–20

[9] K. Bosch. *Notwendige und hinreichende Bedingungen dafür, daß eine Funktion einer homogenen Markoffschen Kette Markoffsch ist.* Z. Wahrscheinlichkeitsth. Geb. **31** (1974–1975) 199–202. MR 0383535 (52 #4416)

[10] M. Boyle. *Putnam's resolving maps in dimension zero.* Ergodic Theory Dyn. Syst. **25**(5) (2005) 1485–1502. MR 2173429 (2006h:37013)

[11] M. Boyle and S. Tuncel. *Infinite-to-one codes and Markov measures.* Trans. Amer. Math. Soc. **285**(2) (1984) 657–684. MR 752497 (86b:28024)

[12] C. J. Burke and M. Rosenblatt. *A Markovian function of a Markov chain.* Ann. Math. Statist. **29** (1958) 1112–1122. MR 0101557 (21 #367)

[13] J.-R. Chazottes and E. Ugalde. *Projection of Markov measures may be Gibbsian.* J. Statist. Phys. **111**(5–6) (2003) 1245–1272. MR 1975928 (2004d:37008)

[14] J.-R. Chazottes and E. Ugalde. *On the preservation of Gibbsianness under symbol amalgamation*. This volume, Chapter 2, 2011

[15] S. W. Dharmadhikari. *Functions of finite Markov chains*. Ann. Math. Statist. **34** (1963) 1022–1032. MR 0152020 (27 #2001a)

[16] S. W. Dharmadhikari. *Sufficient conditions for a stationary process to be a function of a finite Markov chain*. Ann. Math. Statist. **34** (1963) 1033–1041. MR 0152021 (27 #2001b)

[17] S. W. Dharmadhikari. *Exchangeable processes which are functions of stationary Markov chains*. Ann. Math. Statist. **35** (1964) 429–430. MR 0161370 (28 #4577)

[18] S. W. Dharmadhikari. *A characterisation of a class of functions of finite Markov chains*. Ann. Math. Statist. **36** (1965) 524–528. MR 0172333 (30 #2552)

[19] S. W. Dharmadhikari. *Splitting a single state of a stationary process into Markovian states*. Ann. Math. Statist. **39** (1968) 1069–1077. MR 0224154 (36 #7200)

[20] S. W. Dharmadhikari and M. G. Nadkarni. *Some regular and non-regular functions of finite Markov chains*. Ann. Math. Statist. **41** (1970) 207–213. MR 0263161 (41 #7766)

[21] T. Downarowicz and J. Serafin. *Fiber entropy and conditional variational principles in compact non-metrizable spaces*. Fund. Math. **172**(3) (2002) 217–247. MR 1898686 (2003b:37027)

[22] M. H. Ellis. *Lumping states of an irreducible stationary Markov chain*. Unpublished manuscript

[23] Y. Ephraim and N. Merhav. *Hidden Markov processes*. IEEE Trans. Inf. Theory **48**(6) (2002) 1518–1569. Special issue on Shannon theory: perspective, trends, and applications. MR 1909472 (2003f:94024)

[24] M. Fannes, B. Nachtergaele, and L. Slegers. *Functions of Markov processes and algebraic measures*. Rev. Math. Phys. **4**(1) (1992) 39–64. MR 1160137 (93g:82010)

[25] H. Furstenberg. *Stationary Processes and Prediction Theory*. Annals of Mathematics Studies **44**. Princeton University Press, Princeton, NJ, 1960

[26] E. J. Gilbert. *On the identifiability problem for functions of finite Markov chains*. Ann. Math. Statist. **30** (1959) 688–697

[27] L. Gurvits and J. Ledoux. *Markov property for a function of a Markov chain: a linear algebra approach*. Linear Algebra Appl. **404** (2005) 85–117. MR 2149655 (2006g:60108)

[28] O. Häggström. *Is the fuzzy Potts model Gibbsian?* Ann. Inst. H. Poincaré Prob. Statist. **39**(5) (2003) 891–917. MR 1997217 (2005f:82049)

[29] G. Han and B. Marcus. *Analyticity of entropy rate of hidden Markov chains*. IEEE Trans. Inf. Theory **52**(12) (2006) 5251–5266. MR 2300690 (2007m:62008)

[30] G. Han and B. Marcus. *Derivatives of entropy rate in special families of hidden Markov chains*. IEEE Trans. Inf. Theory **53**(7) (2007) 2642–2652. MR 2319402 (2008b:94023)

[31] G. Hansel and D. Perrin. *Rational probability measures*. Theor. Comput. Sci. **65** (1989) 171–188

[32] T. E. Harris. *On chains of infinite order*. Pacific J. Math. **5** (1955) 707–724. MR 0075482 (17,755b)

[33] G. A. Hedlund. *Endomorphisms and automorphisms of the shift dynamical system*. Math. Syst. Theory **3** (1969) 320–375. MR 0259881 #4510

[34] A. Heller. *On stochastic processes derived from Markov chains*. Ann. Math. Statist. **36** (1965) 1286–1291

[35] A. Heller. *Probabilistic automata and stochastic transformations*. Math. Syst. Theory **1** (1967) 197–208. MR 0235926 (38 #4227)

[36] P. W. Holland. *Some properties of an algebraic representation of stochastic processes*. Ann. Math. Statist. **39** (1968) 164–170. MR 0221574 (36 #4626)

[37] Y. Inagaki, T. Fukumura, and H. Matuura. *Some aspects of linear space automata*. Inf. Control **20** (1972) 439–479. MR 0332402 (48 #10729)

[38] R. B. Israel. *Convexity in the Theory of Lattice Gases*. Princeton Series in Physics. With an introduction by A. S. Wightman. Princeton University Press, Princeton, NJ, 1979. MR 517873 (80i:82002)

[39] J. G. Kemeny and J. L. Snell. *Finite Markov Chains*. Undergraduate Texts in Mathematics. Springer, New York, 1976. Reprinting of the 1960 original. MR 0410929 (53 #14670)

[40] B. Kitchens. *An invariant for continuous factors of Markov shifts*. Proc. Amer. Math. Soc. **83** (1981) 825–828. MR 0630029 (82k:28021)

[41] B. Kitchens. *Linear algebra and subshifts of finite type*. In Conf. Modern Analysis and Probability. New Haven, CT, 1982. Contemporary Mathematics **26**, American Mathematical Society, Providence, RI, 1984, pp. 231–248. MR 737405 (85m:28022)

[42] B. Kitchens and S. Tuncel, *Finitary measures for subshifts of finite type and sofic systems*. Mem. Amer. Math. Soc. **58**(338) (1985) iv + 68. MR 818917 (87h:58110)

[43] B. P. Kitchens. *Symbolic Dynamics. One-sided, two-sided and countable state Markov shifts*. Universitext. Springer, Berlin, 1998. MR 1484730 (98k:58079)

[44] S. C. Kleene. *Representation of events in nerve nets and finite automata*. In *Automata Studies* (C. E. Shannon and J. McCarthy, eds.). Princeton University Press, Princeton, NJ, 1956, pp. 3–42

[45] H. Künsch, S. Geman, and A. Kehagias. *Hidden Markov random fields*. Ann. Appl. Prob. **5**(3) (1995) 577–602. MR 1359820 (97a:60070)

[46] F. Ledrappier and P. Walters. *A relativised variational principle for continuous transformations*. J. Lond. Math. Soc. (2) **16**(3) (1977) 568–576. MR 0476995 (57 #16540)

[47] D. Lind and B. Marcus. *An Introduction to Symbolic Dynamics and Coding*. Cambridge University Press, Cambridge, 1995. MR 1369092 (97a:58050)

[48] C. Maes and K. Vande Velde. *The fuzzy Potts model*. J. Phys. A **28**(15) (1995) 4261–4270. MR 1351929 (96i:82022)

[49] B. Marcus, K. Petersen, and S. Williams. *Transmission rates and factors of Markov chains*. Contemp. Math. **26** (1984) 279–293

[50] M. Nasu. *An invariant for bounded-to-one factor maps between transitive sofic subshifts*, Ergodic Theory Dyn. Syst. **5**(1) (1985) 89–105. MR 782790 (86i:28030)

[51] W. Parry. *Intrinsic Markov chains*. Trans. Amer. Math. Soc. **112** (1964) 55–66. MR 0161372 (28 #4579)

[52] W. Parry and S. Tuncel. *Classification Problems in Ergodic Theory*. London Mathematical Society Lecture Note Series **67**. Cambridge University Press, Cambridge, 1982. MR 666871 (84g:28024)

[53] W. Parry and S. Tuncel. *On the stochastic and topological structure of Markov chains*. Bull. Lond. Math. Soc. **14**(1) (1982) 16–27. MR 642417 (84i:28024)

[54] A. Paz. *Introduction to Probabilistic Automata.* Academic Press, New York, 1971. MR 0289222

[55] K. Petersen. *Symbolic Dynamics.* http://www.math.unc.edu/Faculty/petersen/ m261s98.pdf, 1998

[56] K. Petersen. *Ergodic Theory.* Cambridge Studies in Advanced Mathematics **2**. Cambridge University Press, Cambridge, 1989. Corrected reprint of the 1983 original. MR 1073173 (92c:28010)

[57] K. Petersen, A. Quas, and S. Shin. *Measures of maximal relative entropy.* Ergodic Theory Dyn. Syst. **23**(1) (2003) 207–223. MR 1971203 (2004b:37009)

[58] K. Petersen and S. Shin. *On the definition of relative pressure for factor maps on shifts of finite type.* Bull. Lond. Math. Soc. **37**(4) (2005) 601–612. MR 2143740 (2005m:37066)

[59] R. R. Phelps. *Unique equilibrium states.* In Dynamics and Randomness, Santiago, 2000. Nonlinear Phenomena and Complex Systems **7**. Kluwer Academic Dordrecht, 2002, pp. 219–225. MR 1975579 (2004c:28028)

[60] I. F. Putnam. *Lifting factor maps to resolving maps.* Israel J. Math. **146** (2005) 253–280. MR 2151603 (2007i:37020)

[61] M. O. Rabin. *Probabilistic automata.* Inf. Control **6** (1963) 230–245

[62] G. Rubino and B. Sericola. *On weak lumpability in Markov chains.* J. Appl. Prob. **26**(3) (1989) 446–457. MR 1010934 (90j:60069)

[63] G. Rubino and B. Sericola. *A finite characterization of weak lumpable Markov processes. I. The discrete time case.* Stoch. Process. Appl. **38**(2) (1991) 195–204. MR 1119981 (92g: 60092)

[64] A. Schönhuth. *Equations for hidden Markov models.* Preprint (2009)

[65] M. P. Schützenberger. *On the definition of a family of automata.* Inf. Control **4** (1961) 245–270

[66] C. E. Shannon and W. Weaver. *The Mathematical Theory of Communication.* The University of Illinois Press, Urbana, IL, 1949. MR 0032134 (11,258e)

[67] S. Shin. *An example of a factor map without a saturated compensation function.* Ergodic Theory Dyn. Syst. **21**(6) (2001) 1855–1866. MR 1869074 (2002h: 37020)

[68] S. Shin. *Measures that maximize weighted entropy for factor maps between subshifts of finite type.* Ergodic Theory Dyn. Syst. **21**(4) (2001) 1249–1272. MR 1849609 (2002i:37009)

[69] S. Shin. *Relative entropy functions for factor maps between subshifts.* Trans. Amer. Math. Soc. **358**(5) (2006) 2205–2216 (electronic). MR 2197440 (2006i: 37026)

[70] S. Tuncel. *Conditional pressure and coding.* Israel J. Math. **39**(1–2) (1981) 101–112. MR 617293 (82j:28012)

[71] P. Walters. *An Introduction to Ergodic Theory.* Graduate Texts in Mathematics **79**. Springer, New York, 1982. MR 648108 (84e:28017)

[72] P. Walters. *Relative pressure, relative equilibrium states, compensation functions and many-to-one codes between subshifts.* Trans. Amer. Math. Soc. **296**(1) (1986) 1–31. MR 837796 (87j:28028)

[73] B. Weiss. *Subshifts of finite type and sofic systems.* Monatsh. Math. **77** (1973) 462–474. MR 0340556 (49 #5308)

[74] Y. Yayama. *Existence of a measurable saturated compensation function between subshifts and its applications.* Preprint (2009)

[75] J. Yoo. *Measures of maximal relative entropy with full support.* Preprint (2009)

Further reading

A. M. Abdel-Moneim and F. W. Leysieffer. *Weak lumpability in finite Markov chains.* J. Appl. Prob. **19**(3) (1982) 685–691. MR 664854 (84a:60077)

A. M. Abdel-Moneim and F. W. Leysieffer. *Lumpability for nonirreducible finite Markov chains.* J. Appl. Prob. **21**(3) (1984) 567–574. MR 752021 (85k:60089)

R. Ahmad. *An algebraic treatment of Markov processes.* In Trans. Seventh Prague Conf. Information Theory, Statistical Decision Functions, Random Processes and Eighth European Meeting Statisticians, Technical University of Prague, Prague, 1974, Vol. A. Reidel, Dordrecht, 1977, pp. 13–22. MR 0488303 (58 #7854)

M. Arbib. *Realization of stochastic systems.* Ann. Math. Statist. **38** (1967) 927–933 MR 0225606 (37 #1199)

F. Bancilhon. *A geometric model for stochastic automata.* IEEE Trans. Comput. **C-23**(12) (1974) 1290–1299. MR 0406736 (53 #10522)

P. Billingsley. *Probability and Measure.* Wiley, New York, 1995.

D. Blackwell and L. Koopmans. *On the identifiability problem for functions of finite Markov chains.* Ann. Math. Statist. **28** (1957) 1011–1015. MR 0099081 (20 #5525)

P. E. Boudreau. *Functions of finite Markov chains and exponential type processes.* Ann. Math. Statist. **39** (1968) 1020–1029. MR 0224161 (36 #7207)

C. Burke and M. Rosenblatt. *Consolidation of probability matrices.* Bull. Inst. Int. Statist. **36**(3) (1958) 7–8. MR 0120680 (22 #11429)

T. Downarowicz and R.D. Mauldin. *Some remarks on output measures.* Topol. Appl. **152** (2005) 11–25

M. H. Ellis. *The $\bar{d}$-distance between two Markov processes cannot always be attained by a Markov joining.* Israel J. Math. **24**(3–4) (1976) 269–273. MR 0414820 (54 #2912)

R. V. Erickson. *Functions of Markov chains.* Ann. Math. Statist. **41** (1970) 843–850. MR 0264769 (41 #9360)

M. Fox and H. Rubin. *Functions of processes with Markovian states.* Ann. Math. Statist. **39** (1968) 938–946. MR 0232450 (38 #775)

M. Fox and H. Rubin. *Functions of processes with Markovian states. II.* Ann. Math. Statist. **40** (1969) 865–869. MR 0243607 (39 #4928)

M. Fox and H. Rubin. *Functions of processes with Markovian states. III.* Ann. Math. Statist. **41** (1970) 472–479. MR 0258099 (41 #2746)

J. Hachigian and M. Rosenblatt. *Functions of reversible Markov processes that are Markovian.* J. Math. Mech. **11** (1962) 951–960. MR 0145588 (26 #3118)

F. P. Kelly. *Markovian functions of a Markov chain.* Sankhyā Ser. A **44**(3) (1982) 372–379. MR 705461 (85d:60129)

Y. Komota and M. Kimura. *A characterization of the class of structurally stable probabilistic automata. I. Discrete-time case.* Int. J. Syst. Sci. **9**(4) (1978) 369–394. MR 0490599 (58 #9937a)

Y. Komota and M. Kimura. *A characterization of the class of structurally stable probabilistic automata. II. Continuous-time case.* Int. J. Syst. Sci. **9**(4) (1978) 395–424. MR 0490600 (58 #9937b)

Y. Komota and M. Kimura. *On Markov chains generated by Markovian controlled Markov systems: structural stability.* Int. J. Syst. Sci. **12**(7) (1981) 835–854. MR 626281 (83c:93060)

F. W. Leysieffer. *Functions of finite Markov chains.* Ann. Math. Statist. **38** (1967) 206–212. MR 0207043 (34 #6859)

R. W. Madsen. *Decidability of $\alpha(P^k) > 0$ for some k.* J. Appl. Prob. **12** (1975) 333–340. MR 0373011 (51 #9213)

D. L. Neuhoff and P. C. Shields. *Indecomposable finite state channels and primitive approximation.* IEEE Trans. Inf. Theory **28**(1) (1982) 11–18. MR 651096 (83k:94025)

A. Paz. *Word functions of pseudo-Markov chains.* Linear Algebra Appl. **10** (1975) 1–5. MR 0388543 (52 #9379)

J. B. Robertson. *The mixing propoerties of certain processes related to Markov chains.* Math. Syst. Theory **7** (1973) 39–43

J. B. Robertson. *A spectral representation of the states of a measure preserving transformation.* Z. Wahrscheinlichkeitsth. Geb. **27** (1973) 185–194

C. B. Silio, Jr. *An efficient simplex coverability algorithm in E^2 with application to stochastic sequential machines.* IEEE Trans. Comput. **28**(2) (1979) 109–120. MR 519218 (80d:68069)

2

On the preservation of Gibbsianness
under symbol amalgamation

JEAN-RENÉ CHAZOTTES

Centre de Physique Théorique, CNRS École Polytechnique,
91128 Palaisau Cedex, France
E-mail address: jeanrene@cpht.polytechnique.fr

EDGARDO UGALDE

Instituto de Física, Universidad Autónoma de San Luis Potosí,
San Luis de Potosí, S.L.P., 78290 México
E-mail address: ugalde@ifisica.uaslp.mx

Abstract. Starting from the full shift on a finite alphabet A, by mingling some symbols of A, we obtain a new full shift on a smaller alphabet B. This amalgamation defines a factor map from $(A^{\mathbb{N}}, T_A)$ to $(B^{\mathbb{N}}, T_B)$, where T_A and T_B are the respective shift maps. According to the thermodynamic formalism, to each regular function ("potential") $\psi : A^{\mathbb{N}} \to \mathbb{R}$, we can associate a unique Gibbs measure μ_ψ. In this article, we prove that, for a large class of potentials, the pushforward measure $\mu_\psi \circ \pi^{-1}$ is still Gibbsian for a potential $\phi : B^{\mathbb{N}} \to \mathbb{R}$ having a "bit less" regularity than ψ. In the special case where ψ is a "two-symbol" potential, the Gibbs measure μ_ψ is nothing but a Markov measure and the amalgamation π defines a hidden Markov chain. In this particular case, our theorem can be recast by saying that a hidden Markov chain is a Gibbs measure (for a Hölder potential).

1 Introduction

From different viewpoints and under different names, the so-called *hidden Markov measures* have received a lot of attention in the last fifty years [3]. One considers a (stationary) Markov chain $(X_n)_{n \in \mathbb{N}}$ with finite state space A and looks at its "instantaneous" image $Y_n := \pi(X_n)$, where the map π is an amalgamation of the elements of A yielding a smaller state space, say B. It is well known that in general the resulting chain, $(Y_n)_{n \in \mathbb{N}}$, has infinite memory. For concrete examples, see e.g., [1] or the more easily accessible reference [3], where they are recalled.

Entropy of Hidden Markov Processes and Connections to Dynamical Systems: Papers from the Banff International Research Station Workshop, ed. B. Marcus, K. Petersen, and T. Weissman. Published by Cambridge University Press. © Cambridge University Press 2011.

A stationary Markov chain with finite state space A can be equivalently defined as a shift-invariant Markov measure μ on the path space $A^{\mathbb{N}}$ (of infinite sequences of "symbols" from the finite "alphabet" A), where the shift map $T : A^{\mathbb{N}} \to A^{\mathbb{N}}$ is defined by $(Ta)_i = a_{i+1}$. A hidden Markov measure can be therefore seen as the pushforward measure $\mu_{\psi} \circ \pi^{-1}$ on the path space $B^{\mathbb{N}}$ formed by the instantaneous image under the amalgamation π of paths in $A^{\mathbb{N}}$.

In the present article, instead of focusing on shift-invariant Markov measures, we consider a natural generalization of them. Let $\psi : A^{\mathbb{N}} \to \mathbb{R}$ be a "potential"; then, under an appropriate regularity condition on ψ (see more details below), there is a unique so-called Gibbs measure μ_{ψ} associated to it. It is a shift-invariant probability measure on $A^{\mathbb{N}}$ with remarkably nice properties. Each r-step Markov measure falls in this category, since an r-step Markov measure is nothing but a Gibbs measure defined by an $(r+1)$-symbol potential, i.e., a potential ψ such that $\psi(a) = \psi(\tilde{a})$ whenever $a_i = \tilde{a}_i$, $i = 0,\ldots,r$, with r a strictly positive integer.[1] On the other hand, given ψ one can construct a sequence (ψ_r) of $(r+1)$-symbol potentials (uniformly approximating ψ) such that the sequence of associated r-step Markov measures μ_{ψ_r} converges to μ_{ψ} (in the vague or weak* topology, at least).

Now let B be the alphabet obtained from A by amalgamation of some of the symbols of A.[2] The amalgamation defines a surjective (i.e., onto) map $\pi : A \to B$ which extends to $A^{\mathbb{N}}$ in the obvious way. Given a Gibbs measure μ_{ψ} on $A^{\mathbb{N}}$, this map induces a measure $\mu_{\psi} \circ \pi^{-1}$ supported on the full shift $B^{\mathbb{N}}$. The question we address now reads as follows.

Question 1.1. Under which condition is the measure $\mu_{\psi} \circ \pi^{-1}$, supported on the full shift $B^{\mathbb{N}}$, still Gibbsian? In other words, under which conditions on ψ can one build a "nice" potential $\phi : B^{\mathbb{N}} \to \mathbb{R}$ such that $\mu_{\psi} \circ \pi^{-1} = \mu_{\phi}$? In particular, for ψ a two-symbol potential, what is the nature of $\mu_{\psi} \circ \pi^{-1}$?

In this article we make the following answer (made precise below, see Theorems 3.1 and 4.1):

> Under a mild regularity condition on ψ, the pushforward of the Gibbs measure μ_{ψ}, namely $\mu_{\psi} \circ \pi^{-1}$, is Gibbsian as well, and the associated potential ϕ can be computed from ψ. Furthermore, when ψ is a two-symbol potential, the corresponding hidden Markov chain is Gibbsian, and it is associated to a Hölder potential.

A slightly more general problem is the following. Suppose that we do not start with the full shift $A^{\mathbb{N}}$ but with a *subshift of finite type* (henceforth SFT) or a

---

[1] The case $r = 0$ corresponds to product measures (independent and identically distributed (i.i.d.) process).

[2] We assume that B has cardinality at least equal to two.

topological Markov chain X [10]. The image of X is not in general of finite type but it is a *sofic subshift* [10].

Question 1.2. When $X \subseteq A^{\mathbb{N}}$ is an SFT, is the measure $\mu_\psi \circ \pi^{-1}$ still Gibbsian?

Question 1.2 has only received very partial answers to date. We shall comment on that in Section 5.

The present work is motivated, on the one hand, by our previous work in [5] in which we attempted to solve Question 1.2 and were partially successful. On the other hand, it was motivated by [6], where we were interested in approximating Gibbs measures on sofic subshifts by Markov measures on subshifts of finite type. Here we combine ideas and techniques from both [5] and [6], but we need extra work to get more uniformity than previously obtained.

Let us mention two recent works related to ours. In [14], another kind of transformation of the alphabet is considered, and the method employed to prove Gibbsianity is completely different from ours. In [7], the authors study random functions of Markov chains and obtain results about their loss of memory.

The paper is organized as follows. In the next section we give some notation and definitions. In particular, we present the weak* convergence of measures as a projective convergence and we define the notion of Markov approximants of a Gibbs measure. In Section 3, we state Theorem 3.1, which answers Question 1.1 when the starting potential ψ is Hölder continuous (its modulus of continuity decays exponentially to 0). The proof relies on two lemmas, which are proved in Appendices 6.2 and 6.3, respectively. In Section 4, we generalize Theorem 3.1 to a class of potentials with subexponential (strictly subexponential or polynomial) decay of modulus of continuity. We finish in Section 5 by discussing Question 1.2 and giving a conjecture. Appendix 6.1 is devoted to Birkhoff's version of the Perron–Frobenius theorem [16] for positive matrices, our main tool.

2 Background material

2.1 Symbolic dynamics

Let A be a finite set ("alphabet") and $A^{\mathbb{N}}$ be the set of infinite sequences of symbols drawn from A. We define $\mathbb{N}$ to be the set $\{0, 1, 2, \ldots\}$, that is, the set of positive integers plus 0. We denote by $\boldsymbol{a}, \boldsymbol{b}$, etc., elements of $A^{\mathbb{N}}$ and use the notation $\boldsymbol{a}_m^n$ ($m \le n$, $m, n \in \mathbb{N}$) for the word $\boldsymbol{a}_m \boldsymbol{a}_{m+1} \cdots \boldsymbol{a}_{n-1} \boldsymbol{a}_n$ (of length

$n - m + 1$). We endow $A^{\mathbb{N}}$ with the distance

$$d_A(\boldsymbol{a}, \boldsymbol{b}) := \begin{cases} \exp(-\min\{n \geq 0 : a_0^n \neq b_0^n\}) & \text{if } \boldsymbol{a} \neq \boldsymbol{b}, \\ 0 & \text{otherwise.} \end{cases}$$

The resulting metric space $(A^{\mathbb{N}}, d_A)$ is compact.

The *shift transformation* $T : A^{\mathbb{N}} \to A^{\mathbb{N}}$ is defined by $(T\boldsymbol{a})_n = a_{n+1}$ for all $n \in \mathbb{N}$. A *subshift* X of $A^{\mathbb{N}}$ is a closed T-invariant subset of $A^{\mathbb{N}}$.

Given a set of *admissible words* $\mathcal{L} \subset A^\ell$ for some fixed integer $\ell \geq 2$, one defines a *subshift of finite type* $A_{\mathcal{L}} \subset A^{\mathbb{N}}$ by

$$A_{\mathcal{L}} := \{\boldsymbol{a} \in A^{\mathbb{N}} : a_n^{n+\ell-1} \in \mathcal{L}, \ \forall n \in \mathbb{N}\}.$$

A subshift of finite type defined by words in $\mathcal{L} \subset A^2$ is called a *topological Markov chain*. It can be equivalently described by the transition matrix $M : A \times A \to \{0, 1\}$ such that $M(a, b) = \chi_{\mathcal{L}}(ab)$, where $\chi_{\mathcal{L}}$ is the indicator function of the set $\mathcal{L}$. We will use both $A_{\mathcal{L}}$ and A_M to denote the corresponding subshift of finite type.

Note that the "full shift" $(A^{\mathbb{N}}, T)$ can be seen as the subshift of finite type defined by all the words of length ℓ, and we have the identification $A^{\mathbb{N}} \equiv A_{A^\ell}$.

Let $X \subset A^{\mathbb{N}}$ be a subshift. A point $\boldsymbol{a} \in X$ is *periodic* with period $p \geq 1$ if $T^p \boldsymbol{a} = \boldsymbol{a}$, and p is its minimal period if in addition $T^k \boldsymbol{a} \neq \boldsymbol{a}$ whenever $0 < k < p$. We denote by $\mathrm{Per}_p(X)$ the collection of all periodic points with period p in X, and by $\mathrm{Per}(X)$ the collection of all periodic points in X, i.e., $\mathrm{Per}(X) = \bigcup_{p \geq 1} \mathrm{Per}_p(X)$.

Given an arbitrary subshift $X \subset A^{\mathbb{N}}$ and $m \in \mathbb{N}$, the set of all the X-*admissible words of length* $m + 1$ is the set

$$X_m := \{w \in A^{m+1} : \exists\, \boldsymbol{a} \in X, \ w = a_0^m\}.$$

It is a well-known fact that a topological Markov chain A_M is topologically mixing if its transition matrix M is primitive, i.e., if and only if there exists an integer $n \geq 1$ such that $M^n > 0$.[3] In this case, the smallest of such integers is the so-called primitivity index of M.

For a subshift $X \subset A^{\mathbb{N}}$, $w \in X_m$, and $m \in \mathbb{N}$, the set

$$[w] := \{\boldsymbol{a} \in X : a_0^m = w\}$$

is the *cylinder* based on w.

We will use boldfaced symbols $\boldsymbol{a}, \boldsymbol{b}$, etc., not only for infinite sequences but also for finite ones (i.e., for words). The context will make clear whether we deal with a finite or an infinite sequence.

---

[3] On the other hand, if none of the rows or columns of M is identically zero, A_M being topologically mixing implies that M is primitive.

2.2 Thermodynamic formalism

For a subshift $X \subset A^{\mathbb{N}}$, cylinders are clopen sets and generate the Borel σ-algebra. We denote by $\mathcal{M}(X)$ the set of Borel probability measures on X and by $\mathcal{M}_T(X)$ *the subset of T-invariant probability measures* on X. Both are compact convex sets in weak* topology. The weak* topology can be metrized [2] by the distance

$$D(\mu, \nu) := \sum_{m=0}^{\infty} 2^{-(m+1)} \left(\sum_{w \in X_m} |\mu[w] - \nu[w]| \right).$$

It turns out that the following notion of convergence is very convenient in our later calculations.

Definition 2.1. We say that a sequence $(\mu_n)_{n \in \mathbb{N}}$ of probability measures in $\mathcal{M}(X)$ *converges in the projective sense* to a measure $\mu \in \mathcal{M}(X)$ if for all $\epsilon > 0$ and $N > 1$ there exists $N' > 1$ such that

$$\exp(-\epsilon) \leq \frac{\mu_n[w]}{\mu[w]} \leq \exp(\epsilon)$$

for all admissible words w of length $k \leq N$, and for all $n \geq N'$.

It is easy to verify that convergence in the projective sense implies weak* convergence. On the other hand, when all the measures involved share the same support, weak* and projective convergence coincide. Though it is the case in this article, we will speak of projective convergence.

We make the following definitions.

Definition 2.2. $((r+1)$-symbol potentials) A function $\psi : A^{\mathbb{N}} \to \mathbb{R}$ will be called a *potential*. We say that a potential $\psi : A^{\mathbb{N}} \to \mathbb{R}$ is an $(r+1)$-symbol potential if there is an $r \in \mathbb{N}$ such that

$$\psi(\boldsymbol{a}) = \psi(\boldsymbol{b}) \text{ whenever } \boldsymbol{a}_0^r = \boldsymbol{b}_0^r.$$

Of course, we take r to be the smallest integer with this property.

We will say that ψ is *locally constant* if it is an $(r+1)$-symbol potential for some $r \in \mathbb{N}$.

A way of quantifying the regularity of a potential $\psi : A^{\mathbb{N}} \to \mathbb{R}$ is by using its modulus of continuity on cylinders, or variation, defined by

$$\mathrm{var}_n \psi := \sup\{|\psi(\boldsymbol{a}) - \psi(\boldsymbol{b})| : \boldsymbol{a}, \boldsymbol{b} \in A^{\mathbb{N}}, \boldsymbol{a}_0^n = \boldsymbol{b}_0^n\}.$$

A potential ψ is continuous if and only if $\mathrm{var}_n \psi \to 0$ as $n \to \infty$. An $(r+1)$-symbol potential ψ can be alternatively defined by requiring that $\mathrm{var}_n \psi = 0$

whenever $n \geq r$, and thus it is trivially continuous. If there are $C > 0$ and $\varrho \in \,]0, 1[$ such that $\text{var}_n \phi \leq C\varrho^n$ for all $n \geq 0$, then ψ is said to be Hölder continuous.

We will use the notation

$$S_n \psi(a) := \sum_{k=0}^{n-1} \psi \circ T^k(a), \quad n = 1, 2, \ldots.$$

Throughout, we will write

$$x \lesseqgtr y \, C^{\pm 1} \quad \text{for} \quad C^{-1} \leq \frac{x}{y} \leq C$$

for x, y, and C strictly positive numbers. Accordingly, we will use the notation $x \lesseqgtr y \exp(\pm C)$. We also write $x \lesseqgtr y \pm C$ for $-C \leq x - y \leq C$.

We now define the notion of Gibbs measure we will use in the sequel.

Definition 2.3. (Gibbs measures) Let $X \subset A^{\mathbb{N}}$ be a subshift and $\psi : A^{\mathbb{N}} \to \mathbb{R}$ be a potential such that $\psi|_X$ is continuous. A measure $\mu \in \mathcal{M}_T(X)$ is a *Gibbs measure for the potential* ψ if there are constants $C = C(\psi, X) \geq 1$ and $P = P(\psi, X) \in \mathbb{R}$ such that

$$\frac{\mu[a_0^n]}{\exp(S_{n+1}\psi(a) - (n+1)P)} \lesseqgtr C^{\pm 1} \tag{1}$$

for all $n \in \mathbb{N}$ and $a \in X$. We denote by μ_ψ such a measure.

The constant $P = P(\psi, X)$ is the *topological pressure* [9] of X with respect to ψ. It can be obtained, for X a subshift of finite type, as follows:

$$P(\psi, X) = \limsup_{n \to \infty} \frac{1}{n} \log \sum_{a \in \text{Per}_n(X)} \exp(S_n \psi(a)). \tag{2}$$

We will say that *the potential* ψ *is normalized on* X if $P(\psi, X) = 0$. We can always normalize a potential ψ by replacing ψ by $\psi - P(\psi, X)$. This does not affect the associated Gibbs measure μ_ψ.

In the above definition, we allow that $\psi = -\infty$ on $A^{\mathbb{N}} \backslash X$. In other words, ψ is upper semicontinuous on $A^{\mathbb{N}}$.

Remark 2.4. If $\mu \in \mathcal{M}_T(X)$ is such that the sequence $\left(\log(\mu[a_0^n]/\mu[a_1^n])\right)_{n=1}^{\infty}$ converges uniformly in $a \in X$, then the potential $\psi : X \to \mathbb{R}$ given by

$$\psi(a) = \lim_{n \to \infty} \log \left(\frac{\mu[a_0^n]}{\mu[a_1^n]} \right) \tag{3}$$

is continuous on X, and μ is a Gibbs measure with respect to ψ, i.e., $\mu = \mu_\psi$. Furthermore, ψ is such that $P(\psi) = 0$.

Notice that $\mu[a_0^n]/\mu[a_1^n]$ is nothing but the probability under μ of a_0 given a_1^n. Therefore, by the martingale convergence theorem [8], the sequence $\left(\log(\mu[a_0^n]/\mu[a_1^n])\right)_{n=1}^\infty$ converges for μ-a.e. $a \in X$. The uniform convergence is what makes μ a Gibbs measure.

We have the following classical theorem.

Theorem 2.5. [12] *Let $X \subset A^\mathbb{N}$ be a topologically mixing subshift of finite type and $\psi : X \to \mathbb{R}$. If*

$$\sum_{n=0}^\infty \mathrm{var}_n \psi < \infty, \tag{4}$$

then there exists a unique Gibbs measure μ_ψ, i.e., a unique T-invariant probability measure satisfying (1).

Remark 2.6. By this theorem, we have a partial converse to (3) in the sense that there the potential is defined by the measure, while in the theorem it is the potential which defines the measure.

Notice that the uniqueness part of the theorem is granted by the Gibbs inequality (1), since two measures satisfying it have to be absolutely continuous with respect to each other. It is the existence part which is nontrivial.

For a proof of Theorem 2.5, see e.g., [9]. This includes the case of Hölder continuous potentials treated in, e.g., [2, 15].

2.3 Markov measures and Markov approximants

Markov measures can be seen as Gibbs measures. Colloquially, an r-step Markov measure is defined by the property that the probability that $a_n = a \in A$ given a_0^{n-1} depends only on a_{n-r}^{n-1}.[4] What is usually called a Markov measure corresponds to one-step Markov measures. On the full shift, the case $r = 0$ gives product measures. A T-invariant probability measure is an r-step Markov measure if and only if it is the Gibbs measure of an $(r+1)$-symbol potential. Given an $(r+1)$-symbol potential ψ, which we identify as a function on A^{r+1}, one can define the transition matrix $\mathcal{M}_\psi : A^r \times A^r \to \mathbb{R}^+$ such that

$$\mathcal{M}_\psi(v,w) := \begin{cases} \exp(\psi(vw_{r-1})) & \text{if } v_1^{r-1} = w_0^{r-2}, \\ 0 & \text{otherwise.} \end{cases}$$

By vw_{r-1}, we mean the word obtained by concatenation of v and w_{r-1} (the last letter of w).

---

[4] We assume that $r \geq 1$. The case $r = 0$ corresponds to an i.i.d. process, in which case the Gibbsianity is evident.

By the Perron–Frobenius theorem (cf. Appendix 6.1), there exist a right eigenvector $\bar{R}_\psi > 0$ such that $\sum_{a \in A^r} \bar{R}_\psi(a) = 1$, and a left eigenvector $\bar{L}_\psi > 0$ such that $\bar{L}_\psi^\dagger \bar{R}_\psi = 1$, associated to the maximal eigenvalue $0 < \rho_\psi := \max \mathrm{spec}(\mathcal{M}_\psi)$. Then the measure μ defined by

$$\mu[a_0^n] := \bar{L}_\psi(a_0^{r-1}) \, \frac{\prod_{j=0}^{n-r} \mathcal{M}_\psi(a_j^{j+r-1}, a_{j+1}^{j+r})}{\rho_\psi^{n-r+1}} \, \bar{R}_\psi(a_{n-r+1}^n) \tag{5}$$

for each $a \in A^{\mathbb{N}}$ and $n \in \mathbb{N}$ such that $n \geq r$ is easily seen to be a T-invariant probability measure satisfying (1) with

$$P = \log(\rho_\psi) \quad \text{and} \quad C = \rho_\psi^r \, e^{-r\|\psi\|} \, \frac{\max\{\bar{L}_\psi(w)\bar{R}_\psi(w') : w, w' \in A^r\}}{\min\{\bar{L}_\psi(w)\bar{R}_\psi(w') : w, w' \in A^r\}},$$

where $\|\psi\| := \sup\{|\psi(a)| : a \in A^{\mathbb{N}}\}$. Therefore, $\mu = \mu_\psi$ is the unique Gibbs measure associated to the $(r+1)$-symbol potential ψ.

2.4 Markov and locally constant approximants

Given a continuous $\psi : A^{\mathbb{N}} \to \mathbb{R}$, one can uniformly approximate it by a sequence of $(r+1)$-symbol potentials ψ_r, $r = 1, 2, \ldots$, in such a way that $\|\psi - \psi_r\| \leq \mathrm{var}_r(\psi)$, which goes to 0 as r goes to ∞. The ψ_r are not defined in a unique way but this does not matter since the associated r-step Markov measures μ_{ψ_r} approximate the same Gibbs measure μ_ψ. We can choose $\psi_r(a) := \max\{\psi(b) : b \in [a_0^r]\}$ for instance.

The potential ψ_r will be called the $(r+1)$-symbol approximant of ψ and the associated r-step Markov measure μ_{ψ_r} will be the *rth Markov approximant* of μ_ψ. It is well known (and not difficult to prove) that μ_{ψ_r} converges in the weak* topology to μ_ψ.

3 Main result

The next theorem answers Question 1.1 when ψ is Hölder continuous (Theorem 3.1). For the sake of simplicity, we discuss the generalization of that theorem to a class of less regular potentials (i.e., $\mathrm{var}_n(\psi)$ decreases subexponentially or polynomially) in Section 4.

Amalgamation map. Let A, B be two finite alphabets, with $\mathrm{Card}(A) > \mathrm{Card}(B)$, and $\pi : A \to B$ be a surjective map ("amalgamation") which extends to the map $\pi : A^{\mathbb{N}} \to B^{\mathbb{N}}$ (we use the same letter for both) such that $(\pi a)_n = \pi(a_n)$

for all $n \in \mathbb{N}$. The map π is continuous and shift commuting, i.e., it is a factor map from $A^{\mathbb{N}}$ onto $B^{\mathbb{N}}$.

Theorem 3.1. *Let* $\pi : A^{\mathbb{N}} \to B^{\mathbb{N}}$ *be the amalgamation map just defined and* $\psi : A^{\mathbb{N}} \to \mathbb{R}$ *be a Hölder continuous potential. Then the measure* $\mu_\psi \circ \pi^{-1}$ *is a Gibbs measure with support* $B^{\mathbb{N}}$ *for a potential* $\phi : B^{\mathbb{N}} \to \mathbb{R}$ *such that*

$$\mathrm{var}_n(\phi) \leq \mathcal{D} \exp(-c\sqrt{n})$$

for some $c, \mathcal{D} > 0$, *and all* $n \in \mathbb{N}$.

Furthermore, this potential $\phi : B^{\mathbb{N}} \to \mathbb{R}$ *is normalized and it is given by*

$$\phi(\boldsymbol{b}) = \lim_{r \to \infty} \lim_{n \to \infty} \log \left(\frac{\mu_{\psi_r} \circ \pi^{-1}[\boldsymbol{b}_0^n]}{\mu_{\psi_r} \circ \pi^{-1}[\boldsymbol{b}_1^n]} \right), \qquad (6)$$

where ψ_r *is the* $(r+1)$-*symbol approximant of* ψ.

If ψ *is locally constant, then, for all* n,

$$\mathrm{var}_n(\phi) \leq C \vartheta^n,$$

where $\vartheta \in \,]0, 1[, \, C > 0$.

The case of locally constant potentials in the theorem can be rephrased as follows: when μ_ψ is an r-step Markov measure, with $r > 0$, the pushforward measure $\mu_\psi \circ \pi^{-1}$, i.e., the hidden Markov measure, is a Gibbs measure for a Hölder continuous potential ϕ given by

$$\phi(\boldsymbol{b}) = \lim_{n \to \infty} \log \left(\frac{\mu_\psi \circ \pi^{-1}[\boldsymbol{b}_0^n]}{\mu_\psi \circ \pi^{-1}[\boldsymbol{b}_1^n]} \right). \qquad (7)$$

The case $r = 0$ is trivial: the Gibbs measure is simply a product measure and its pushforward is also a product measure.

The proof of Theorem 3.1 relies on the following two lemmas, whose proofs are deferred to Appendices 6.2 and 6.3.

Lemma 3.2. (Amalgamation for $(r+1)$-symbol potentials) *The measure* $\mu_{\psi_r} \circ \pi^{-1}$, *with* $r > 0$, *is a Gibbs measure for the potential* $\phi_r : B^{\mathbb{N}} \to \mathbb{R}$ *obtained as the following limit:*

$$\phi_r(\boldsymbol{b}) := \lim_{n \to \infty} \log \left(\frac{\mu_{\psi_r} \circ \pi^{-1}[\boldsymbol{b}_0^n]}{\mu_{\psi_r} \circ \pi^{-1}[\boldsymbol{b}_1^n]} \right). \qquad (8)$$

Furthermore, there are constants $C > 0$ and $\theta \in [0, 1[$ such that, for any positive integer $n > r$ and for any $\boldsymbol{b} \in B^{\mathbb{N}}$, we have

$$\left| \phi_r(\boldsymbol{b}) - \log\left(\frac{\mu_{\psi_r} \circ \pi^{-1}[\boldsymbol{b}_0^n]}{\mu_{\psi_r} \circ \pi^{-1}[\boldsymbol{b}_1^n]} \right) \right| \leq C r^2 \theta^{n/r}. \tag{9}$$

Lemma 3.3. (Projective convergence of Markov approximants) *The sequence of measures (μ_{ψ_r}) converges in the projective sense to the Gibbs measure μ_ψ associated to the potential ψ.*

Furthermore, for all $n, r > 0$, and $w \in A^n$, we have

$$\mu_{\psi_r}[w] \lesssim \mu_\psi[w] \exp(\pm\epsilon_{r,n}),$$

where

$$\epsilon_{r,n} := D \sum_{s=r}^{\infty} ((n + (s+1)(s+2))\mathrm{var}_s \psi + s\theta^s) \tag{10}$$

for adequate constants $D > 0$ and $\theta \in [0, 1[$ (the same θ as in Lemma 3.2).

With the two previous lemmas at hand, we can proceed to the proof of Theorem 3.1.

Proof of Theorem 3.1. We start by proving that the sequence $(\mu_{\psi_r} \circ \pi^{-1})_r$ converges in the projective sense to $\mu_\psi \circ \pi^{-1}$.

On the one hand, Lemma 3.2 tells us that the measure $\nu_r := \mu_{\psi_r} \circ \pi^{-1}$ is Gibbsian for the potential $\phi_r : B^{\mathbb{N}} \to \mathbb{R}$ given by

$$\phi_r(\boldsymbol{b}) = \lim_{n \to \infty} \log\left(\frac{\nu_r[\boldsymbol{b}_0^n]}{\nu_r[\boldsymbol{b}_1^n]} \right).$$

On the other hand, Lemma 3.3 ensures that for each $n, r > 0$ with $n \geq r$, and each $v \in A^n$, we have $\mu_{\psi_r}[v] \lesssim \mu_\psi[v] \exp(\pm\epsilon_{r,n})$, where $\epsilon_{r,n}$ is defined as in (10). From this, it follows that for each $w \in B^n$ we have

$$\nu_r[w] := \sum_{v \in A^n : \pi v = w} \mu_{\psi_r}[v]$$

$$\lesssim \exp(\pm\epsilon_{r,n}) \sum_{v \in A^n : \pi v = w} \mu_\psi[v]$$

$$\lesssim \exp(\pm\epsilon_{r,n}) \mu_\psi \circ \pi^{-1}[w]. \tag{11}$$

Otherwise said, the sequence of approximants $(\nu_r \equiv \mu_{\psi_r} \circ \pi^{-1})_r$ converges in the projective sense to the induced measure $\mu_\psi \circ \pi^{-1}$, and the speed of convergence is the same for both the factor and the original system.

Now we prove that the pushforward measure $\nu := \mu_\psi \circ \pi^{-1}$ is a Gibbs measure.

According to Lemma 3.2 and (11), for any $\boldsymbol{b} \in B^{\mathbb{N}}$, and $n, r > 0$ with $n \geq r$, we have

$$\left| \phi_r(\boldsymbol{b}) - \log\left(\frac{\nu[\boldsymbol{b}_0^n]}{\nu[\boldsymbol{b}_1^n]}\right) \right| \leq 2\epsilon_{r,n} + C\, r^2 \theta^{n/r}. \tag{12}$$

Let us take, for each $r > 0$, $n = n(r) := r^2$, and let $r^* > 0$ be such that both $s \mapsto s^2\theta^s$ and $s \mapsto \epsilon_{s,s^2}$ define decreasing functions in $[r^*, \infty)$. Hence, using the triangle inequality, we obtain

$$|\phi_r(\boldsymbol{b}) - \phi_{r'}(\boldsymbol{b})| \leq 2(2\epsilon_{r,r^2} + C\, r^2\theta^r)$$

for all $r^* \leq r < r'$, and for any $\boldsymbol{b} \in B^{\mathbb{N}}$. This proves uniform convergence of the sequence of potentials $(\phi_r)_r$. The limit is the continuous function $\phi : B^{\mathbb{N}} \to \mathbb{R}$ defined by

$$\phi(\boldsymbol{b}) := \lim_{n \to \infty} \log\left(\frac{\nu[\boldsymbol{b}_0^n]}{\nu[\boldsymbol{b}_1^n]}\right).$$

If we verify that ϕ satisfies condition (4), then, according to the observation following Theorem 2.5, this will prove that $\nu \equiv \mu_\psi \circ \pi^{-1}$ is the unique Gibbs measure for ϕ. From (12), it follows that

$$|\phi(\boldsymbol{b}) - \phi(\tilde{\boldsymbol{b}})| \leq \left|\phi(\boldsymbol{b}) - \phi_r(\boldsymbol{b})\right| + \left|\phi_r(\boldsymbol{b}) - \log\left(\frac{\nu[\boldsymbol{b}_0^n]}{\nu[\boldsymbol{b}_1^n]}\right)\right|$$

$$+ \left|\phi_r(\tilde{\boldsymbol{b}}) - \log\left(\frac{\nu[\boldsymbol{b}_0^n]}{\nu[\boldsymbol{b}_1^n]}\right)\right| + |\phi_r(\tilde{\boldsymbol{b}}) - \phi(\tilde{\boldsymbol{b}})|$$

$$\leq 4(2\epsilon_{r,r^2} + C\, r^2\theta^r) + 2(2\epsilon_{r,n} + C\, r^2 \theta^{n/r})$$

for all $\boldsymbol{b}, \tilde{\boldsymbol{b}} \in B^{\mathbb{N}}$ such that $\tilde{\boldsymbol{b}} \in [\boldsymbol{b}_0^n]$, and every $n > r \geq r^*$.

Since ψ is Hölder continuous and

$$\epsilon_{r,r^2} := D \sum_{s=r}^{\infty} ((r^2 + (s+1)(s+2))\mathrm{var}_s\psi + s\theta^s),$$

then there exist $C > 0$ and $\varrho \in [\theta, 1[$ (remember that $\theta \in [0, 1[$) such that $\max(\epsilon_{r,r^2}, r^2\theta^r) \leq C\varrho^r$. We take again $n = n(r) = r^2$ and obtain, for all $n \in \mathbb{N}$,

$$\mathrm{var}_n\phi \leq \mathcal{D}\exp(-c\sqrt{n}),$$

with a convenient $\mathcal{D} \geq 6C(2 + C)$, and $c = -\log(\varrho)$.

The case of a locally constant ψ is the immediate consequence of Lemma 3.2 and one has $\vartheta = \theta^{1/r}$.

The theorem is now proved. □

Remark 3.4. The competition between the terms $\epsilon_{r,n}$ and $\theta^{n/r}$ in the upper bound of $\mathrm{var}_n \phi$ leads to a subexponential bound, namely $\mathrm{var}_n \phi \leq \mathcal{D} \exp(-cn^{\delta/(1+\delta)})$, for any $\delta > 0$. We made the choice $\delta = 1$.

4 Generalization to less regular potentials

In this section, we go beyond Hölder continuous potentials and look at potentials ψ such that $\mathrm{var}_r(\psi)$ decreases slower than exponentially. Besides the fact that $\sum_r \mathrm{var}_r \psi < \infty$ is always assumed, the only place where a finer control in the decrease of $\mathrm{var}_r(\psi)$ is required is inside the proof of Lemma 3.3. There, the projective convergence of the Markov approximants depends on the fact that

$$\epsilon_{r,n} := D \sum_{s=r}^{\infty} ((n+(s+1)(s+2))\mathrm{var}_s \psi + s\theta^s) \to 0 \quad \text{when } r \to \infty$$

for each $n > 0$. Furthermore, the variation of the induced potential, $\mathrm{var}_n \phi$, is upper bounded by a linear combination of $\epsilon_{\sqrt{n},n}$ and $n\theta^{n/r}$. After this consideration, we can generalize Theorem 3.1 as follows.

Theorem 4.1. *Let* $\pi : A^{\mathbb{N}} \to B^{\mathbb{N}}$ *be the amalgamation map just defined and* $\psi : A^{\mathbb{N}} \to \mathbb{R}$ *be such that* $\sum_{s=0}^{\infty} s^2 \mathrm{var}_s \psi < \infty$. *Then the measure* $\mu_\psi \circ \pi^{-1}$ *is a Gibbs measure with support* $B^{\mathbb{N}}$ *for a normalized potential* $\phi : B^{\mathbb{N}} \to \mathbb{R}$ *defined by the limit*

$$\phi(\boldsymbol{b}) = \lim_{r\to\infty} \lim_{n\to\infty} \log \left(\frac{\mu_{\psi_r} \circ \pi^{-1}[\boldsymbol{b}_0^n]}{\mu_{\psi_r} \circ \pi^{-1}[\boldsymbol{b}_1^n]} \right),$$

where ψ_r *is the* $(r+1)$-*symbol approximant of* ψ.

If $\mathrm{var}_n \psi$ *is subexponentially decreasing, i.e., if* $\mathrm{var}_n \psi \leq C \exp(-cn^\gamma)$ *for some* $c, C > 0$ *and* $\gamma \in \,]0, 1[$, *then there are constants* $D > C$ *and* $0 < d < c$ *such that*

$$\mathrm{var}_n(\phi) \leq D \exp\left(-d\, n^{\gamma/(1+\gamma)}\right)$$

for all $n \in \mathbb{N}$.

If $\mathrm{var}_n \psi$ *is polynomially decreasing, i.e., if* $\mathrm{var}_n \psi \leq Cn^{-q}$ *for some* $C > 0$
and $q > 3$, *then for all* $\epsilon \in (0, q-3)$ *there is a constant* $D > C$ *such that*

$$\mathrm{var}_n(\phi) \leq D \frac{1}{n^{q-2-\epsilon}}$$

for all $n \in \mathbb{N}$.

Remark 4.2. As mentioned above, the n variation of the induced potential is
upper bounded by a linear combination of $\epsilon_{r,n}$ and $r^2 \theta^{n/r}$. We have to optimize
the choice of the function $r \mapsto n(r)$ in such a way that $n/r \to \infty$ when $r \to \infty$,
and that the resulting n variation of ψ has the fastest possible decrease. In
the subexponential case, $\mathrm{var}_n \psi \leq C \exp(-cn^\gamma)$, the optimal choice turns out to
be $n(r) = r^{1+\gamma}$, while in the polynomially decreasing case, $\mathrm{var}_n \psi \leq Cn^{-q}$, the
optimal choice is $n(r) = r^{(q-1)/(q-1-\epsilon)}$. This gives a bound in $n^{-q+2+\epsilon}$.

5 Comments and open questions

In our previous work [5], we made two restrictive assumptions, namely that
ψ is a locally constant potential and the image of the starting SFT under the
amalgamation map π is still an SFT (in general it is a sofic subshift). In that
setting, we could prove, under sufficient conditions, that $\mu_\psi \circ \pi^{-1}$ is a Gibbs
measure for a Hölder continuous potential ϕ. We also exhibited an example
showing that one of our sufficient conditions turns out to be necessary in that
otherwise the induced potential ϕ is not defined everywhere.

We conjecture the following: let $\pi : A \to B$ be an amalgamation map as above,
$X \subset A^{\mathbb{N}}$ an SFT, and $Y \subset B^{\mathbb{N}}$ the resulting sofic subshift. Then the pushforward
measure of a Gibbs measure for a Hölder continuous potential is a "weak"
Gibbs measure μ_ϕ in that (1) does not hold for every a but for almost all a
(with respect to μ_ϕ).

6 Appendix

6.1 Preliminary result: Birkhoff's refinement of the Perron–Frobenius theorem

Let E, E' be finite sets and $M : E \times E' \to \mathbb{R}^+$ be a row-allowable nonnegative
matrix, i.e., a matrix such that $Mx > 0$ whenever $x > 0$. Let us define the set

$$\Delta_E := \left\{ x \in {]0, 1[}^E : |x|_1 := \sum_{e \in E} x(e) = 1 \right\},$$

and similarly $\Delta_{E'}$. We supply Δ_E with the distance

$$\delta_E(x,y) := \max_{e,f \in E} \log \frac{x(e)y(f)}{x(f)y(e)}.$$

On $\Delta_{E'}$ we define $\delta_{E'}$ accordingly. Let us now define

$$\tau(M) := \frac{1 - \sqrt{\Phi(M)}}{1 + \sqrt{\Phi(M)}},$$

where

$$\Phi(M) := \begin{cases} \min\limits_{e,f \in E, e',f' \in E'} \dfrac{M(e,e')M(f,f')}{M(e,f')M(f,e')} & \text{if } M > 0, \\ 0 & \text{otherwise.} \end{cases}$$

Here $M > 0$ means that all entries of M are strictly positive.

Theorem 6.1. (After Birkhoff [13]) *Let $M : E \times E' \to \mathbb{R}^+$ be row allowable, and $F_M : \Delta_{E'} \to \Delta_E$ be such that*

$$F_M x := \frac{M x}{|M x|_1} \quad \text{for each } x \in \Delta_{E'}.$$

Then, for all $x, y \in \Delta_{E'}$, we have

$$\delta_E(F_M x, F_M y) \le \tau(M) \delta_{E'}(x,y).$$

We have $\tau(M) < 1$ if and only if $M > 0$.

For a proof of this important result, see [4] for instance. It can also be deduced from the proof of a similar theorem concerning square matrices, which can be found in [16]. As a corollary of this result, we obtain the following form of the Perron–Frobenius theorem.

Corollary 6.2. (Enhanced Perron–Frobenius theorem) *Suppose that $M : E \times E \to \mathbb{R}^+$ is primitive, i.e., there exists $\ell \in \mathbb{N}$ such that $M^\ell > 0$. Then its maximal eigenvalue ρ_M is simple and it has a unique right eigenvector $\bar{R}_M \in \Delta_E$ and a unique left eigenvector $\bar{L}_M$ satisfying $\bar{L}_M^\dagger \bar{R}_M = 1$. Furthermore, for each $x \in \Delta_E$ and each $n \in \mathbb{N}$, we have*

$$M^n x \lesseqgtr (\bar{L}_M^\dagger x) \rho_M^n \bar{R}_M \exp\left(\pm \frac{\ell \delta_E(x, F_M x)}{1 - \tau} \tau^{\lfloor n/\ell \rfloor} \right),$$

with $\tau := \tau(M^\ell) < 1$.

Proof. Let us first remark that $F_{M^\ell} = F_M^\ell$. Since $M^\ell > 0$, Theorem 6.1 and the contraction mapping theorem [11] imply the existence of a unique fixed point $x_M = F_M x_M \in \Delta_E$ such that

$$\delta_E(F_M^n x, x_M) \leq \sum_{k=0}^{\infty} \delta_E\left(F_M^{n+k\ell} x, F_M^{n+(k+1)\ell} x\right)$$

$$\leq \frac{\delta_E(x, F_M^\ell x)\tau^{\lfloor n/\ell \rfloor}}{1-\tau} \leq \frac{\ell\delta_E(x, F_M x)\tau^{\lfloor n/\ell \rfloor}}{1-\tau}$$

for each $n \in \mathbb{N}$ and $x \in \Delta_E$. From the definition of projective distance, it follows that, for each $x \in \Delta_E$ and $n \in \mathbb{N}$, there is a constant $C = C(x, n)$ such that

$$M^n x \lessgtr C(x, n) x_M \exp\left(\pm \frac{\ell\delta_E(x, F_M x)}{1-\tau}\tau^{\lfloor n/\ell \rfloor}\right). \tag{13}$$

Let us now prove that $x_M \equiv \bar{R}_M \in \Delta_E$ is the unique positive right eigenvector associated to the maximum eigenvalue $\rho_M := \max \operatorname{spec}(M)$. Indeed, since $F_M x_M = x_M$, $M x_M = \lambda x_M$ for some $\lambda > 0$. Now, if $M y = \lambda y$ for some $y \in \mathbb{C}^E$, and taking into account that M is a real matrix, then $y = az$ for some $a \in \mathbb{C}$ and $z \in \Delta_E$. Therefore, λ is a simple eigenvalue. It follows from Theorem 6.1 and the contraction mapping theorem that $z = \bar{R}_M$ is the associated eigenvector.

Consider the map $x \mapsto \min_{e \in E}(Mx)(e)/x(e)$ on Δ_E, and extend it to $\operatorname{clos}(\Delta_E)$ (the closure is taken with respect to the Euclidean distance) by allowing values in the extended reals $\bar{\mathbb{R}} := \mathbb{R} \cup \{\infty\}$.[5] The resulting transformation is upper semicontinuous, and therefore there exists $x_0 \in \operatorname{clos}(\Delta_E)$ attaining the supremum, i.e., such that

$$\rho := \sup_{x \in \Delta_E} \min_{e \in E} \frac{(Mx)(e)}{x(e)} = \min_{e \in E} \frac{(Mx_0)(e)}{x_0(e)}.$$

This supremum is an eigenvalue, and the point where it is attained is its corresponding positive eigenvector. Indeed, if $Mx_0 \neq \rho x_0$, i.e., if $(Mx_0)(e) > \rho x_0(e)$ for some $e \in E$, then $M^{\ell+1}x_0 > \rho M^\ell x_0$, which implies that $\rho < \sup_{x \in \Delta_E} \min_{e \in E}(Mx)(e)/x(e)$. Therefore, x_0 is a nonnegative eigenvector for M, but, since $M^\ell x_0 = \rho^\ell x_0 > 0$, then necessarily $x_0 = \bar{R}_M$ and $\lambda = \rho$.

Finally, if $0 \neq y \in \mathbb{C}^E$ is a right eigenvector of M, associated to another eigenvalue $\lambda' \in \mathbb{C}$, then

$$|\lambda'|\,|y| = |My| \leq M|y|,$$

[5] Here we are following a standard argument, which can be found in [16] for instance.

where $|z|$ denotes the coordinatewise absolute value of the vector $z \in \mathbb{C}^E$, and the inequality holds at each coordinate. If $|\lambda'| < \min_{e \in E}(|(My)(e)|)/(|y(e)|)$, we can find a vector $y^+ \in \Delta_E$ by slightly changing $|y|$ at coordinates $e \in E$ where $y(e) = 0$ and then normalizing, so that $|\lambda'| \leq \min_{e \in E}(My^+)(e)/y^+(e)$. If on the contrary $|\lambda'| = \min_{e \in E}|(My)(e)|/|y(e)|$, then $M^{\ell+1}|y| \geq |\lambda'|M^\ell|y|$, and normalizing $M^\ell|y|$ we obtain $y^+ \in \Delta_E$ in such a way that $|\lambda'| \leq \min_{e \in E}(My^+)(e)/y^+(e)$. We conclude that

$$|\lambda'| \leq \sup_{x \in \Delta_E} \min_{e \in E} \frac{(Mx)(e)}{x(e)} := \rho$$

for each $\lambda' \in \text{spec}(M)$; therefore, $\rho \equiv \rho_M := \max \text{spec}(M)$.

It remains to prove that in (13) we have $C(x,n) = (\bar{L}_M^\dagger x)\rho_M^n$, where $\bar{L}_M > 0$ is the left eigenvector associated to ρ_M, normalized so that $\bar{L}_M^\dagger \bar{R}_M = 1$. For this note that, by multiplying (13) at left by $\bar{L}_M$, we obtain

$$\rho_M^n(\bar{L}_M^\dagger x) \lesseqgtr (\bar{L}_M^\dagger \bar{R}_M)\, C(x,n) \exp\left(\pm \frac{\ell \delta_E(x, F_M x)}{1-\tau} \tau^{\lfloor n/\ell \rfloor}\right);$$

hence, $C(x,n) = \rho_M^n(\bar{L}_M^\dagger x)/(\bar{L}_M^\dagger \bar{R}_M)$, and the proof is finished. $\square$

6.2 Proof of Lemma 3.2

6.2.1 The right eigenvector

Notice that the transition matrix $\mathcal{M}_r := \mathcal{M}_{\psi_r}$ is primitive with primitivity index r; hence, according to Corollary 6.2,

$$\mathcal{M}_r^n x \lesseqgtr (\bar{L}_r^\dagger x)\rho_r^n \bar{R}_r \exp\left(\pm \frac{r \delta_{A^r}(x, F_r x)}{1-\tau} \tau^{\lfloor n/r \rfloor}\right)$$

for each $x \in \Delta_E$ and $n \in \mathbb{N}$. Here ρ_r denotes the maximal eigenvalue of $\mathcal{M}_r$, $\bar{R}_r \in \Delta_{A^r}$ its unique right eigenvector in the simplex, $\bar{L}_r$ its unique associated left eigenvector satisfying $\bar{L}_r^\dagger \bar{R}_r = 1$, and $\tau := \tau(\mathcal{M}_r^r)$ the contraction coefficient associated to the positive matrix $\mathcal{M}_r^r$.

Let us now obtain an explicit upper bound for τ and for the distance $\delta_{A^r}(x, F_r x)$ for particular values of $x \in \Delta_E$. First,

$$\Phi(\mathcal{M}_r^r) \geq \min_{u,v,u',u'',v',v'' \in A^r} \frac{\mathcal{M}_r(u,u')\mathcal{M}_r(v,v'')}{\mathcal{M}_r(u,v')\mathcal{M}_r(v,u'')} \geq \exp\left(-2\sum_{k=0}^r \text{var}_k \psi\right) > 0.$$

Therefore, $\tau \leq 1 - e^{-\sum_{k=0}^r \text{var}_k \psi}$ and $(1-\tau)^{-1} \leq e^{\sum_{k=0}^r \text{var}_k \psi}$.

Let $s_\psi := \sum_{k=0}^{\infty} \mathrm{var}_k \psi$ and $\theta := 1 - e^{-s_\psi}$. With this, and taking into account the upper bound for τ and $(1-\tau)^{-1}$, we obtain

$$\mathcal{M}_r^n \mathbf{r} \lesssim (\bar{l}_{,}^\dagger \mathbf{r}) \rho_r^n \bar{R}_r \, \exp\left(\pm r\, \delta_{A^r}(\mathbf{x}, F_r \mathbf{x})\, e^{s_\psi} \theta^{\lfloor n/r \rfloor}\right). \tag{14}$$

On the other hand, for $\bar{u} := (1/(\mathrm{Card}(A^r)), \ldots, 1/(\mathrm{Card}(A^r)))^\dagger \in \Delta_{A^r}$, we have

$$\delta_{A^r}(\bar{u}, F_r \bar{u}) := \max_{w,w' \in A^r} \log\left(\frac{\bar{u}(w')(\mathcal{M}_r \bar{u})(w)}{\bar{u}(w)(\mathcal{M}_r \bar{u})(w')}\right)$$

$$\leq r \log(\mathrm{Card}(A)) + 2\|\psi\| < r\, C_0,$$

where $C_0 := 2\,(\log(\mathrm{Card}(A)) + \|\psi\|)$. Therefore, by taking $\mathbf{x} = \bar{u}$ and $n = r^2$ in (14), we finally obtain

$$\bar{R}_r(\mathbf{u}) \lesssim \frac{\sum_{a \in \mathrm{Per}_{r^2}(A^\mathbb{N}) \cap [u]} e^{S_{r^2-r-1}\psi_r(a)}}{\rho_r^{r^2} |\bar{L}_r|_1} \, e^{\pm C_0\, r^2 \exp(s_\psi)\theta^r}. \tag{15}$$

6.2.2 Ansatz for the induced potential

To each word $w \in B^r$ we associate the simplex

$$\Delta_w := \left\{ \mathbf{x} \in (0,1)^{E_w} : |\mathbf{x}|_1 := \sum_{v \in E_w} \mathbf{x}_v = 1 \right\},$$

where $E_w := \{v \in A^r : \pi v = w\}$.

Let $\mathcal{M}_r, \rho_r, \bar{L}_r := \bar{L}_{\psi_r}$, and $\bar{R}_r := \bar{R}_{\psi_r}$ be as above, and define, for each $w \in B^r$, the restriction $\bar{L}_{r,w} := \bar{L}_r|_{E_w} \in (0,\infty)^{E_w}$. Define $\bar{R}_{r,w}$ in the analogous way, and for each $w \in B^{r+1}$ let $\mathcal{M}_{r,w}$ be the restriction of $\mathcal{M}_r$ to the coordinates in $E_{w_0^{r-1}} \times E_{w_1^r}$. Using this, and taking into account (5), which applies to our $(r+1)$-symbol potential ψ_r, we derive the matrix expression

$$\nu_r[\mathbf{b}_0^n] \equiv \sum_{\pi a_0^n = b_0^n} \mu_{\psi_r}[\mathbf{a}_0^n] = \bar{L}_{r,\mathbf{b}_0^{r-1}}^\dagger \left(\frac{\prod_{j=0}^{n-r} \mathcal{M}_{r,\mathbf{b}_j^{j+r}}}{\rho_r^{n-r+1}}\right) \bar{R}_{r,\mathbf{b}_{n-r+1}^n}$$

for the induced measure $\nu_r := \mu_{\psi_r} \circ \pi^{-1}$. It follows from this that

$$\log\left(\frac{\nu_r[\mathbf{b}_0^n]}{\nu_r[\mathbf{b}_1^n]}\right) = \log\left(\frac{\left(\bar{L}_{r,\mathbf{b}_0^{r-1}}\right)^\dagger \prod_{j=0}^{n-r} \mathcal{M}_{r,\mathbf{b}_j^{j+r}} \bar{R}_{r,\mathbf{b}_{n-r+1}^n}}{\left(\bar{L}_{r,\mathbf{b}_1^r}\right)^\dagger \prod_{j=1}^{n-r} \mathcal{M}_{r,\mathbf{b}_j^{j+r}} \bar{R}_{r,\mathbf{b}_{n-r+1}^n}}\right) - \log(\rho_r). \tag{16}$$

For each $w \in A^{r+s}$, with $s \geq 1$, let $\mathcal{M}_{r,w} := \prod_{j=0}^{s-1} \mathcal{M}_{r,w_j^{j+r}}$, and define the transformation $F_{r,w} : \Delta_{w_s^{r+s-1}} \to \Delta_{w_0^{r-1}}$ such that

$$F_{r,w}x = \frac{\mathcal{M}_{r,w}x}{|\mathcal{M}_{r,w}x|_1}.$$

For each $\boldsymbol{b} \in B^{\mathbb{N}}$ and $s,t \in \mathbb{N}$, let

$$x_{r,\boldsymbol{b}_{s+1}^{s+t+r}} := F_{r,\boldsymbol{b}_{s+1}^{s+r+1}} \circ \cdots \circ F_{r,\boldsymbol{b}_{s+t}^{s+t+r}} \left(\bar{R}_{r,\boldsymbol{b}_{s+t+1}^{s+t+r}} / |\bar{R}_{r,\boldsymbol{b}_{s+t+1}^{s+t+r}}|_1 \right) \in \Delta_{\boldsymbol{b}_{s+1}^{s+r}}.$$

By convention, $x_{r,\boldsymbol{b}_{s+1}^{s+r}} \equiv \bar{R}_{r,\boldsymbol{b}_{s+1}^{s+r}} / |\bar{R}_{r,\boldsymbol{b}_{s+1}^{s+r}}|_1 \in \Delta_{\boldsymbol{b}_{s+1}^{s+r}}$. Using this notation, and after adequate renormalization, (16) becomes

$$\log\left(\frac{\nu_r[\boldsymbol{b}_0^n]}{\nu_r[\boldsymbol{b}_1^n]} \right) = \log\left(\frac{\left(\bar{L}_{r,\boldsymbol{b}_0^{r-1}}\right)^{\dagger} \mathcal{M}_{r,\boldsymbol{b}_0^r} x_{r,\boldsymbol{b}_1^n}}{\left(\bar{L}_{r,\boldsymbol{b}_1^r}\right)^{\dagger} x_{r,\boldsymbol{b}_1^n}} \right) - \log(\rho_r). \quad (17)$$

6.2.3 Convergence of the inhomogeneous product

Let us now prove the convergence of the sequence $(x_{\boldsymbol{b}_1^n})_{n \geq r}$. For this, notice that

$$x_{r,\boldsymbol{b}_1^n} := F_{r,\boldsymbol{b}_1^{r+1}} \circ \cdots \circ F_{r,\boldsymbol{b}_{n-r}^n} x_{r,\boldsymbol{b}_{n-r+1}^n}$$

$$= F_{r,\boldsymbol{b}_1^n} x_{r,\boldsymbol{b}_{n-r+1}^n} = F_{r,\boldsymbol{b}_1^{2r}} \circ F_{r,\boldsymbol{b}_{r+1}^{3r}} \circ \cdots \circ F_{r,\boldsymbol{b}_{(k-1)r+1}^{(k+1)r}} x_{r,\boldsymbol{b}_{kr+1}^n},$$

where $k := \lfloor n/r \rfloor - 1$. Now, since $\mathcal{M}_{r,w} > 0$ for each $w \in B^{2r}$, Theorem 6.1 ensures that the associated transformation $F_{r,w} : \Delta_{w_r^{2r-1}} \to \Delta_{w_0^{r-1}}$ is a contraction with contraction coefficient $\tau_w = (1 - \sqrt{\Phi_w})/(1 + \sqrt{\Phi_w})$, where

$$\Phi_w := \min_{v,u \in E_{w_0^{r-1}}, v',u' \in E_{w_r^{2r-1}}} \frac{\mathcal{M}_{r,w}(v,v')\mathcal{M}_{r,w}(u,u')}{\mathcal{M}_{r,w}(v,u')\mathcal{M}_{r,w}(u,v')}$$

$$\geq \min_{v,u,v',u',v'',u'' \in A^r} \frac{\mathcal{M}_{r,w}(v,v')\mathcal{M}_{r,w}(u,u'')}{\mathcal{M}_{r,w}(v,u')\mathcal{M}_{r,w}(u,v'')}$$

$$\geq \exp\left(-2\sum_{k=0}^r \mathrm{var}_k \psi \right) \geq e^{-2s_\psi} > 0. \quad (18)$$

Recall that $s_\psi = \sum_{k=0}^\infty \mathrm{var}_k \psi$. From (18), we obtain a uniform upper bound for the contraction coefficients, $\tau_w \leq \theta := 1 - \exp(-s_\psi) < 1$, which allows us to establish the uniform convergence of the sequence $(x_{b_1^n})_{n \geq r}$ with respect to $b \in B^{\mathbb{N}}$. Indeed, for $b \in B^{\mathbb{N}}$ fixed and $m > n$, we have

$$\delta_{E_{b_1^r}}(x_{b_1^n}, x_{b_1^m}) \leq \theta^k \, \delta_{E_{b_{kr+1}^{(k+1)r}}}(x_{r,b_{kr+1}^n}, x_{r,b_{kr+1}^m}), \tag{19}$$

where $k := \lfloor n/r \rfloor - 1$. On the other hand,

$$\delta_{E_{b_{kr+1}^{(k+1)r}}}\!\left(x_{r,b_{kr+1}^n}, x_{r,b_{kr+1}^m}\right) \leq \sum_{j=0}^{k'} \delta_{E_{b_{kr+1}^{(k+1)r}}}\!\left(x_{r,b_{kr+1}^{n+jr}}, x_{r,b_{kr+1}^{n+(j+1)r}}\right)$$
$$+ \delta_{E_{b_{kr+1}^{(k+1)r}}}\!\left(x_{r,b_{kr+1}^{n+(k'+1)r}}, x_{r,b_{kr+1}^m}\right),$$

where $k' := \lfloor (m-n)/r \rfloor - 1$. By convention, when $k' = -1$, the summation on the right-hand side is zero. Then, since all the matrices $\mathcal{M}_{r,w}$ are row allowable and positive for $w \in B^{2r}$, we have

$$\delta_{E_{b_{kr+1}^{(k+1)r}}}\!\left(x_{r,b_{kr+1}^n}, x_{r,b_{kr+1}^m}\right) \leq T_1 + T_2 + T_3, \tag{20}$$

where

$$T_1 := \delta_{E_{b_{n-r+1}^n}}\!\left(x_{r,b_{n-r+1}^n}, F_{r,b_{n-r+1}^{n+r}} x_{r,b_{n+1}^{n+r}}\right), \tag{21}$$

$$T_2 := \sum_{j=1}^{k'} \theta^j \, \delta_{E_{b_{n+(j-1)r+1}^{n+jr}}}\!\left(x_{r,b_{n+(j-1)r+1}^{n+jr}}, F_{r,b_{n+(j-1)r+1}^{n+(j+1)r}} x_{r,b_{n+jr+1}^{n+(j+1)r}}\right), \tag{22}$$

and

$$T_3 := \theta^{k'} \, \delta_{E_{b_{n+k'r+1}^{n+(k'+1)r}}}\!\left(x_{r,b_{n+k'r+1}^{n+(k'+1)r}}, F_{r,b_{n+k'r+1}^m} x_{r,b_{m-r+1}^m}\right). \tag{23}$$

Once again, we conclude that $T_2 = 0$ if $k' = -1$.

Now, for each $w, w' \in B^r$, and $v \in B^s$ with $r < s < 2r$, and such that $v_0^{r-1} = w$, $v_{s-r+1}^s = w'$, we have

$$\delta_w(x_{r,w}, F_{r,v} x_{r,w'}) = \max_{u,u' \in E_w} \log \left(\frac{x_{r,w}(u)\,(F_{r,v} x_{r,w'})(u')}{x_{r,w}(u')\,(F_{r,v} x_{r,w'})(u)} \right)$$
$$\leq \max_{u,u' \in E_w} \log \left(\frac{\bar{R}_r(u)\,(\mathcal{M}_{r,v}\bar{R}_{r,w'})(u')}{\bar{R}_r(u')\,(\mathcal{M}_{r,v}\bar{R}_{r,w'})(u)} \right).$$

Hence, using the estimate for the right eigenvectors given in (15), it follows that

$$\delta_w(\boldsymbol{x}_{r,w}, F_{r,v}\boldsymbol{x}_{r,w'})$$

$$\leq \max_{u,u' \in A^r} \log \left(\frac{\displaystyle\sum_{a \in \mathrm{Per}_{r,2}(A^{\mathbb{N}})} e^{S_{r2-r-1}\psi_r(a)} \quad \displaystyle\sum_{a \in \mathrm{Per}_{r,2+s-r}(A^{\mathbb{N}})} e^{S_{r2+s-2r-1}\psi_r(a)}}{\displaystyle\min_{a \in \mathrm{Per}_{r,2}(A^{\mathbb{N}})} e^{S_{r2-r-1}\psi_r(a)} \quad \displaystyle\min_{a \in \mathrm{Per}_{r,2+s-r}(A^{\mathbb{N}})} e^{S_{r2+s-2r-1}\psi_r(a)}} \right)$$

$$+ 2r^2 C_0 e^{s_\psi}\theta^r \leq 2r(r+1)C_0(e^{s_\psi}\theta^r + 1),$$

with $C_0 = 2(\log(\mathrm{Card}(A)) + \|\psi\|)$ and $\theta = 1 - \exp(-s_\psi)$ as in (15). Using this upper bound in (21), (22), and (23), we obtain from (20)

$$\delta_{E_{b_{kr+1}}^{(k+1)r}}\left(\boldsymbol{x}_{r,b_{kr+1}^n}, \boldsymbol{x}_{r,b_{kr+1}^m}\right) \leq 2r(r+1)C_0(e^{s_\psi}\theta^r + 1)\left(\theta^k + \frac{1}{1-\theta}\right)$$

and, with this, (19) becomes

$$\delta_{E_{b_1^r}}\left(\boldsymbol{x}_{b_1^n}, \boldsymbol{x}_{b_1^m}\right) \leq 2r(r+1)C_0(e^{s_\psi}\theta^r + 1)\theta^{\lfloor n/r \rfloor - 1}\left(\theta^{\lfloor n/r \rfloor - 1} + \frac{1}{1-\theta}\right), \quad (24)$$

which holds for all $\boldsymbol{b} \in B^{\mathbb{N}}$ and $r < n < m$. Hence, $(\boldsymbol{x}_{b_1^n})_{n \geq r}$ is a Cauchy sequence in complete space Δ_{b^r}, and the existence of the limit $\boldsymbol{x}_{b_1^\infty} := \lim_{m \to \infty} \boldsymbol{x}_{b_1^m}$ is ensured for each $\boldsymbol{b} \in B^{\mathbb{N}}$. Furthermore, from (24) it follows that

$$\delta_{E_{b_1^r}}(\boldsymbol{x}_{b_1^n}, \boldsymbol{x}_{b_1^\infty}) \leq 2r(r+1)C_0(e^{s_\psi}\theta^r + 1)\theta^{\lfloor n/r \rfloor - 1}\left(\theta^{\lfloor n/r \rfloor - 1} + \frac{1}{1-\theta}\right) \leq C_1 r^2 \theta^{n/r},$$
$$(25)$$

with $C_1 := 4C_0(1 + e^{s_\psi}\theta)/(\theta^2(1-\theta))$.

6.2.4 The induced potential and the Gibbs inequality
Taking (25), it follows that the limit

$$\phi_r(\boldsymbol{b}) = \lim_{n \to \infty} \log\left(\frac{\nu_r[\boldsymbol{b}_0^n]}{\nu_r[\boldsymbol{b}_1^n]}\right) = \log\left(\frac{\left(\bar{L}_{r,b_0^{r-1}}\right)^\dagger \mathcal{M}_{r,b_0^r}\boldsymbol{x}_{r,b_1^\infty}}{\left(\bar{L}_{r,b_1^r}\right)^\dagger \boldsymbol{x}_{r,b_1^\infty}}\right) - \log(\rho_r) \quad (26)$$

exists for each $\boldsymbol{b} \in B^{\mathbb{N}}$, and defines a continuous function $\boldsymbol{b} \mapsto \phi_r(\boldsymbol{b})$. This proves that the limit (8) in the statement of the lemma does exist. It remains to find an upper bound to its modulus of continuity.

Inequality (25), and the fact that $|x_{b_1^n}|_1 = |x_{b_1^\infty}|_1 = 1$, imply that

$$x_{b_1^n} \lessgtr x_{b_1^\infty} \exp\left(\pm C_1 r^2 \theta^{n/r}\right)$$

for all $b \in B^{\mathbb{N}}$ and $n > r$. With this, and taking into account (17) and (26), it follows that

$$\left| \phi_r(b) - \log\left(\frac{\nu_r[b_0^n]}{\nu_r[b_1^n]}\right) \right| \leq \left| \log\left(\frac{\left(\bar{L}_{r,b_0^{r-1}}\right)^\dagger \mathcal{M}_{r,b_0^r} x_{r,b_1^\infty}}{\left(\bar{L}_{r,b_0^r}\right)^\dagger \mathcal{M}_{r,b_0^r} x_{r,b_1^n}}\right) \right.$$

$$\left. - \log\left(\frac{\left(\bar{L}_{r,b_1^{r-1}}\right)^\dagger x_{r,b_1^n}}{\left(\bar{L}_{r,b_1^r}\right)^\dagger x_{r,b_1^\infty}}\right) \right|$$

$$\leq C r^2 \theta^{n'/r}$$

for all $b \in B^{\mathbb{N}}, n > r$, and $C := 2C_1 = 8C_0(1 + e^{s\psi}\theta)/(\theta^2(1-\theta))$. This proves (9) in the statement of the lemma.

From this, it can be easily deduced that $\nu_r \equiv \mu_{\psi_r} \circ \pi^{-1}$ satisfies the Gibbs inequality (1) with potential ϕ_r and constants $P(\phi_r, B^{\mathbb{N}}) = 0$ and

$$C(\phi_r, B^{\mathbb{N}}) = \max_{b \in B^{\mathbb{N}}} \left(\frac{\exp(S_{r^2}\phi_r(b))}{\nu_r[b_0^{r^2}]}, \frac{\nu_r[b_0^{r^2}]}{\exp(S_{r^2}\phi_r(b))}\right) \exp\left(\frac{C r^2 \theta^r}{1-\theta^{1/r}}\right).$$

This proves the first statement of the lemma, the proof of which is now complete. □

Remark 6.3. As mentioned above (see (2)), the topological pressure of ψ is given by

$$P(\psi) = P(\psi, A^{\mathbb{N}}) = \lim_{n \to \infty} \frac{1}{n} \log\left(\sum_{a \in \mathrm{Per}_n(A^{\mathbb{N}})} e^{S_n\psi(a)}\right).$$

Since $\psi \lessgtr \psi_r \pm \mathrm{var}_r \psi$, we get

$$\log(\rho_r) = \lim_{n \to \infty} \frac{1}{n} \log(\mathrm{Trace}(\mathcal{M}_r^n))$$

$$\leqslant \lim_{n \to \infty} \frac{1}{n} \log \left(\sum_{a \in \mathrm{Per}_n(A^{\mathbb{N}})} e^{S_n \psi(a)} \right) \pm \mathrm{var}_r \psi$$

$$\leqslant P(\psi) \pm \mathrm{var}_r \psi$$

for each $r \in \mathbb{N}$.

6.3 Proof of Lemma 3.3

6.3.1 Periodic approximations

Each Markov approximant μ_{ψ_r} can be seen as the limit of measures supported on periodic points as follows. Fix $n, r \in \mathbb{N}$ with $n \geq r$, and $w \in A^n$. Then, for each $p > r + n$, we have

$$\mathcal{P}_r^{(p)}[w] := \frac{\sum_{a \in \mathrm{Per}_p(A^{\mathbb{N}}) \cap [w]} e^{S_p \psi_r(a)}}{\sum_{a \in \mathrm{Per}_p(A^{\mathbb{N}})} e^{S_p \psi_r(a)}}.$$

We can rewrite the above equation as

$$\mathcal{P}_r^{(p)}[w] = \frac{\left(\prod_{s=0}^{n-r-1} M_r(w_s^{s+r-1}, w_{s+1}^{s+r}) \right) \bar{e}_{w_{n-r}^{n-1}}^{\dagger} M_r^{p-n+r} \bar{e}_{w_0^{r-1}}}{\sum_{\zeta \in A^r} \bar{e}_{\zeta}^{\dagger} \mathcal{M}_r^p \bar{e}_{\zeta}}$$

$$= \frac{\left(\prod_{s=0}^{n-r-1} M_r(w_s^{s+r-1}, w_{s+1}^{s+r}) \right) \bar{e}_{w_{n-r}^{n-1}}^{\dagger} M_r^{p-n} (M_r^r \bar{e}_{w_0^{r-1}})}{\sum_{\zeta \in A^r} \bar{e}_{\zeta}^{\dagger} \mathcal{M}_r^{p-n} (\mathcal{M}_r^n \bar{e}_{\zeta})},$$

with $\mathcal{M}_r$ as above, and $\bar{e}_{\zeta} \in \{0, 1\}^{A^r}$ the vector with 1 at coordinate ζ and zeros everywhere else. Now, since $\mathcal{M}_r^k \bar{e}_{\zeta} > 0$ for each $k \geq r$ and $\zeta \in A^r$, Corollary 6.2 applies, and using (5) we obtain

$$\mathcal{P}_r^{(p)}[w] \lesssim \frac{\bar{L}_r^\dagger(\mathcal{M}_r^r \bar{e}_{w_0^{r-1}}) \left(\prod_{s=0}^{n-r-1} \mathcal{M}_r(w_s^{s+r-1}, w_{s+1}^{s+r})\right) \bar{e}_{w_{n-r}^{n-1}}^\dagger \bar{R}_r}{\sum_{\zeta \in A^r} \bar{L}_r^\dagger(\mathcal{M}_r^n \bar{e}_{w'}) \bar{e}_\zeta^\dagger \bar{R}_r} \, e^{\pm \frac{r D_0}{1-\tau} \tau^{\frac{p-n}{r} - 2}}$$

$$\lesssim \frac{\bar{L}_r^\dagger(w_0^{r-1}) \left(\prod_{s=0}^{n-r-1} \mathcal{M}_r(w_s^{s+r-1}, w_{s+1}^{s+r})\right) \bar{R}_r(w_{n-r}^{n-1})}{\rho_r^{n-r}} \, e^{\pm \frac{r D_0}{1-\tau} \tau^{\frac{p-n}{r} - 2}}$$

$$\lesssim \mu_{\psi_r}[w] \exp\left(\pm \frac{r D_0}{1-\tau} \tau^{\frac{p-n}{r} - 2}\right),$$

with $\bar{R}_r, \bar{L}_r, \rho_r$, and $\tau := \tau(\mathcal{M}_r^r)$ as before, and

$$D_0 := 2 \max_{\zeta \in A^r} \delta_{A^r}(\mathcal{M}_r^r \bar{e}_\zeta, \mathcal{M}_r^{r+1} \bar{e}_\zeta).$$

Since $\tau = (1 - \sqrt{\Phi})/(1 + \sqrt{\Phi})$, with

$$\Phi := \min_{v, u, v', u' \in A^r} \frac{\mathcal{M}_{r,w}(v, v') \mathcal{M}_{r,w}(u, u')}{\mathcal{M}_{r,w}(v, u') \mathcal{M}_{r,w}(u, v')} \geq \exp\left(-2 \sum_{k=0}^r \mathrm{var}_k \psi\right) > 0,$$

$\tau \leq \theta := 1 - \exp(-s_\psi)$ and $(1 - \tau)^{-1} \leq \exp(s_\psi)$, with $s_\psi := \sum_{k=0}^\infty \mathrm{var}_k \psi$ as in Lemma 3.2. On the other hand,

$$D_0 \leq 2 \max_{w', u, u' \in A^r} \log\left(\frac{\mathcal{M}_r^r(u, w') \mathcal{M}_r^{r+1}(u', w')}{\mathcal{M}_r^{r+1}(u, w') \mathcal{M}_r^r(u', w')}\right)$$

$$\leq 4 \left(\log(\mathrm{Card}(A)) + \sum_{k=1}^r \mathrm{var}_k \psi + \|\psi\|\right)$$

$$\leq 4 \left(\log(\mathrm{Card}(A)) + s_\psi + \|\psi\|\right) =: D_1.$$

Using this explicit bound just obtained, we deduce the inequalities

$$\mu_{\psi_r}[w] \lesssim \frac{\sum_{a \in \mathrm{Per}_p(A^{\mathbb{N}}) \cap [w]} e^{S_p \psi_r(a)}}{\sum_{a \in \mathrm{Per}_p(A^{\mathbb{N}})} e^{S_p \psi_r(a)}} \, \exp\left(\pm D_1 \, r \, e^{s_\psi} \theta^{\frac{p-n}{r} - 2}\right)$$

for each $w \in A^n$, and all $p > n + r$. It is easy to check that these inequalities extend to each $w \in \bigcup_{k=1}^n A^k$, and we finally obtain

$$\mu_{\psi_r}[w] \lesssim \frac{\sum_{a \in \mathrm{Per}_p(A^{\mathbb{N}}) \cap [w]} e^{S_p \psi_r(a)}}{\sum_{a \in \mathrm{Per}_p(A^{\mathbb{N}})} e^{S_p \psi_r(a)}} \, \exp\left(\pm D_1 \, r \, e^{s_\psi} \theta^{\frac{p - \max(n,r)}{r} - 2}\right) \tag{27}$$

for all $r, n \in \mathbb{N}$, and $w \in A^n$.

6.3.2 Telescopic product

Let us now compare two consecutive Markov approximants. Fix $n, r > 0$, and $w \in A^n$. Then, for each $p > n + r + 1$, (27) ensure that

$$
\frac{\mu_{\psi_r}[w]}{\mu_{\psi_{r+1}}[w]} \lessgtr \frac{\sum_{a \in \mathrm{Per}_p(A^{\mathbb{N}}) \cap [w]} e^{S_p \psi_r(a)}}{\sum_{a \in \mathrm{Per}_p(A^{\mathbb{N}}) \cap [w]} e^{S_p \psi_{r+1}(a)}} \frac{\sum_{a \in \mathrm{Per}_p(A^{\mathbb{N}})} e^{S_p \psi_{r+1}(a)}}{\sum_{a \in \mathrm{Per}_p(A^{\mathbb{N}})} e^{S_p \psi_r(a)}} \exp\left(\pm C r \theta^{\frac{p-q}{r}-2}\right)
$$

with $q = \max(r+1, n)$ and $C := 2 e^{s \psi} D_1$. Since $\psi_{r+1} \lessgtr \psi_r \pm \mathrm{var}_{r+1} \psi$, we have

$$
\frac{\mu_{\psi_r}[w]}{\mu_{\psi_{r+1}}[w]} \lessgtr \exp\left(\pm \left(2p \, \mathrm{var}_{r+1} \psi + C r \theta^{\frac{p-q}{r}-2}\right)\right)
$$

$$
\lessgtr \exp\left(\pm \left(2p \, \mathrm{var}_{r+1} \psi + C r \theta^{\frac{p-r-n-1}{r}-2}\right)\right)
$$

for all $r \in \mathbb{N}$, $w \in \bigcup_{k=1}^{n} A^k$, and $p > n + r + 1$. Let $p = (r+1)(r+2) + n - 1$; then, for each $r' > r \in \mathbb{N}$ and $w \in \bigcup_{k=1}^{n} A^k$, we have

$$
\frac{\mu_{\psi_r}[w]}{\mu_{\psi_{r'}}[w]} \lessgtr \exp\left(\pm D \sum_{s=r}^{\infty} \left((n + (s+1)(s+2)) \, \mathrm{var}_s \psi + s \theta^s\right)\right),
$$

with $D := \max(2, C)$. Since ψ is Hölder continuous and $\theta \in (0, 1)$,

$$
\epsilon_{r,n} := D \sum_{s=r}^{\infty} \left((n + (s+1)(s+2)) \mathrm{var}_s \psi + s \theta^s\right) \to 0 \quad \text{when } r \to \infty
$$

for each $n, r \in \mathbb{N}$. We conclude that $\mu[w] := \lim_{r \to \infty} \mu_{\psi_r}[w]$ exists for each $w \in \bigcup_{k=0}^{\infty} A^k$, and we have

$$
\frac{\mu_{\psi_r}[w]}{\mu[w]} \lessgtr \exp\left(\pm D \sum_{s=r}^{\infty} \left((|w| + (s+1)(s+2)) \mathrm{var}_s \psi + s \theta^s\right)\right)
$$

for every $r \in \mathbb{N}$ and $w \in \bigcup_{k=1}^{\infty} A^k$.

6.3.3 The limit $\lim_{r \to \infty} \mu_{\psi_r}$ is the Gibbs measure μ_ψ

It only remains to prove that μ such that $\mu[w] := \lim_{r \to \infty} \mu_{\psi_r}$ coincides with the original Gibbs measure μ_ψ. Note first that μ so defined is T-invariant. Indeed, it is the weak* limit of the sequence $(\mu_{\psi_r})_{r \geq 1}$ of T-invariant Markov approximants; it is a T-invariant probability measure as well.

Now, replacing ψ_r by $\psi \pm \mathrm{var}_r\psi$, and making $p = (r+1)(r+2) + n - 1$ in (27), it follows that

$$\mu\lceil w\rceil \lesssim \mu_{\psi_r}\lceil w\rceil \exp(\pm\epsilon_{r,n})$$

$$\lesssim \frac{\sum_{a\in\mathrm{Per}_p(A^{\mathbb{N}})\cap[w]} e^{S_p\psi(a)}}{\sum_{a\in\mathrm{Per}_p(A^{\mathbb{N}})} e^{S_p\psi(a)}} \exp(\pm 2\epsilon_{r,n})$$

for every $w \in \bigcup_{k=1}^{n} A^k$. By taking $n = r^2$, we obtain

$$\mu[w] \lesssim \frac{\sum_{a\in\mathrm{Per}_{(2r+1)(r+1)}(A^{\mathbb{N}})\cap[w]} e^{S_{(2r+1)(r+1)}\psi(a)}}{\sum_{a\in\mathrm{Per}_{(2r+1)(r+1)}(A^{\mathbb{N}})} e^{S_{(2r+1)(r+1)}\psi(a)}} \exp(\pm 2\epsilon_{r,r^2}) \qquad (28)$$

for each $r \in \mathbb{N}$ and $w \in \bigcup_{k=1}^{r^2} A^k$. On the other hand, the Gibbs measure μ_ψ, whose existence is ensured by the fact that $\sum_r \mathrm{var}_r\psi < \infty$, is such that

$$\mu_\psi[w] \lesssim C^{\pm 1} \frac{\sum_{a\in\mathrm{Per}_p(A^{\mathbb{N}})\cap[w]} e^{S_p\psi(a)}}{\sum_{a\in\mathrm{Per}_p(A^{\mathbb{N}})} e^{S_p\psi(a)}}$$

for each $w \in A^k$ with $k \leq p$. Since $\epsilon_{r,r^2} \to 0$ when $r \to \infty$, it follows from this and (28) that μ is absolutely continuous with respect to μ_ψ. The ergodic decomposition theorem [13] implies that μ_ψ is the only ergodic measure entering in the decomposition of the invariant measure μ; therefore, $\mu = \mu_\psi$.

Acknowledgment. We have greatly benefited from the careful reading of an anonymous referee whose valuable comments helped us to improve the article.

References

[1] D. Blackwell. *The entropy of functions of finite–state Markov chains.* In Trans. First Prague Conf. Information Theory, Statistical Decision Functions, Random Processes, 1957, pp. 13–20

[2] R. Bowen. *Equilibrium States and the Ergodic Theory of Anosov Diffeomorphisms,* 2nd edn. Lecture Notes in Mathematics **470**. Springer, Berlin, 2008.

[3] M. Boyle and K. Petersen. *Hidden Markov processes in the context of symbolic dynamics.* This volume, Chapter 1, 2011

[4] J. E. Caroll. *Birkhoff's contraction coefficient.* Linear Algebra Appl. **389** (2004) 227–234

[5] J.-R. Chazottes and E. Ugalde. *Projection of Markov measures may be Gibbsian.* J. Statist. Phys. **111**(5–6) (2003) 1245–1272

[6] J.-R. Chazottes, L. Ramírez, and E. Ugalde. *Finite type approximations to Gibbs measures on sofic subshifts.* Nonlinearity **18**(1) (2005) 445–465

[7] P. Collet and F. Leonardi. *Loss of memory of random functions of Markov chains and Lyapunov exponents*. Preprint (2009). http://arxiv.org/abs/0908.0077

[8] R. Durrett. *Probability: Theory and Examples*, 4th edn. Cambridge University Press, Cambridge, 2010

[9] G. Keller. *Equilibrium States in Ergodic Theory*. London Mathematical Society Student Texts **42**. Cambridge University Press, New York, 1998

[10] B. Kitchens. *Symbolic Dynamics*. Springer, Berlin, 1998

[11] A. N. Kolmogorov and S. V. Fomin. *Introductory Real Analysis*. Dover, Rochester, NY, 1975

[12] O. E. Lanford III and D. Ruelle. *Observables at infinity and states with short range correlations in statistical mechanics*. Commun. Math. Phys. **13** (1969) 194–215

[13] K. E. Petersen. *Ergodic Theory*. Cambridge University Press, Cambridge, 1990

[14] F. Redig and F. Wang. *Transformations of one-dimensional Gibbs measures with infinite-range interaction*. Preprint (2010). http://arxiv.org/abs/1002.4796

[15] D. Ruelle. *Thermodynamic Formalism. The Mathematical Structures of Equilibrium Statistical Mechanics*, 2nd edn. Cambridge Mathematical Library. Cambridge University Press, Cambridge, 2004

[16] E. Seneta. *Non-negative Matrices and Markov Chains*, 2nd edn. Springer, New York, 2006

3

A note on a complex Hilbert metric with application to domain of analyticity for entropy rate of hidden Markov processes

GUANGYUE HAN*

Department of Mathematics, University of Hong Kong, Pokfulam Road, Pokfulam, Hong Kong
E-mail address: ghan@hku.hk

BRIAN MARCUS

Department of Mathematics, University of British Columbia, Vancouver, BC V6T 1Z2, Canada
E-mail address: marcus@math.ubc.ca

YUVAL PERES

Department of Statistics, University of California at Berkeley, Berkeley, CA 94720-3860, USA
E-mail address: peres@stat.berkeley.edu

Abstract. In this article, we show that small complex perturbations of positive matrices are contractions, with respect to a complex version of the Hilbert metric, on a neighborhood of the interior of the real simplex within the complex simplex. We show that this metric can be used to obtain estimates of the domain of analyticity of the entropy rate for a hidden Markov process when the underlying Markov chain has strictly positive transition probabilities.

1 Introduction

The purpose of this article is twofold. First, in Section 2, we introduce a new complex version of the Hilbert metric on the standard real simplex. This metric is defined on a complex neighborhood of the interior of the standard real simplex, within the standard complex simplex. We show that if the neighborhood is sufficiently small, then any sufficiently small complex perturbation of a strictly positive square matrix acts as a contraction, with respect to this metric. While this article was nearing completion, we were informed of a different complex

Entropy of Hidden Markov Processes and Connections to Dynamical Systems: Papers from the Banff International Research Station Workshop, ed. B. Marcus, K. Petersen, and T. Weissman. Published by Cambridge University Press. © Cambridge University Press 2011.

Hilbert metric, which was recently introduced. We briefly discuss the relation between this metric [3] and our metric in Remark 2.7.

Secondly, we show how one can use a complex Hilbert metric to obtain lower estimates of the domain of analyticity of the entropy rate for a hidden Markov process when the underlying Markov chain has strictly positive transition probabilities. The domain of analyticity is important because it specifies an explicit region where a Taylor series converges to the entropy rate and also gives an explicit estimate of the rate of convergence of the Taylor approximation.

In principle, an estimate of the domain can be obtained by examining the proof of analyticity in [5]. That proof was based on a contraction mapping argument, using the fact that the real Euclidean metric is equivalent to the real Hilbert metric. However, in the course of transforming the Euclidean metric to the Hilbert metric, the setup is changed in a way that makes it difficult to keep track of the domain of analyticity. In Section 3.1, we revisit certain aspects of the proof and outline how to modify it using a complex Hilbert metric; this yields a more direct estimate. In Section 3.2, we illustrate this with a small example.

We remark that the entropy rate of a hidden Markov process can be interpreted as a top Lyapunov exponent for a random matrix product [6]. In principle, a complex Hilbert metric can be used, more generally, to estimate the domain of analyticity of the top Lyapunov exponent for certain random matrix products; see [7, 8].

2 Contraction mapping and a complex Hilbert metric

We begin with a review of the real Hilbert metric. Let B be a positive integer, and let W be the standard simplex in B-dimensional real Euclidean space:

$$W = \left\{ w = (w_1, w_2, \ldots, w_B) \in \mathbb{R}^B : w_i \geq 0, \sum_i w_i = 1 \right\},$$

and let W° denote its interior, consisting of the vectors with positive coordinates. For any two vectors $v, w \in W^\circ$, the Hilbert metric [10] is defined as

$$d_H(w, v) = \max_{i,j} \log \left(\frac{w_i/w_j}{v_i/v_j} \right). \tag{1}$$

For a $B \times B$ strictly positive matrix $T = (t_{ij})$, the mapping f_T induced by T on W is defined by $f_T(w) = wT/(wT\mathbf{1})$, where $\mathbf{1}$ is the all-ones vector. It is well known that f_T is a contraction mapping under the Hilbert metric [10]. The contraction coefficient of T, which is also called the Birkhoff coefficient, is given by

$$\tau(T) = \sup_{v \neq w} \frac{d_H(vT, wT)}{d_H(v, w)} = \frac{1 - \sqrt{\phi(T)}}{1 + \sqrt{\phi(T)}}, \qquad (2)$$

where $\phi(T) = \min_{i,j,k,l}(t_{ik}t_{jl}/t_{jk}t_{il})$. This result extends to the case where T has all columns strictly positive or all zero and at least one strictly positive column (then, in the definition of $\phi(T)$, consider only k, l corresponding to strictly positive columns).

Let $W_{\mathbb{C}}$ denote the complex version of W, i.e., $W_{\mathbb{C}}$ denotes the complex simplex comprising the vectors

$$\left\{ w = (w_1, w_2, \ldots, w_B) \in \mathbb{C}^B : \sum_i w_i = 1 \right\}.$$

Let $W_{\mathbb{C}}^+ = \{v \in W_{\mathbb{C}} : \mathcal{R}(v_i/v_j) > 0 \text{ for all } i, j\}$. For $v, w \in W_{\mathbb{C}}^+$, let

$$d_H(v, w) = \max_{i,j} \left| \log \left(\frac{w_i/w_j}{v_i/v_j} \right) \right|, \qquad (3)$$

where log is taken as the principal branch of the complex $\log(\cdot)$ function (i.e., the branch whose branch cut is the negative real axis). Since the principal branch of log is additive on the right half-plane, d_H is a metric on $W_{\mathbb{C}}^+$, which we call a *complex Hilbert metric*.

We will show that any sufficiently small perturbation of a strictly positive matrix is a contraction, with respect to d_H, on a sufficiently small complex neighborhood of W°. We begin with the following very simple lemma.

Lemma 2.1. *Let $n \geq 2$. For any fixed $z_1, z_2, \ldots, z_n, z \in \mathbb{C}$, and fixed $t > 0$, we have*

$$\sup_{t_1, \ldots, t_n \geq 0, \, t_1 + t_2 + \cdots + t_n = t} |t_1 z_1 + t_2 z_2 + \cdots + t_n z_n + z| = \max_{i=1,\ldots,n} |t z_i + z|.$$

Proof. The convex hull of $z_1, z_2, \ldots, z_n$ is a solid polygon, taking the form

$$\{(t_1/t)z_1 + (t_2/t)z_2 + \cdots + (t_n/t)z_n : t_1, t_2, \ldots, t_n \geq 0, t_1 + t_2 + \cdots + t_n = t\}.$$

By convexity, the distance from any point in this solid polygon to the point $(-1/t)z$ will achieve the maximum at one of the extreme points, namely

$$\sup_{t_1, \ldots, t_n \geq 0, t_1 + t_2 + \cdots + t_n = t} |(t_1/t)z_1 + (t_2/t)z_2 + \cdots + (t_n/t)z_n - (-1/t)z|$$

$$= \max_{i=1,\ldots,n} |z_i - (-1/t)z|.$$

The lemma then immediately follows. □

The following lemma is implied by the proof of Lemma 2.1 of [2]; we give a proof for completeness.

Lemma 2.2. *For fixed* $a_1, a_2, \ldots, a_B > 0 \in \mathbb{R}$ *and fixed* $x_1, x_2, \ldots, x_B > 0 \in \mathbb{R}$, *define*

$$D_n = \frac{a_n x_n}{\sum_{m=1}^{B} a_m x_m} - \frac{x_n}{\sum_{m=1}^{B} x_m}.$$

Let $T_0 = \{n : D_n \geq 0\}$ *and* $T_1 = \{n : D_n < 0\}$. *Then we have*

$$\sum_{n \in T_0} D_n = \sum_{n \in T_1} |D_n| \leq \frac{1 - \sqrt{a/A}}{1 + \sqrt{a/A}},$$

where $a = \min\{a_1, a_2, \ldots, a_B\}$ *and* $A = \max\{a_1, a_2, \ldots, a_B\}$.

Proof. It immediately follows from $\sum_{n=1}^{B} D_n = 0$ and the definitions of T_0 and T_1 that

$$\sum_{n \in T_0} D_n = \sum_{n \in T_1} |D_n|.$$

Now

$$\sum_{n \in T_0} D_n = \sum_{n \in T_0} \left(\frac{a_n x_n}{\sum_{m \in T_0} a_m x_m + \sum_{m \in T_1} a_m x_m} - \frac{x_n}{\sum_{m \in T_0} x_m + \sum_{m \in T_1} x_m} \right)$$

$$\leq \sum_{n \in T_0} \left(\frac{A x_n}{A \sum_{m \in T_0} x_m + a \sum_{m \in T_1} x_m} - \frac{x_n}{\sum_{m \in T_0} x_m + \sum_{m \in T_1} x_m} \right).$$

Let

$$z = \frac{\sum_{m \in T_1} x_m}{\sum_{m \in T_0} x_m};$$

we then have

$$\sum_{n \in T_0} D_n \leq \frac{1}{1 + (a/A)z} - \frac{1}{1 + z} = f(z).$$

Simple calculus shows that $f(z)$ will be bounded above by $\frac{1 - \sqrt{a/A}}{1 + \sqrt{a/A}}$ on $[0, \infty)$. This establishes the lemma. □

Let $B_\delta(W^\circ)$ denote the neighborhood of radius δ about W°, contained in $W_{\mathbb{C}}^+$, measured in the Hilbert metric:

$$B_\delta(W^\circ) = \{v \in W_{\mathbb{C}}^+ : \exists u \in W^\circ \text{ such that } d_H(u,v) < \delta\},$$

While we will state our result in terms of $B_\delta(W^\circ)$, our proof will make use of a slightly different neighborhood:

$$W_{\mathbb{C}}^\circ(\delta) = \{v = (v_1, v_2, \ldots, v_B) \in W_{\mathbb{C}} : \exists u \in W^\circ, |v_i - u_i| \leq \delta u_i, i = 1, 2, \ldots, B\}.$$

The neighborhoods $W_{\mathbb{C}}^\circ(\delta)$ and $B_\delta(W^\circ)$ are equivalent in the following sense.

Lemma 2.3. *For some $L > 0$ and sufficiently small $\delta > 0$,*

(1) $W_{\mathbb{C}}^\circ(\delta) \subseteq B_{L\delta}(W^\circ)$,
(2) $B_\delta(W^\circ) \subseteq W_{\mathbb{C}}^\circ(L\delta)$.

Proof. (1) Let $v \in W_{\mathbb{C}}^\circ(\delta)$. Then there exists $u \in W^\circ$ such that for each i, $|v_i/u_i - 1| \leq \delta$. Thus,

$$|v_i/u_i| \geq 1 - \delta \quad \text{for all } i \tag{4}$$

and

$$|v_i/u_i - v_j/u_j| \leq 2\delta \quad \text{for all } i,j. \tag{5}$$

Dividing (5) by $|v_j/u_j|$ and using (4), we see that for each i,j,

$$\left| \frac{v_i/u_i}{v_j/u_j} \right| - 1 \leq \frac{2\delta}{1-\delta} < 4\delta, \quad \text{for } \delta < \frac{1}{2}.$$

This implies that there is a constant $L > 0$ such that for sufficiently small δ, $v \in W_{\mathbb{C}}^+$ and $d_H(u,v) < L\delta$.

(2) Let $v \in B_\delta(W^\circ)$. Then $v \in W_{\mathbb{C}}$ and there exists $u \in W^\circ$ such that

$$d_H(v,u) = \max_{i,j} \left| \log\left(\frac{v_i/u_i}{v_j/u_j} \right) \right| < \delta.$$

It follows that for some $L > 0$,

$$\max_{i,j} \left| \frac{v_i/u_i}{v_j/u_j} - 1 \right| < L\delta.$$

Let $\alpha_j = v_j/u_j$. Then, for all i,j,

$$|v_i - \alpha_j u_i| \le L\delta |\alpha_j| u_i,$$

and so

$$|1 - \alpha_j| = \left| \sum_{i=1}^{n} v_i - \alpha_j u_i \right| \le \sum_{i=1}^{n} |v_i - \alpha_j u_i| \le L\delta |\alpha_j| \sum_{i=1}^{n} u_i = L\delta |\alpha_j|.$$

It follows that $|v_j - u_j| \le L\delta |v_j|$, and so $|v_j| \le u_j/(1 - L\delta)$, and so

$$|v_j - u_j| \le \frac{L\delta}{1 - L\delta} u_j \le 2L\delta u_j,$$

if δ is sufficiently small. Part (2) is then established by replacing L by $2L$. $\qquad\square$

We consider complex matrices $\hat{T} = (\hat{t}_{ij})$ which are perturbations of a strictly positive matrix $T = (t_{ij})$. For such a matrix T and $r > 0$, let $B_T(r)$ denote the set of all complex matrices $\hat{T}$ such that, for all i,j,

$$|t_{ij} - \hat{t}_{ij}| \le r.$$

With the aid of the above lemmas, we shall prove the following theorem.

Theorem 2.4. *Let T be a strictly positive matrix. There exist $r, \delta > 0$ such that whenever $\hat{T} \in B_T(r)$, $f_{\hat{T}}$ is a contraction mapping on $B_\delta(W^\circ)$, with respect to the complex Hilbert metric.*

Proof. For $\hat{x}, \hat{y} \in W_{\mathbb{C}}$, $\hat{x} \ne \hat{y}$, and i,j, let

$$L_{ij} = \frac{\log(\sum_m \hat{x}_m \hat{T}_{mi} / \sum_m \hat{x}_m \hat{T}_{mj}) - \log(\sum_m \hat{y}_m \hat{T}_{mi} / \hat{y}_m \hat{T}_{mj})}{\max_{k,l} \left| \log(\hat{x}_k/\hat{y}_k) - \log(\hat{x}_l/\hat{y}_l) \right|}.$$

Note that

$$\frac{d_H(\hat{x}\hat{T}, \hat{y}\hat{T})}{d_H(\hat{x}, \hat{y})} = \max_{i,j} |L_{ij}|.$$

It suffices to prove that there exists $0 < \rho < 1$ such that for sufficiently small $r, \delta > 0$, $\hat{x}, \hat{y} \in B_\delta(W^\circ)$ $(\hat{x} \ne \hat{y})$, $\hat{T} \in B_T(r)$, and any i,j,

$$|L_{ij}| < \rho.$$

For each m, let $\hat{c}_m = \log(\hat{x}_m/\hat{y}_m)$; then $\hat{x}_m = \hat{y}_m e^{\hat{c}_m}$. Choose $p \neq q$ such that

$$|\hat{c}_p - \hat{c}_q| = \max_{k,l} |\hat{c}_k - \hat{c}_l|.$$

Hence,

$$L_{ij} = \frac{\log(\sum_m \hat{y}_m e^{\hat{c}_m - \hat{c}_q} \hat{T}_{mi} / \sum_m \hat{y}_m e^{\hat{c}_m - \hat{c}_q} \hat{T}_{mj}) - \log(\sum_m \hat{y}_m \hat{T}_{mi} / \hat{y}_m \hat{T}_{mj})}{|\hat{c}_p - \hat{c}_q|}.$$

Define

$$F(t) = \log\left(\sum_m \hat{y}_m e^{(\hat{c}_m - \hat{c}_q)t} \hat{T}_{mi} \Big/ \sum_m \hat{y}_m e^{(\hat{c}_m - \hat{c}_q)t} \hat{T}_{mj} \right).$$

Since

$$|F(1) - F(0)| = \left| \int_0^1 F'(t)\, dt \right| \leq \max_{\xi \in [0,1]} |F'(\xi)|,$$

we have

$$|L_{ij}| = \frac{|F(1) - F(0)|}{|\hat{c}_p - \hat{c}_q|} \leq \frac{\max_{\xi \in [0,1]} |F'(\xi)|}{|\hat{c}_p - \hat{c}_q|}. \tag{6}$$

Note that $F'(\xi)$ takes the following form:

$$F'(\xi) = \frac{\sum_m (\hat{c}_m - \hat{c}_q) \hat{y}_m e^{(\hat{c}_m - \hat{c}_q)\xi} \hat{T}_{mi}}{\sum_m \hat{y}_m e^{(\hat{c}_m - \hat{c}_q)\xi} \hat{T}_{mi}} - \frac{\sum_m (\hat{c}_m - \hat{c}_q) \hat{y}_m e^{(\hat{c}_m - \hat{c}_q)\xi} \hat{T}_{mj}}{\sum_m \hat{y}_m e^{(\hat{c}_m - \hat{c}_q)\xi} \hat{T}_{mj}}.$$

Now, for all m, let $\hat{a}_m = \hat{T}_{mi}/\hat{T}_{mj}$. Then

$$\frac{F'(\xi)}{|\hat{c}_p - \hat{c}_q|} = \sum_n \frac{\hat{c}_n - \hat{c}_q}{|\hat{c}_p - \hat{c}_q|} \left(\frac{\hat{y}_n e^{(\hat{c}_n - \hat{c}_q)\xi} \hat{a}_n \hat{T}_{nj}}{\sum_m \hat{y}_m e^{(\hat{c}_m - \hat{c}_q)\xi} \hat{a}_m \hat{T}_{mj}} - \frac{\hat{y}_n e^{(\hat{c}_n - \hat{c}_q)\xi} \hat{T}_{nj}}{\sum_m \hat{y}_m e^{(\hat{c}_m - \hat{c}_q)\xi} \hat{T}_{mj}} \right)$$

$$= \sum_n \frac{\hat{c}_n - \hat{c}_q}{|\hat{c}_p - \hat{c}_q|} \hat{D}_n, \tag{7}$$

where $\hat{D}_n$ denotes the quantity in parentheses in the middle expression.

If $\hat{x}, \hat{y} \in B_\delta(W^\circ)$, there exist $x, y \in W^\circ$ such that for all k, $|\hat{x}_k - x_k| \leq L\delta x_k$ and $|\hat{y}_k - y_k| \leq L\delta y_k$, where L is as in Lemma 2.3(2).

Let $a_m = T_{mi}/T_{mj}$, $c_m = \log(x_m/y_m)$, and let D_n denote the unperturbed version of $\hat{D}_n$:

$$D_n = \frac{y_n e^{(c_n - c_q)\xi} a_n T_{nj}}{\sum_m y_m e^{(c_m - c_q)\xi} a_m T_{mj}} - \frac{y_n e^{(c_n - c_q)\xi} T_{nj}}{\sum_m y_m e^{(c_m - c_q)\xi} T_{mj}}. \tag{8}$$

By Lemma 2.2, we have

$$\sum_{n \in \mathcal{T}_0} D_n = \sum_{n \in \mathcal{T}_1} |D_n| \le \max_{k,l} \frac{1 - \sqrt{a_k/a_l}}{1 + \sqrt{a_k/a_l}} \le \tau(T), \tag{9}$$

where $\mathcal{T}_0 = \{n : D_n \ge 0\}$ and $\mathcal{T}_1 = \{n : D_n < 0\}$.

Now, for some universal constant K_0,

$$\left| \sum_n \frac{\hat{c}_n - \hat{c}_q}{|\hat{c}_p - \hat{c}_q|} \hat{D}_n - \sum_n \frac{\hat{c}_n - \hat{c}_q}{|\hat{c}_p - \hat{c}_q|} D_n \right| < K_0(L\delta + r). \tag{10}$$

Applying Lemma 2.1 twice, we conclude that there exist $n_0 \in \mathcal{T}_0, n_1 \in \mathcal{T}_1$ such that

$$\left| \sum_n \frac{\hat{c}_n - \hat{c}_q}{|\hat{c}_p - \hat{c}_q|} D_n \right| \le \left| \frac{\hat{c}_{n_0} - \hat{c}_q}{|\hat{c}_p - \hat{c}_q|} \sum_{n \in \mathcal{T}_0} D_n + \sum_{n \in \mathcal{T}_1} \frac{\hat{c}_n - \hat{c}_q}{|\hat{c}_p - \hat{c}_q|} D_n \right|$$

$$\le \left| \frac{\hat{c}_{n_0} - \hat{c}_q}{|\hat{c}_p - \hat{c}_q|} \sum_{n \in \mathcal{T}_0} D_n - \frac{\hat{c}_{n_1} - \hat{c}_q}{|\hat{c}_p - \hat{c}_q|} \sum_{n \in \mathcal{T}_1} |D_n| \right|.$$

Then together with (6), (7), (10), (9), and the fact that $|\hat{c}_{n_1} - \hat{c}_{n_0}| \le |\hat{c}_p - \hat{c}_q|$, we obtain that for sufficiently small $r, \delta > 0$, $|L_{ij}|$ is upper bounded by some $\rho < 1$, as desired. □

Remark 2.5. One can further choose $r, \delta > 0$ such that when $\hat{T} \in B_T(r)$, $f_{\hat{T}}(W_{\mathbb{C}}^\circ(\delta)) \subset W_{\mathbb{C}}^\circ(\delta)$. Consider a compact subset $N \subset W^\circ$ such that $f_T(W) \subset N$. Let $N(R)$ denote the Euclidean R-neighborhood of N in $W_{\mathbb{C}}$. The proof of Theorem 2.4 implies that when $T > 0$ or ($T \ge 0$ and $\sup_{x,y \in N, 0 \le \xi \le 1} \sum_{n \in \mathcal{T}_0} D_n < 1$ (here D_n is defined in (8))), there exist $r, R > 0$ such that when $\hat{T} \in B_T(r)$, $f_{\hat{T}}$ is a contraction mapping on $N(R)$ under the complex Hilbert metric.

Example 2.6. Consider a 2×2 strictly positive matrix

$$T = \begin{bmatrix} a & c \\ b & d \end{bmatrix}.$$

If we parameterize the interior of the simplex W° by $(0,\infty)$: $w = (x,y) \mapsto x/y$, then, letting $z = x/y$, we have $f_T(z) = (az + b)/(cz + d)$; the domain of this mapping naturally extends from $(0,\infty)$ to the open right half complex plane H, and the complex Hilbert metric becomes simply $d_H(z_1, z_2) = |\log(z_1/z_2)|$. This metric is simply the image, via the exponential map, of the Euclidean metric on the strip $\{z \in \mathbb{C} : |\mathcal{I}(z)| < \pi/2\}$.

One can show that f_T is a contraction on all of H with contraction coefficient

$$\tau(T) = \frac{1 - bc/ad}{1 + bc/ad}$$

(assuming that $\det(T) \geq 0$; otherwise, the last expression is replaced by $(1 - ad/bc)/(1 + ad/bc)$). To see this, for any $z, w \in H$, consider

$$L = \left| \frac{\log(f_T(z)) - \log(f_T(w))}{\log(z) - \log(w)} \right|.$$

With the change of variables $u = \log(z), v = \log(w)$, we have

$$L = \left| \frac{\log(f_T(e^u)) - \log(f_T(e^v))}{u - v} \right| = \left| \int_0^1 e^{v+t(u-v)} \frac{f_T'(e^{v+t(u-v)})}{f_T(e^{v+t(u-v)})} dt \right|,$$

which implies that

$$L \leq \sup_{z \in H} \left| \frac{z f_T'(z)}{f_T(z)} \right|.$$

A simple computation shows that

$$\frac{z f_T'(z)}{f_T(z)} = \frac{ad - bc}{acz + (ad + bc) + bd/z}. \tag{11}$$

To see that the supremum is $(1 - bc/ad)/(1 + bc/ad)$, first note that since $ad - bc \geq 0$ and $a, b, c, d > 0$, the absolute value of the quantity on the right-hand side of (11) is maximized by minimizing $|acz + bd/z|$; since the only solutions to $acz + bd/z = 0$ are $z = \pm i\sqrt{bd/ac}$, one sees that the supremum is obtained by substituting $z = \pm i\sqrt{bd/ac}$ into (11), and this shows that the supremum is indeed $(1 - bc/ad)/(1 + bc/ad)$.

Note that this contraction coefficient on H is strictly larger (i.e., worse) than the contraction coefficient on $[0,\infty)$: $(1 - \sqrt{bc/ad})/(1 + \sqrt{bc/ad})$.

When

$$\hat{T} = \begin{bmatrix} \hat{a} & \hat{c} \\ \hat{b} & \hat{d} \end{bmatrix}$$

is a sufficiently small complex perturbation of T, then $f_{\hat{T}}(H) \subseteq H$ and one obtains

$$\tau(\hat{T}) = \sup_{z \in H} \left| \frac{z f'_{\hat{T}}(z)}{f_{\hat{T}}(z)} \right| = \sup_{z \in H} \left| \frac{\hat{a}\hat{d} - \hat{b}\hat{c}}{\hat{a}\hat{c}z + (\hat{a}\hat{d} + \hat{b}\hat{c}) + \hat{b}\hat{d}/z} \right|,$$

which will approximate $(1 - bc/ad)/(1 + bc/ad)$, and so $f_{\hat{T}}$ will still be a contraction on H.

Remark 2.7. While this article was nearing completion, we were informed that alternative complex Hilbert metrics, based on the Poincaré metric in the right half complex plane, were recently introduced in Rugh [9] and Dubois [3]. Contractiveness with respect to these metrics is proven in great generality and yields far-reaching consequences for complex Perron–Frobenius theory. The proofs of contractiveness in these papers seem rather different from the calculus approach in our article.

The complex Hilbert metric, which we call d_P, used in [3] (see equation (3.23)) is explicit and natural, but slightly more complicated than our complex Hilbert metric; for $v, w \in W_{\mathbb{C}}^+$,

$$d_P(w, v) = \log \frac{\max_{i,j}(|\overline{w_i}v_j + \overline{w_j}v_i| + |w_i v_j - w_j v_i|)(2\mathcal{R}(\overline{w_i}w_j))^{-1}}{\min_{i,j}(|\overline{w_i}v_j + \overline{w_j}v_i| - |w_i v_j - w_j v_i|)(2\mathcal{R}(\overline{w_i}w_j))^{-1}}, \quad (12)$$

where $\bar{z}$ denotes complex conjugate, $\mathcal{R}(z)$ denotes real part, and log is the ordinary real logarithm. In the two-dimensional case, it can be verified that, if one transforms $w = (w_1, w_2)$ and $v = (v_1, v_2)$ to $z_1 = w_2/w_1$ and $z_2 = w_2/w_1$, then d_P reduces to the Poincaré metric on H:

$$d_P(z_1, z_2) = \log \frac{|z_1 + \bar{z}_2| + |z_1 - z_2|}{|z_1 + \bar{z}_2| - |z_1 - z_2|}.$$

Using the infinitesimal form for the Poincaré metric (as a Riemannian metric on H), one checks that, in the 2×2 case, the Lipschitz constant for a complex matrix $\hat{T}$ such that $f_{\hat{T}}(H) \subseteq H$ is

$$\sup_{z \in H} \left| \frac{\mathcal{R}(z) f'_{\hat{T}}(z)}{\mathcal{R}(f_{\hat{T}}(z))} \right|, \quad (13)$$

in contrast to

$$\sup_{z \in H} \left| \frac{z f'_{\hat{T}}(z)}{f_{\hat{T}}(z)} \right| \tag{14}$$

for our complex Hilbert metric (as in Example 2.6 above).

While we have not analyzed in detail the differences between these metrics, there are a few things that can be said in the 2×2 case:

- $f_{\hat{T}}$ is a contraction with respect to d_P on H whenever it maps H into its interior; this follows from standard complex analysis (Section IX.3 of [4]), and Dubois [3] proves an analogue of this for the metric d_P (12) in higher dimensions. However, this does not hold for d_H.

- When $\hat{T} = T$ is strictly positive, the contraction coefficient, with respect to d_P, is always at least as good as (i.e., at most) the contraction coefficient with respect to d_H. This can be seen as follows.

  First, recall that any fractional linear transformation T can be expressed as the composition of transitions, dilations, and inversions. In the case where T is strictly positive, the translations are by positive real numbers and the dilations are by real numbers; see page 65 of [4]. Using the infinitesimal forms (13), (14), our assertion would follow from

$$\left| \frac{\mathcal{R}(z)}{z} \right| \le \left| \frac{\mathcal{R}(f_T(z))}{f_T(z)} \right| \quad \text{for all } z \in H. \tag{15}$$

  This is true indeed: it is easy to see that in fact we get equality in (15) for inversions and dilations by real numbers, and we get strict inequality in (15) for translations by positive real numbers.

- When $\hat{T}$ is a complex perturbation of a strictly positive transformation T, (15) (with T replaced by $\hat{T}$) need not hold; in fact, for perturbations $\hat{T}$ of T on the order of 1% and $z = x + yi \in H$, with $|y|/x$ on the order of 1%, the contraction coefficient with respect to d_H may be slightly smaller than that with respect to d_P. The reason is that, in this case, the dilations may be complex (nonreal) and for such a dilation the inequality (15) may be reversed. Examples of this can be randomly generated in MATLAB. For example, if

$$\hat{T} = \begin{bmatrix} 0.012890500224 + 0.000128905002i & 0.310402226067 + 0.003104022260i \\ 0.779079247486 - 0.007790792474i & 0.307296084921 - 0.003072960849i \end{bmatrix}$$

and $z = 0.926678310631 - 0.009266783106i$, then the contraction coefficent of d_H is approximately 0.664396 and that of d_P is approximately 0.664599. For larger perturbations, the differences in contraction coefficient can be greater. The relative strength of contraction of d_H, d_P seems to be heavily dependent on specific choices of $\hat{T}$ and z.

- For any point z, other than 0, of the imaginary axis, the metric d_H can be extended to a neighborhood with respect to which any sufficiently small complex perturbation $\hat{T}$ of a strictly positive matrix acts as a contraction; on the other hand, there is no way to do this with d_P since it blows up as one approaches the imaginary axis.

- Also, on a small punctured neighborhood of 0, if we replace d_H by the metric $d(z_1, z_2) = |\log(z_1) - \log(z_2)|$, then a small complex perturbation $\hat{T}$ of a strictly positive matrix still acts as a contraction.

In the next section, we use d_H for estimates of the domain of analyticity of the entropy rate of a hidden Markov process. Alternatively, d_P could be used; however, it appears to be computationally easier to use d_H for the estimation.

3 Domain of analyticity of entropy rate of hidden Markov processes

3.1 Background

For $m, n \in \mathbb{Z}$ with $m \leq n$, we denote a sequence of symbols $y_m, y_{m+1}, \ldots, y_n$ by y_m^n. Consider a stationary stochastic process Y with a finite set of states $\mathcal{I} = \{1, 2, \ldots, B\}$ and distribution $p(y_m^n)$. Denote the conditional distributions by $p(y_{n+1}|y_m^n)$. The entropy rate of Y is defined as

$$H(Y) = \lim_{n \to \infty} -E_p(\log(p(y_0|y_{-n}^{-1}))),$$

where E_p denotes expectation with respect to the distribution p.

Let Y be a stationary first-order Markov chain with

$$\Delta(i,j) = p(y_1 = j | y_0 = i).$$

In this section, we only consider the case when Δ is strictly positive.

A *hidden Markov process* (HMP) Z is a process of the form $Z = \Phi(Y)$, where Φ is a function defined on $\mathcal{I} = \{1, 2, \ldots, B\}$ with values in $\mathcal{J} = \{1, 2, \ldots, A\}$.

Recall that W is the B-dimensional real simplex and $W_\mathbb{C}$ is the complex version of W. For $a \in \mathcal{J}$, let $\mathcal{I}(a)$ denote the set of all indices $i \in \mathcal{I}$ with $\Phi(i) = a$.

Let

$$W_a = \{w \in W : w_i = 0 \text{ whenever } i \notin \mathcal{I}(a)\}$$

and

$$W_{a,\mathbb{C}} = \{w \in W_{\mathbb{C}} : w_i = 0 \text{ whenever } i \notin \mathcal{I}(a)\}.$$

Let Δ_a denote the $B \times B$ matrix such that $\Delta_a(i,j) = \Delta(i,j)$ for $j \in \mathcal{I}(a)$, and $\Delta_a(i,j) = 0$ for $j \notin \mathcal{I}(a)$ (i.e., Δ_a is formed from Δ by "zeroing out" the columns corresponding to indices that are not in $\mathcal{I}(a)$. For $a \in \mathcal{J}$, define the scalar-valued and vector-valued functions r_a and f_a on W by

$$r_a(w) = w \Delta_a \mathbf{1}$$

and

$$f_a(w) = w \Delta_a / r_a(w).$$

Note that f_a defines the action of the matrix Δ_a on the simplex W. For any fixed n and z_{-n}^0 and for $i = -n, -n+1, \ldots,$ define

$$x_i = x_i(z_{-n}^i) = p(y_i = \cdot \,|\, z_i, z_{i-1}, \ldots, z_{-n}) \tag{16}$$

(here $\cdot$ represents the states of the Markov chain Y); then, from Blackwell [1], we have that $\{x_i\}$ satisfies the random dynamical iteration

$$x_{i+1} = f_{z_{i+1}}(x_i), \tag{17}$$

starting with

$$x_{-n-1} = p(y_{-n-1} = \cdot), \tag{18}$$

where $p(y_{-n-1} = \cdot)$ is the stationary distribution for the underlying Markov chain. One checks that $p(z_{i+1}|z_{-n}^i)$ can be recovered from this dynamical system; more specifically, we have

$$p(z_{i+1}|z_{-n}^i) = r_{z_{i+1}}(x_i).$$

If the entries of $\Delta = \Delta^{\vec{\varepsilon}}$ are analytically parameterized by a real variable vector $\vec{\varepsilon} \in \mathbb{R}^k$ (k is a positive integer), then we obtain a family $Z = Z^{\vec{\varepsilon}}$ and corresponding $\Delta_a = \Delta_a^{\vec{\varepsilon}}$, $f_a = f_a^{\vec{\varepsilon}}$, etc.

The following result was proven in [5].

Theorem 3.1. *Suppose that the entries of $\Delta = \Delta^{\vec{\varepsilon}}$ are analytically parameterized by a real variable vector $\vec{\varepsilon}$. If at $\vec{\varepsilon} = \vec{\varepsilon}_0$, Δ is strictly positive, then $H(Z) = H(Z^{\vec{\varepsilon}})$ is a real analytic function of $\vec{\varepsilon}$ at $\vec{\varepsilon}_0$.*

In [5] this result is stated in greater generality, allowing some entries of Δ to be zero. The proof is based on an analysis of the action of perturbations of f_a on neighborhoods of $\hat{W}_b \overset{\triangle}{=} f_b(W)$, with respect to the Euclidean metric. The proof assumes that each f_a is a contraction on each $\hat{W}_b$. While this need not hold, one can arrange for this to be true by replacing the original system with a higher-power system: namely, one replaces the original alphabet $\mathcal{J}$ with $\mathcal{J}^n$ for some n and replaces the mappings $\{f_a : a \in \mathcal{J}\}$ with $\{f_{a_0} \circ f_{a_1} \circ \cdots \circ f_{a_{n-1}} : a_0 a_1 \cdots a_{n-1} \in \mathcal{J}^n\}$. The existence of such an n follows from (a) the equivalence of the (real) Hilbert metric and the Euclidean metric on each $\hat{W}_b$ (Proposition 2.1 of [5]) and (b) the contractiveness of each f_a with respect to the (real) Hilbert metric. However, in the course of this replacement, one easily loses track of the domain of analyticity.

When at $\vec{\varepsilon} = \vec{\varepsilon}_0$, Δ is strictly positive, an alternative is to directly use a complex Hilbert metric, as follows. For each $a \in \mathcal{J}$, we can define a complex Hilbert metric $d_{a,H}$ on $W_{a,\mathbb{C}}^{\circ}$ as follows: for $w, v \in W_{a,\mathbb{C}}^{\circ}$,

$$d_{a,H}(w,v) = d_H(w_{\mathcal{I}(a)}, v_{\mathcal{I}(a)}) = \max_{i,j \in \mathcal{I}(a)} \left| \log \left(\frac{w_i/w_j}{v_i/v_j} \right) \right|. \tag{19}$$

Theorem 2.4 implies that for each $a, b \in \mathcal{J}$, sufficiently small perturbations of f_a are contractions on sufficiently small complex neighborhoods of $\hat{W}_b$ in $W_{b,\mathbb{C}}$; see Remark 2.5 (note that while Δ_a is not strictly positive, f_a maps into W_a and so as a mapping from W_b to W_a it can be regarded as the induced mapping of a strictly positive matrix). For complex $\vec{\varepsilon}$ close to $\vec{\varepsilon}_0$, $f_a = f_a^{\vec{\varepsilon}}$ is sufficiently close to $f_a^{\vec{\varepsilon}_0}$ to guarantee that $f_a^{\vec{\varepsilon}}$ is a contraction.

Let $\Omega_{a,H}(R)$ denote the neighborhood of diameter R, measured in the complex Hilbert metric, of $\hat{W}_a$ in $W_{a,\mathbb{C}}$. Let $B_{\vec{\varepsilon}_0}(r)$ denote the complex r-neighborhood of $\vec{\varepsilon}_0$ in $\mathbb{C}^k$.

Following the proof of Theorem 3.1 (especially pages 5254–5255 of [5]), one obtains a lower bound $r > 0$ on the domain of analyticity if there exist $R > 0$ and $0 < \rho < 1$ satisfying the following conditions.

(1) For any $a, z \in \mathcal{A}$ and any $\vec{\varepsilon} \in B_r(\vec{\varepsilon}_0)$, $f_z^{\vec{\varepsilon}}$ is a contraction, with respect to the complex Hilbert metric, on $\Omega_{a,H}(R)$:

$$\sup_{x \neq y \in \Omega_{a,H}(R)} \left| \frac{d_{z,H}(f_z^{\vec{\varepsilon}}(x), f_z^{\vec{\varepsilon}}(y))}{d_{a,H}(x,y)} \right| \leq \rho < 1.$$

(2) For any $\vec{\varepsilon} \in B_r(\vec{\varepsilon}_0)$, any $x \in \cup_a \hat{W}_a$, and any $z \in \mathcal{A}$,

$$d_{z,H}(f_z^{\vec{\varepsilon}}(x), f_z^{\vec{\varepsilon}_0}(x)) \leq R(1-\rho)$$

and

$$d_{z,H}(f_z^{\vec{\varepsilon}}(\pi(\varepsilon)), f_z^{\vec{\varepsilon}_0}(\pi(\varepsilon_0))) \leq R(1-\rho)$$

(where $\pi(\varepsilon)$ denotes the stationary vector for the Markov chain defined by $\Delta^{\vec{\varepsilon}}$).

(3) For any $x \in \Omega_{a,H}(R)$ and $\vec{\varepsilon} \in B_r(\vec{\varepsilon}_0)$,

$$\sum_a |r_a^{\vec{\varepsilon}}(x)| \leq 1/\rho.$$

The existence of r, R, ρ follows from Theorem 3.1. In fact, we can choose ρ to be any positive number such that $\max_{a \in \mathcal{A}} \tau(\Delta_a) < \rho < 1$, and small r, R to satisfy Condition 1, then smaller r, R, if necessary, to further satisfy Conditions 2 and 3.

Let $\Omega_{a,E}(R)$ denote the neighborhood of diameter R, measured in the Euclidean metric, of $\hat{W}_a$ in $W_{a,\mathbb{C}}$. To facilitate the computation, at the expense of obtaining a smaller lower bound, it may be easier to use $\Omega_{a,E}(R)$ instead of $\Omega_{a,H}(R)$; then, the conditions above are replaced with the following conditions.

(1′) Condition 1 above with $\Omega_{a,H}(R)$ replaced by $\Omega_{a,E}(R)$ (the map f_z^ε is still required to be a contraction under the complex *Hilbert* metric).

(2′) Condition 2 above with R on the right-hand side of the inequalities replaced by R/K, where

$$K = \sup_{x \neq y \in \Omega_{a,E}(R),a} \left| \frac{d_{a,E}(x,y)}{d_{a,H}(x,y)} \right|;$$

note that for R sufficiently small, $0 < K < \infty$ since $d_{a,H}$ and $d_{a,E}$ are equivalent metrics (this in turn follows from the fact that the Euclidean metric and (real) Hilbert metric are equivalent on any compact subset of the interior of the real simplex).

(3′) Condition 3 above with $\Omega_{a,H}(R)$ replaced by $\Omega_{a,E}(R)$.

3.2 Example for domain of analyticity

In the following, we consider hidden Markov processes obtained by passing binary Markov chains through binary symmetric channels with cross-over probability ε. Suppose that the Markov chain is defined by a 2×2 stochastic matrix $\Pi = [\pi_{ij}]$. From now through the end of this section, we assume that

- $\det(\Pi) > 0$,
- all $\pi_{ij} > 0$, and
- $0 < \varepsilon < 1/2$.

We remark that the condition $\det(\Pi) > 0$ is purely for convenience.

Strictly speaking, the underlying Markov process of the resulting hidden Markov process is given by a four-state matrix (the states are the ordered pairs of a state of Π and a noise state (0 for "noise off" and 1 for "noise on"); see page 5255 of [5]). However, the information contained in each f_a can be reduced to an equivalent map induced by a 2×2 matrix and then reduced to an equivalent function of a single variable as in Example 2.6. We describe this as follows.

Let $a_i = p(z_1^i, y_i = 0)$ and $b_i = p(z_1^i, y_i = 1)$. The pair (a_i, b_i) satisfies the following dynamical system:

$$(a_i, b_i) = (a_{i-1}, b_{i-1}) \begin{bmatrix} p_E(z_i)\pi_{00} & p_E(z_i)\pi_{10} \\ p_E(\bar{z}_i)\pi_{01} & p_E(\bar{z}_i)\pi_{11} \end{bmatrix},$$

where $p_E(0) = \varepsilon$ and $p_E(1) = 1 - \varepsilon$.

Similar to Example 2.6, let $x_i = a_i/b_i$; we then have a dynamical system with just one variable:

$$x_{i+1} = f_{z_{i+1}}^{\varepsilon}(x_i),$$

where

$$f_z^{\varepsilon}(x) = \frac{p_E(z)}{p_E(\bar{z})} \frac{\pi_{00}x + \pi_{10}}{\pi_{01}x + \pi_{11}}, \qquad z = 0, 1$$

starting with

$$x_0 = \pi_{10}/\pi_{01}, \tag{20}$$

which comes from the stationary vector of Π.

It can be shown that

$$p^{\varepsilon}(z_i = 0 | z_1^{i-1}) = r_0^{\varepsilon}(x_{i-1}), \qquad p^{\varepsilon}(z_i = 1 | z_1^{i-1}) = r_1^{\varepsilon}(x_{i-1}),$$

where

$$r_0^{\varepsilon}(x) = \frac{((1-\varepsilon)\pi_{00} + \varepsilon\pi_{01})x + ((1-\varepsilon)\pi_{10} + \varepsilon\pi_{11})}{x+1} \tag{21}$$

and

$$r_1^\varepsilon(x) = \frac{(\varepsilon\pi_{00} + (1-\varepsilon)\pi_{01})x + (\varepsilon\pi_{10} + (1-\varepsilon)\pi_{11})}{x+1}. \tag{22}$$

Now let $\Omega(R)$ denote the complex R-neighborhood (in the Euclidean metric) of the interval

$$S = [S_1, S_2] = \left[\frac{\varepsilon_0\pi_{10}}{(1-\varepsilon_0)\pi_{11}}, \frac{(1-\varepsilon_0)\pi_{00}}{\varepsilon_0\pi_{01}} \right];$$

this interval is the union of $f_0^{\varepsilon_0}([0,\infty])$ and $f_1^{\varepsilon_0}([0,\infty])$; again let $B_{\varepsilon_0}(r)$ denote the complex r-neighborhood of a given cross-over probability $\varepsilon_0 > 0$.

The sufficient conditions (1′), (2′), and (3′) in Section 3.1 are guaranteed by the following. There exist $R > 0, r > 0, 0 < \rho < 1$ such that the following hold.

(1″) For any z, $f_z^\varepsilon(x)$ is a contraction on $\Omega(R)$ under the complex Hilbert metric,

$$\sup_{x \neq y \in \Omega(R)} \left| \frac{\log f_z^\varepsilon(x) - \log f_z^\varepsilon(y)}{\log x - \log y} \right| \leq \rho < 1.$$

Note that here

$$\log f_z^\varepsilon(x) - \log f_z^\varepsilon(y) = \log \frac{\pi_{00}x + \pi_{10}}{\pi_{01}x + \pi_{11}} - \log \frac{\pi_{00}y + \pi_{10}}{\pi_{01}y + \pi_{11}}.$$

(2″) For any $\varepsilon \in B_{\varepsilon_0}(r)$, any $x \in S$, and any z,

$$|\log f_z^\varepsilon(x) - \log f_z^{\varepsilon_0}(x)| \leq (R/K)(1-\rho),$$

where

$$K = \sup_{x \neq y \in \Omega(R)} \left| \frac{x-y}{\log x - \log y} \right| = \sup_{x \in \Omega(R)} |x| = S_2 + R$$

(note that here the second condition in (2′) is vacuous since by (20) x_0 does not depend on ε).

(3″) For any $x \in \Omega(R)$ and $\varepsilon \in B_{\varepsilon_0}(r)$,

$$|r_0^\varepsilon(x)| + |r_1^\varepsilon(x)| \leq 1/\rho.$$

By considering extreme cases, the above conditions can be further relaxed to:

(1‴)

$$0 < \frac{\pi_{00}\pi_{11} - \pi_{10}\pi_{01}}{\pi_{01}\pi_{00}(S_1 - R) + \pi_{01}\pi_{10} + \pi_{11}\pi_{00} + \pi_{11}\pi_{10}/(S_2 + R)} \leq \rho$$

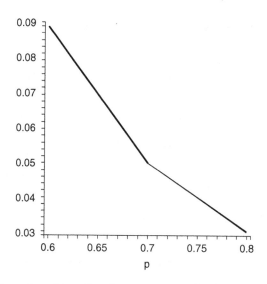

Figure 1 Lower bound on radius of convergence as a function of p.

(here we applied the mean value theorem to give an upper bound on
$|\log((\pi_{00}x+\pi_{10})/(\pi_{01}x+\pi_{11}))-\log((\pi_{00}y+\pi_{10})/(\pi_{01}y+\pi_{11}))|$);
(2''')

$$0<\frac{r}{\varepsilon_0-r}+\frac{r}{1-\varepsilon_0-r}\leq(R/(S_2+R))(1-\rho)$$

(here we applied the mean value theorem to give an upper bound on
$|\log((1-\varepsilon)/\varepsilon)-\log((1-\varepsilon_0)/\varepsilon_0)|$);
(3''')

$$0<\frac{((1-\varepsilon_0+r)\pi_{00}+(\varepsilon_0+r)\pi_{01})(S_2+R)+((1-\varepsilon_0+r)\pi_{10}+(\varepsilon_0+r)\pi_{11})}{S_1-R+1}$$
$$+\frac{((\varepsilon_0+r)\pi_{00}+(1-\varepsilon_0+r)\pi_{01})(S_2+R)+((\varepsilon_0+r)\pi_{10}+(1-\varepsilon_0+r)\pi_{11})}{S_1-R+1}$$
$$\leq1/\rho.$$

In other words, choose r, R, and ρ to satisfy the conditions (1'''), (2'''), and (3''').
Then the entropy rate is an analytic function of ε on $|\varepsilon-\varepsilon_0|<r$.

Consider the symmetric case: $\pi_{00}=\pi_{11}=p$ and $\pi_{01}=\pi_{10}=1-p$. We plot
lower bounds on the radius of convergence of $H(Z)$ (as a function of ε at
$\varepsilon_0=0.4$) against p in Figure 1. For a fixed p, the lower bound is obtained by
randomly generating many 3-tuples (r, R, ρ) and taking the maximal r from the
3-tuples which satisfies the conditions.

Acknowledgments. We thank Albert Chau for helpful discussions on Riemannian metrics in H.

The first author gratefully acknowledges the support of Research Grants Council of the Hong Kong Special Administrative Region, China under grant No. HKU 701708P.

References

[1] D. Blackwell. *The entropy of functions of finite-state Markov chains.* In Trans. First Prague Conf. Information Theory, Statistical Decision Functions, Random Processes, 1957, pp. 13–20

[2] R. Cavazos-Cadena. *An alternative derivation of Birkhoff's formula for the contraction coefficient of a positive matrix.* Linear Algebra Appl. **375** (2003) 291–297

[3] L. Dubois. *Projective metrics and contraction principles for complex cones.* J. Lond. Math. Soc. **79**(3) (2009) 719–737

[4] T. Gamelin. *Complex Analysis.* Springer, New York, 2001

[5] G. Han and B. Marcus. *Analyticity of entropy rate of hidden Markov chains.* IEEE Trans. Inf. Theory **52**(12) (2006) 5251–5266.

[6] T. Holliday. A. Goldsmith, and P. Glynn, *Capacity of finite state channels based on Lyapunov exponents of random matrices.* IEEE Trans. Inf. Theory **52**(8) (2006) 3509–3532

[7] Y. Peres. *Analytic Dependence of Lyapunov Exponents on Transition Probabilities.* Lecture Notes in Mathematics **1486**. Springer, Berlin, 1991, pp. 131–148

[8] Y. Peres. *Domains of analytic continuation for the top Lyapunov exponent.* Ann. Inst. H. Poincaré Prob. Statist. **28**(1) (1992) 131–148

[9] H. H. Rugh. *Cones and gauges in complex spaces: spectral gaps and complex Perron–Frobenius theory.* Ann. Math. **171**(3) (2010) 1707–1752

[10] E. Seneta. *Non-negative Matrices and Markov Chains.* Springer Series in Statistics. Springer, Berlin, 1980

4

Bounds on the entropy rate of binary hidden Markov processes

ERIK ORDENTLICH

Hewlett-Packard Laboratories, 1501 Page Mill Road, MS 1181,
Palo Alto, CA 94304, USA
E-mail address: erik.ordentlich@hp.com

TSACHY WEISSMAN

Information Systems Laboratory, Department of Electrical Engineering,
Stanford University, Packard 256, Stanford, CA 94305, USA
E-mail address: tsachy@stanford.edu

Abstract. Let $\{X_i\}$ be a stationary finite-alphabet Markov chain and $\{Z_i\}$ denote its noisy version when corrupted by a discrete memoryless channel. Let $P(X_i \in \cdot | Z^i_{-\infty})$ denote the conditional distribution of X_i given all past and present noisy observations, a simplex-valued random variable. We present an approach to bounding the entropy rate of $\{Z_i\}$ by approximating the distribution of this simplex-valued random variable. This approximation is facilitated by the construction and study of a Markov process whose stationary distribution determines the distribution of $P(X_i \in \cdot | Z^i_{-\infty})$, while being more tractable than the latter. The bounds are particularly meaningful in situations where the support of $P(X_i \in \cdot | Z^i_{-\infty})$ is significantly smaller than the whole simplex. To illustrate its efficacy, we specialize this approach to the case of a binary symmetric channel corrupted binary Markov chain. The bounds obtained are sufficiently tight to characterize the behavior of the entropy rate in asymptotic regimes that exhibit a "concentration of the support". Examples include the "high signal-to-noise ratio", "low signal-to-noise ratio", "rare spikes", and "weak dependence" regimes. Our analysis also gives rise to a deterministic algorithm for approximating the entropy rate, achieving the best known precision–complexity tradeoff for certain subsets of the process parameter space.

1 Introduction

1.1 The problem

Let $\{X_i\}$ be a stationary Markov chain and $\{Z_i\}$ denote its noisy version when corrupted by a discrete memoryless channel (DMC). The components of these

Entropy of Hidden Markov Processes and Connections to Dynamical Systems: Papers from the
Banff International Research Station Workshop, ed. B. Marcus, K. Petersen, and T. Weissman.
Published by Cambridge University Press. © Cambridge University Press 2011.

processes take values, respectively, in the finite alphabets $\mathcal{X}$ and $\mathcal{Z}$. We let $\mathcal{K}$ denote the transition kernel of the Markov chain, i.e., the $|\mathcal{X}| \times |\mathcal{X}|$ matrix with entries

$$\mathcal{K}(x, x') = P(X_{i+1} = x' | X_i = x). \tag{1}$$

Let $\mathcal{C}$ denote the channel transition matrix, i.e., the $|\mathcal{X}| \times |\mathcal{Z}|$ matrix with entries

$$\mathcal{C}(x, z) = P(Z_i = z | X_i = x). \tag{2}$$

The process $\{Z_i\}$ is known as a hidden Markov process (HMP). Its distribution and, a fortiori, its entropy rate which we denote by $\overline{H}(Z)$, are completely determined by the pair $(\mathcal{K}, \mathcal{C})$. However, the explicit form of $\overline{H}(Z)$ as a function of this pair is unknown, and is our interest in this work.

1.2 Motivation

Hidden Markov processes arise naturally in many contexts, both as information sources and as noise (cf. [9] and references therein). Their entropy rate naturally arises in data compression and communications:

- Lossless compression: how many bits per source symbol are required to losslessly encode an HMP?
- Lossy compression: assume that $\mathcal{X} = \mathcal{Z}$ and that addition and subtraction of elements in this alphabet are well defined. Assume further that the DMC relating $\{X_i\}$ to $\{Z_i\}$ is an additive noise channel with a distribution which is maximum entropy [5, Chapter 12] with respect to the per-letter additive distortion criterion $d(x, y) = \rho(x - y)$ for some nonnegative, real-valued function ρ. For example, if the DMC is symmetric, leaving the input symbol unchanged with a certain probability $p > 1/|\mathcal{X}|$, and flipping equiprobably to each of the remaining symbols, the corresponding distortion measure is Hamming loss. For this setting, the rate distortion function satisfies the Shannon lower bound [14, 13, 2, 34, 35], so is explicitly given by

$$R(D) = \overline{H}(Z) - \phi(D), \tag{3}$$

where $\phi(D)$ is the "single-letter" maximum-entropy function defined by

$$\phi(D) = \max\{H(X) : E\rho(X) \le D\},$$

with ρ as above and the maximum being over all $\mathcal{X}$-valued random variables X. Since $\phi(D)$ is readily obtainable in closed form, evaluation of the rate distortion function for the HMP reduces, by (3), to evaluation of its entropy rate.

- Channel coding: consider an additive noise channel of the form

$$Y_i = U_i + Z_i, \tag{4}$$

where $\{U_i\}$ is the transmitted channel input, $\{Y_i\}$ is the received channel output, the noise process $\{Z_i\}$ is the above-described HMP, and all process components are $|\mathcal{Z}|$-valued, with addition in (4) being in the finite-field mod-$|\mathcal{Z}|$ sense. For example, in the binary case this is the "Gilbert–Elliot" or "burst-noise" channel [10, 8, 25]. It is easy to show that the capacity of the channel in (4), for the case of no input constraints, is achieved by an independent and identically distributed (i.i.d.) uniform distribution on the input, implying that its capacity is given by

$$C = \log_2 |\mathcal{Z}| - \overline{H}(Z). \tag{5}$$

Evidently, key questions in lossless compression, lossy compression (3), and channel coding (5) reduce to finding the entropy rate $\overline{H}(Z)$.

1.3 On the hardness of determining $\overline{H}(Z)$

Let $\mathcal{M}(\mathcal{X})$ denote the simplex of distributions on $\mathcal{X}$ and β_i be the $\mathcal{M}(\mathcal{X})$-valued random variable defined by

$$\beta_i(x) = P(X_i = x | Z_{-\infty}^i),$$

where $\beta_i(x)$ denotes the xth component of β_i. We denote this by

$$\beta_i = P(X_i \in \cdot | Z_{-\infty}^i).$$

We refer to $\{\beta_i\}$ as the "belief process", as it represents the "belief" of an observer of the HMP regarding the value of the underlying state. Conditional independence of X_{i+1} and $Z_{-\infty}^i$ given X_i implies that $P(X_{i+1} \in \cdot | Z_{-\infty}^i) = \beta_i \cdot \mathcal{K}$, in turn implying, by the memorylessness of the noise, that

$$P(Z_{i+1} \in \cdot | Z_{-\infty}^i) = \beta_i \cdot \mathcal{K} \cdot \mathcal{C}, \tag{6}$$

where $\mathcal{K}$ and $\mathcal{C}$ are, respectively, the Markov and channel transition matrices defined in (1) and (2) (viewing elements of $\mathcal{M}(\mathcal{X})$ as row vectors). With $H(Q)$

denoting the entropy of a distribution Q on $\mathcal{Z}$,

$$H(Q) = \sum_{z \in \mathcal{Z}} Q(z) \log_2 \frac{1}{Q(z)},$$

we obtain

$$\overline{H}(Z) = H(Z_{i+1} | Z_{-\infty}^i) = EH(P(Z_{i+1} \in \cdot | Z_{-\infty}^i)) = EH(\beta_i \cdot \mathcal{K} \cdot \mathcal{C}). \tag{7}$$

Evidently, the distribution of β_i holds the key to the value of the entropy rate. This distribution, however, shown by Blackwell in [4] (cf. also [29, Claim 1]) to satisfy an integral equation, remains elusive to date even for the simplest HMPs.

Another perspective by which the hardness of the problem can be appreciated is that developed in [21, 20] of Lyapunov exponents. Standard recursion for HMPs yields [9]

$$P(Z^n = z^n) = \mu_s^T \left[\prod_{i=1}^n [\mathcal{K} \odot \mathcal{C}(\cdot, z_i)^T] \right] \mathbf{1},$$

where μ_s is the stationary distribution of the underlying Markov chain (represented as a column vector), $\mathcal{K} \odot \mathcal{C}(\cdot, z_i)^T$ denotes the $|\mathcal{X}| \times |\mathcal{X}|$ matrix whose xth row is given by the componentwise multiplication of the xth row of $\mathcal{K}$ by the row vector whose x'th component is $\mathcal{C}(x', z_i)$, and $\mathbf{1}$ denotes the "all-ones" column vector. The Shannon–McMillan–Breiman theorem [5] implies then that, with probability one,

$$\overline{H}(Z) = \lim_{n \to \infty} -\frac{1}{n} \log_2 \mu_s^T \left[\prod_{i=1}^n [K \odot C(\cdot, Z_i)^T] \right] \mathbf{1}$$

$$= \lim_{n \to \infty} -\frac{1}{n} \log_2 \left\| \prod_{i=1}^n [K \odot C(\cdot, Z_i)^T] \right\|, \tag{8}$$

where $\|\cdot\|$ denotes any matrix norm. In other words, up to sign, $\overline{H}(Z)$ is the (top) Lyapunov exponent (cf., e.g., [30]) associated with the square-matrix-valued process $\{K \odot C(\cdot, Z_i)^T\}_{i \geq 1}$. Characterization of the Lyapunov exponent is an open question, even in the simplest cases of finite-valued i.i.d. matrices [30, 1]. In our case, $\{K \odot C(\cdot, Z_i)^T\}_{i \geq 1}$ is not even a Markov process.

Yet another perspective on the problem is that of statistical physics. For simplicity, consider the HMP given by a binary symmetric channel (BSC) corrupted symmetric binary Markov source. The distribution of Z^n can be put in the form [36, 24]

$$P(Z^n) = \sum_{x^n} P(x^n) P(Z^n | x^n) = \sum_{x^n} P(x_1) \prod_{i=1}^{n-1} \mathcal{K}(x_i, x_{i+1}) \prod_{i=1}^{n} \mathcal{C}(x_i, Z_i)$$

$$= c_1 c_2^n \sum_{\tau^n} \exp\left(J \sum_{i=1}^{n-1} \tau_i \tau_{i+1} + K \sum_{i=1}^{n} \tau_i \sigma_i \right), \tag{9}$$

where the last equality is obtained by a change of variables $\tau_i = (-1)^{X_i}$ and $\sigma_i = (-1)^{Z_i}$, and c_1, c_2, J, K are explicit functions of the process parameters. Characterization of the entropy rate reduces then to that of the limit, in n, of

$$\frac{1}{n} E \log_2 \sum_{\tau^n} \exp\left(J \sum_{i=1}^{n-1} \tau_i \tau_{i+1} + K \sum_{i=1}^{n} \tau_i \sigma_i \right), \tag{10}$$

which is the expected density of the logarithm of the partition function associated with the Gibbs measure for (random) energy levels $E(\tau^n) = -\sum_{i=1}^{n-1} \tau_i \tau_{i+1} - K/J \sum_{i=1}^{n} \tau_i \sigma_i$ at temperature $1/J$. The asymptotic value, for large n, of this expected density is an open problem in statistical physics *even* for the case where the energy levels are i.i.d. [33].

1.4 Existing results

Given the hardness of the problem, the predominant approach to the study of the entropy rate, until relatively recently, has been one of approximation (cf. [25, 21, 7] and references therein). Indeed, what we refer to as the "Cover and Thomas" bounds

$$H(Z_0 | Z_{-n}^{-1}, X_{-n-1}) \leq \overline{H}(Z) \leq H(Z_0 | Z_{-n}^{-1}) \tag{11}$$

hold for every n, becoming arbitrarily tight with increasing n [5, Section 4.5]. We shall discuss these bounds in more detail in Section 5, where we suggest an alternative deterministic scheme for approximating the entropy rate. Another approach for approximating the entropy rate is via (8), which implies that simulating the HMP and evaluating $-(1/n) \log_2 \left\| \prod_{i=1}^{n} [K \odot C(\cdot, Z_i)^{\mathsf{T}}] \right\|$ gives an estimate, which, for large n, becomes arbitrarily precise with probability arbitrarily close to one (cf. [21] and references therein).

Useful as these techniques may be from a numerical standpoint, they lack the capacity to resolve basic questions regarding the dependence of the entropy rate on the Markov transition kernel and the channel parameters. First steps toward the resolution of such questions were taken in [21], where continuity of the entropy rate in the parameters was established. Significant progress in this direction was made by Han and Marcus in a recent series of papers [15, 16, 17, 18], which not only established smoothness, but also characterized conditions for differentiability and analyticity of the entropy rate in the transition parameters.

Expansions of the entropy rate for the BSC-corrupted binary Markov chain in the "high signal-to-noise ratio (SNR)" regime, where the channel cross-over probability is small, have been obtained initially in [22, 29, 36, 27]. Initial results on the behavior in various additional asymptotic regimes such as "rare spikes", "rare bursts", "low SNR", and "almost memoryless" were obtained in [29]. We expand on and strengthen this line of results in subsequent sections by incorporating into the approach of [29] finer properties of the distribution of the belief process β_i, as summarized in the next subsection. More recent refinements and extensions were obtained in [15, 16, 17].

1.5 Our approach

An immediate consequence of (7) is the following observation.

Observation 1.1.

$$\min_{\beta \in S} H\left(\beta \cdot \mathcal{K} \cdot \mathcal{C}\right) \leq \overline{H}(Z) \leq \max_{\beta \in S} H\left(\beta \cdot \mathcal{K} \cdot \mathcal{C}\right),$$

where S denotes the support of β_i.

Trivial as this observation may seem, it was shown in [29] to lead to useful bounds in cases where bounds on the support set S are obtainable, and these bounds are significantly smaller than the whole simplex $\mathcal{M}(\mathcal{X})$.

The bounds of Observation 1.1, which depend on the distribution of β_i through its support only, can be refined by covering S using several disjoint sets and considering also the probabilities of these sets.

Observation 1.2. For any countable collection $\{I_k\}$ of pairwise-disjoint sets $I_k \subseteq \mathcal{M}(\mathcal{X})$ covering S (i.e., for which $S \subseteq \bigcup_k I_k$),

$$\sum_k P(\beta_i \in I_k) \inf_{\beta \in I_k} H\left(\beta \cdot \mathcal{K} \cdot \mathcal{C}\right) \leq \overline{H}(Z) \leq \sum_k P(\beta_i \in I_k) \sup_{\beta \in I_k} H\left(\beta \cdot \mathcal{K} \cdot \mathcal{C}\right). \quad (12)$$

Since the distribution of β_i is unknown, $P(\beta_i \in I_k)$ will also be unknown in general. However, for certain choices of $\{I_k\}$, and in certain regions of the space of parameters governing the HMP, the bounds in (12) can be either explicitly evaluated or closely bounded. This is done by constructing a Markov process which is more tractable than the $\{\beta_i\}$ process. The stationary distribution of this process is directly and simply related to the distribution of β_i. The fraction of times that the process visits the set I_k, for appropriately chosen I_k, is computable, a computation that can then be directly translated to give the value of $P(\beta_i \in I_k)$.

For concreteness and simplicity in illustrating the idea, we concentrate on the case where $\{Z_i\}$ is a BSC-corrupted binary Markov chain. In this context, the two new ingredients of our approach relative to [29] involve covering the support of the belief process by multiple disjoint intervals I_k (as opposed to only two intervals in [29]) and the construction of an alternative, more tractable, Markov process as a tool for analyzing the probabilities $P(\beta_i \in I_k)$ of the belief process falling into these intervals. The incorporation of these finer properties of β_i is shown to lead to tighter characterizations of $\overline{H}(Z)$, in various asymptotic regimes, than were obtained in [29]. The alternative Markov process is also leveraged to obtain and analyze the aforementioned deterministic algorithm for numerically approximating $\overline{H}(Z)$. Several of the above results have appeared in preliminary form in our previous conference papers [27, 28].

We remark that while our approach is based, in part, on finite coverings of the support $\mathcal{S}$ of the belief process, little is known about $\mathcal{S}$, as a whole, beyond the observations in [4]. It is shown therein, via examples, that $\mathcal{S}$, in general, can be a finite set, a countable set, or an uncountable set with Lebesgue measure 0 (in a strong sense made precise in [4]). It is in fact conjectured in [4] that if the distribution of β_i is continuous (e.g., no point masses), it will be singular with respect to Lebesgue measure (again, in a strong sense made precise in [4]).

1.6 Remaining content

In Section 2, we start with a concrete description of the problem setting, and the evolution of the log-likelihood process (equivalent to the belief process but in a more convenient form) for the case of the BSC-corrupted Markov chain. We then detail the construction of an alternative Markov process, and its relationship to the original log-likelihood (and, therefore, belief) process. Section 3 focuses on the case of a symmetric Markov chain, and details the form the bounds in (12) assume for this case, in terms of the alternative Markov process. Using these bounds, we then derive the behavior of the entropy rate in various asymptotic regimes. Section 4 follows a similar development for the case where the underlying binary Markov chain is not necessarily symmetric. In Section 5 we describe a deterministic algorithm, inspired by the alternative

Markov process, for approximating the entropy rate. We show that its guaranteed precision–complexity tradeoff is the best among all known deterministic schemes for approximation of the entropy rate, for certain subsets of the parameter space. This algorithm was preliminarily presented in [28] for the symmetric Markov chain case. More recently, a similar approach was taken in [23], again for the symmetric Markov chain case, but the details of the resulting algorithm and its analysis are different. Section 6 contains a summary of the paper, along with a discussion of some related directions.

2 The BSC-corrupted binary Markov chain

2.1 Setup and some notation

Assume henceforth the case $\mathcal{X} = \mathcal{Z} = \{0, 1\}$, where the Markov transition matrix and the channel matrix are, respectively,

$$\mathcal{K} = \begin{pmatrix} 1 - \pi_{01} & \pi_{01} \\ \pi_{10} & 1 - \pi_{10} \end{pmatrix}, \quad \mathcal{C} = \begin{pmatrix} 1 - \delta & \delta \\ \delta & 1 - \delta \end{pmatrix}. \tag{13}$$

Without loss of generality, we assume that $\delta \leq 1/2$ and $\pi_{01} \leq \pi_{10}$. To avoid trivialities, we also assume below that:

(1) $\pi_{01} + \pi_{10} \neq 1$ (otherwise the state process is i.i.d.);
(2) either $\pi_{01} \in (0, 1)$ or $\pi_{10} \in (0, 1)$ (otherwise the state process is essentially (up to its initial state) deterministic and $\overline{H}(Z) = h_b(\delta)$).

For positive-valued functions f and g, $f(\varepsilon) \sim g(\varepsilon)$ will stand for $\lim_{\varepsilon \downarrow 0} \frac{f(\varepsilon)}{g(\varepsilon)} = 1$ and $f(\varepsilon) \overset{\sim}{<} g(\varepsilon)$ will stand for $\limsup_{\varepsilon \downarrow 0} \frac{f(\varepsilon)}{g(\varepsilon)} \leq 1$; $f(\varepsilon) = O(g(\varepsilon))$ will stand for $\limsup_{\varepsilon \downarrow 0} \frac{f(\varepsilon)}{g(\varepsilon)} < \infty$ and $f(\varepsilon) = \Omega(g(\varepsilon))$ will stand for $\liminf_{\varepsilon \downarrow 0} \frac{f(\varepsilon)}{g(\varepsilon)} > 0$; $f(\varepsilon) \asymp g(\varepsilon)$ will stand for the statement that both $f(\varepsilon) = O(g(\varepsilon))$ and $f(\varepsilon) = \Omega(g(\varepsilon))$ hold. If R is a random variable, $\mathcal{L}(R)$ will denote its law. Similarly, if A is an event, $\mathcal{L}(R|A)$ will denote the law of R conditioned on A. Also, for R, S random variables and $0 \leq \alpha \leq 1$, $\alpha \mathcal{L}(R) + (1 - \alpha)\mathcal{L}(S)$ will denote the law of $BR + (1 - B)S$, for $B \sim \text{Bernoulli}(\alpha)$ that is independent of R and S. We extend this interpretation to the combination of more than two laws, namely to $\sum_i \alpha_i \mathcal{L}(R_i)$ for $\alpha_i \geq 0$ summing to 1 and R_i random variables, in the obvious way. Throughout the article, $\log_2$ and $\log$ will denote the base-2 and natural logarithms, respectively, and all entropies and entropy rates are expressed in bits (underlying $\log_2$).

2.2 Evolution of the log-likelihood

The standard forward recursions [9] are readily shown (cf., e.g., [29]) to assume the form

$$\frac{\beta_i(0)}{1-\beta_i(0)} = \left[\frac{1-\delta}{\delta}\right]^{1-2Z_i} g\left(\frac{\beta_{i-1}(0)}{1-\beta_{i-1}(0)}\right), \tag{14}$$

where

$$g(x) = \frac{x(1-\pi_{01})+\pi_{10}}{x\pi_{01}+(1-\pi_{10})}. \tag{15}$$

Equivalently, this can be expressed as

$$l_i = (2Z_i - 1)\log\left[\frac{1-\delta}{\delta}\right] + f(l_{i-1}), \tag{16}$$

where $l_i = \log\frac{\beta_i(1)}{1-\beta_i(1)}$ and

$$f(x) = \log\frac{\pi_{01}+e^x(1-\pi_{10})}{(1-\pi_{01})+e^x\pi_{10}}. \tag{17}$$

It follows from (7) that, in terms of the log-likelihood process, the entropy rate is given by

$$\overline{H}(Z) = Eh_b\left([\beta_i(1)(1-\pi_{10})+(1-\beta_i(1))\pi_{01}]*\delta\right)$$
$$= Eh_b\left(\left[\frac{e^{l_i}}{1+e^{l_i}}(1-\pi_{10})+\frac{1}{1+e^{l_i}}\pi_{01}\right]*\delta\right), \tag{18}$$

where $*$ denotes binary convolution defined by $p*q = (1-p)q+(1-q)p$ and

$$h_b(x) = -x\log_2 x - (1-x)\log_2(1-x)$$

is the binary entropy function.

In the sequel, we shall make use of the following properties of the function f in (17). First, note that

$$f'(x) = \frac{e^x(1-\pi_{01}-\pi_{10})}{(1-\pi_{01})\pi_{01}+e^{2x}(1-\pi_{10})\pi_{10}+e^x(1-\pi_{01}-\pi_{10}+2\pi_{01}\pi_{10})}, \tag{19}$$

which has the sign of the numerator, so f is either strictly increasing or strictly decreasing according to whether $\pi_{01}+\pi_{10} < 1$ or $\pi_{01}+\pi_{10} > 1$. Another important property of f is its contractiveness. To be sure, note that

$$\sup_x |f'(x)| = \frac{|1 - \pi_{01} - \pi_{10}|}{\min_{y>0}\left[(1-\pi_{01})\pi_{01}y^{-1}+(1-\pi_{10})\pi_{10}y+(1-\pi_{01}-\pi_{10}+2\pi_{01}\pi_{10})\right]}$$

$$= \frac{|1 - \pi_{01} - \pi_{10}|}{2\sqrt{(1-\pi_{01})\pi_{01}(1-\pi_{10})\pi_{10}} + (1-\pi_{01}-\pi_{10}+2\pi_{01}\pi_{10})}$$

$$\overset{\triangle}{=} c(\pi_{01}, \pi_{10}), \tag{20}$$

where for the second equality we have used the elementary fact that for nonnegative a, b, c

$$\min_{y>0}\left[ay^{-1} + by + c\right] = 2\sqrt{ab} + c.$$

This implies that f is contractive since $c(\pi_{01}, \pi_{10}) < 1$ (the denominator is obviously greater than the numerator if $\pi_{01} + \pi_{10} < 1$ and is invariant under the transformation $(\pi_{01}, \pi_{10}) \to (1 - \pi_{01}, 1 - \pi_{10})$).

When specialized to the symmetric case $\pi_{10} = \pi_{01} = \pi$, we obtain the evolution

$$l_i = (2Z_i - 1)\log\left[\frac{1-\delta}{\delta}\right] + f(l_{i-1}), \tag{21}$$

where $f(x) = \log\frac{e^x(1-\pi)+\pi}{e^x\pi+(1-\pi)}$. Specializing (18) for this case gives

$$\overline{H}(Z) = Eh_b\left(\frac{e^{l_i}}{1+e^{l_i}} * \pi * \delta\right). \tag{22}$$

In this symmetric case, (20) becomes

$$\sup_x |f'(x)| = 1 - 2\pi. \tag{23}$$

Thus, as discussed in the introduction, the distribution of β_i (or, equivalently, of l_i) is key to the evaluation of the entropy rate. Although $\{\beta_i\}$ was shown to be a Markov process by Blackwell [4], its analysis turns out to be quite elusive. In what follows, we construct another, more tractable, Markov process, whose stationary distribution is closely related to (and determines) the distribution of β_i.

2.3 An alternative Markov process

In this subsection, we construct a Markov process, which, as a process, is more tractable than the log-likelihood process $\{l_i\}$, but whose stationary distribution is closely and simply related to that of l_i. The benefit is that the entropy rate, which was expressed as the expectation in (18), or (22) in the symmetric case, will be expressible as a similar expectation involving the new process. Our alternative process is closely related to the joint state/belief process (X_i, β_i) studied in [11] and [32]. The former reference derived conditions for the geometric ergodicity (exponential convergence to the stationary distribution) of the joint process and the latter reference applied these conditions to give a simple proof of Birch's result [3] on the exponential decay (in n) of the difference between the upper and lower bounds on the entropy rate in (11) above.

2.3.1 The symmetric case

To illustrate the idea behind the construction of the alternative Markov process in its simplest form, we start with the symmetric case where $\pi_{10} = \pi_{01} = \pi < 1/2$. There is no loss of generality in assuming that $\pi < 1/2$ since the argument in [29, Section 4-C] implies that the entropy rate when the Markov chain is symmetric with transition probability $1 - \pi$ is the same as when it is π.

Theorem 2.1. *Consider the first-order Markov process* $\{Y_i\}_{i \geq 0}$ *formed by letting* $Y_0 = Y$ *and* $\{Y_i\}_{i \geq 1}$ *evolve according to*

$$Y_i = r_i \log \frac{1-\delta}{\delta} + s_i f(Y_{i-1}), \qquad (24)$$

where $\{r_i\}$ *and* $\{s_i\}$ *are independent i.i.d. sequences, independent of* Y, *with*

$$r_i = \begin{cases} -1 & w.p.\ \delta, \\ 1 & w.p.\ 1-\delta, \end{cases} \quad s_i = \begin{cases} -1 & w.p.\ \pi, \\ 1 & w.p.\ 1-\pi. \end{cases} \qquad (25)$$

In this theorem, "w.p." stands for "with probability". Then

(1) [existence and uniqueness of the stationary distribution] *there exists a unique (in distribution) random variable* Y *under which* $\{Y_i\}_{i \geq 0}$ *is stationary;*

(2) [connection to the original process] $\mathcal{L}(Y) = \mathcal{L}(l_i | X_i = 1)$.

Proof. It is evident from (24) and (25) that a distribution on Y is a stationary distribution for the process $\{Y_i\}$ if and only if it satisfies

$$\mathcal{L}(Y) = \pi\delta \cdot \mathcal{L}\left(-\log\frac{1-\delta}{\delta} - f(Y)\right) + (1-\pi)\delta \cdot \mathcal{L}\left(-\log\frac{1-\delta}{\delta} + f(Y)\right)$$
$$+ \pi(1-\delta) \cdot \mathcal{L}\left(\log\frac{1-\delta}{\delta} - f(Y)\right)$$
$$+ (1-\pi)(1-\delta) \cdot \mathcal{L}\left(\log\frac{1-\delta}{\delta} + f(Y)\right). \tag{26}$$

To prove uniqueness, assume first that there exists a distribution on Y satisfying (26), and let $\{\tilde{Y}_i\}$ denote the stationary process evolving according to (24), initiated at time 0 with an arbitrary stationary distribution (and arbitrarily jointly distributed with Y). Then, due to (23),

$$|\tilde{Y}_i - Y_i| = |f(\tilde{Y}_{i-1}) - f(Y_{i-1})| \le (1-2\pi)|\tilde{Y}_{i-1} - Y_{i-1}|,$$

so

$$|\tilde{Y}_i - Y_i| \le (1-2\pi)^i|\tilde{Y}_0 - Y_1|$$

and, in particular,

$$|\tilde{Y}_i - Y_i| \longrightarrow 0$$

as $i \to \infty$ (for all sample paths). This implies, when combined with the stationarity of both processes, that $\tilde{Y}_0 \stackrel{d}{=} Y_0$. To prove existence, as well as the second assertion of the theorem, it will suffice to establish the fact that taking $\mathcal{L}(Y) = \mathcal{L}(l_i|X_i = 1)$ satisfies (26). To see this, note first that for all $\alpha \in \mathbb{R}$,

$$P(f(l_{i-1}) \le \alpha|X_i = 1) = \sum_j P(f(l_{i-1}) \le \alpha, X_{i-1} = j|X_i = 1)$$
$$= \sum_j P(X_{i-1} = j|X_i = 1)P(f(l_{i-1}) \le \alpha|X_{i-1} = j)$$
$$= \pi P(f(l_{i-1}) \le \alpha|X_{i-1} = 0) + (1-\pi)P(f(l_{i-1})$$
$$\le \alpha|X_{i-1} = 1)$$
$$= \pi P(f(-l_{i-1}) \le \alpha|X_{i-1} = 1) + (1-\pi)P(f(l_{i-1})$$
$$\le \alpha|X_{i-1} = 1),$$

the last equality following since, by symmetry, $\mathcal{L}(l_{i-1}|X_{i-1} = 0) = \mathcal{L}(-l_{i-1}|X_{i-1} = 1)$. The strict increasing monotonicity of $f(\cdot)$ implies then that

$$\mathcal{L}(l_{i-1}|X_i = 1) = \pi\mathcal{L}(-l_i|X_i = 1) + (1-\pi)\mathcal{L}(l_i|X_i = 1). \tag{27}$$

Now, *conditioned on the event* $X_i = 1$, the two summands on the right-hand side of (16) are *independent*, with the first being distributed as

$$(2Z_i - 1) \log \left[\frac{1-\delta}{\delta} \right] = \begin{cases} \log \frac{1-\delta}{\delta} & \text{w.p. } 1-\delta, \\ -\log \frac{1-\delta}{\delta} & \text{w.p. } \delta. \end{cases} \tag{28}$$

Combined with (27), this implies that the distribution $\mathcal{L}(l_i|X_i = 1)$ satisfies (26). □

Henceforth, when referring to the process defined in Theorem 2.1, we assume that it was initiated by the stationary distribution.

Corollary 2.2. *For the process constructed in Theorem 2.1,*

$$\overline{H}(Z) = Eh_b \left(\frac{e^{Y_i}}{1+e^{Y_i}} * \pi * \delta \right). \tag{29}$$

Proof. We have

$$\overline{H}(Z) = \frac{1}{2} E \left[h_b \left(\frac{e^{l_i}}{1+e^{l_i}} * \pi * \delta \right) \bigg| X_i = 1 \right] + \frac{1}{2} E \left[h_b \left(\frac{e^{l_i}}{1+e^{l_i}} * \pi * \delta \right) \bigg| X_i = 0 \right]$$

$$= Eh_b \left(\frac{e^{Y_i}}{1+e^{Y_i}} * \pi * \delta \right), \tag{30}$$

the first equality following from (22) and the second from the second item of Theorem 2.1 and the facts that $\mathcal{L}(l_i|X_i = 1) = \mathcal{L}(-l_i|X_i = 0)$ and that

$$h_b \left(\frac{e^y}{1+e^y} * \pi * \delta \right) = h_b \left(\frac{1}{2} - \frac{e^y}{1+e^y} * \pi * \delta \right) = h_b \left(\frac{e^{-y}}{1+e^{-y}} * \pi * \delta \right)$$

for all y. □

The bottom line is that we have transformed the calculation of the entropy rate into an expectation of a simple function of the variable Y_i. It will be seen that the benefit in doing so is that information on the distribution of Y_i, which translates via Corollary 2.2 to bounds on the entropy rate, can be inferred by studying the dynamics of the process $\{Y_i\}$.

2.3.2 The non-symmetric case

In the symmetric case, it sufficed to construct one process with real-valued components whose stationary distribution is $\mathcal{L}(l_i|X_i = 1)$, since this immediately conveyed also $\mathcal{L}(l_i|X_i = 0)$ as, by symmetry, $\mathcal{L}(l_i|X_i = 1) = \mathcal{L}(-l_i|X_i = 0)$. In the nonsymmetric case, we have the following theorem.

Theorem 2.3. *Define* $\{(Y_i, U_i)\}_{i \geq 0}$, *a Markov process with state space* $\mathbb{R}^2$, *by letting* $(Y_0, U_0) = (Y, U)$ *and, for* $i \geq 1$,

$$Y_i = r_i \log \frac{1-\delta}{\delta} + s_i f(U_{i-1}) + (1 - s_i) f(Y_{i-1}) \tag{31}$$

and

$$U_i = q_i \log \frac{1-\delta}{\delta} + (1 - t_i) f(U_{i-1}) + t_i f(Y_{i-1}), \tag{32}$$

where $\{q_i\}$, $\{r_i\}$, $\{s_i\}$, $\{t_i\}$ *are independent i.i.d. sequences, independent of* (Y, U), *with*

$$q_i = \begin{cases} 1 & w.p.\ \delta, \\ -1 & w.p.\ 1-\delta, \end{cases} \quad r_i = \begin{cases} -1 & w.p.\ \delta, \\ 1 & w.p.\ 1-\delta, \end{cases} \tag{33}$$

and $s_i \sim \text{Bernoulli}(\pi_{10})$, $t_i \sim \text{Bernoulli}(\pi_{01})$. *Then*

(1) [existence of a marginally stationary distribution] *there exists a distribution on the pair* (Y, U) *under which* $\{(Y_i, U_i)\}_{i \geq 0}$ *is marginally stationary in the sense that, for all* $i \geq 0$, $\mathcal{L}(Y_i) = \mathcal{L}(Y)$ *and* $\mathcal{L}(U_i) = \mathcal{L}(U)$);
(2) [uniqueness of marginals and connection to the original process] *any distribution on* (Y, U) *giving rise to a process which is marginally stationary in the sense of the previous item satisfies* $\mathcal{L}(Y) = \mathcal{L}(l_i | X_i = 1)$ *and* $\mathcal{L}(U) = \mathcal{L}(l_i | X_i = 0)$.

Remark. We will refer to a distribution on (Y, U) that gives rise to a process $\{(Y_i, U_i)\}$ which is marginally stationary in the above sense as a "marginally stationary distribution". It is evident from the evolution equations (31) and (32) that $\mathcal{L}(Y_i)$ and $\mathcal{L}(U_i)$ depend on $\mathcal{L}(Y_{i-1}, U_{i-1})$ only through the marginal distributions $\mathcal{L}(Y_{i-1})$ and $\mathcal{L}(U_{i-1})$. Thus, if a distribution on the pair (Y, U) gives rise to a marginally stationary process, then any other distribution with the same marginal distributions of U and Y has the same property. Conversely, the second item of the theorem implies that *all* distributions on (Y, U) giving rise to a marginally stationary process will share the same marginals, which, respectively, are given by $\mathcal{L}(l_i | X_i = 1)$ and $\mathcal{L}(l_i | X_i = 0)$. In particular, the marginal stationary distributions are unique.

Proof of Theorem 2.3. Conditioned on the event $X_i = 1$, the two summands on the right-hand side of (16) are independent with

$$(2Z_i - 1) \log \left[\frac{1-\delta}{\delta} \right] = \begin{cases} \log \frac{1-\delta}{\delta} & \text{w.p. } 1 - \delta, \\ -\log \frac{1-\delta}{\delta} & \text{w.p. } \delta. \end{cases} \tag{34}$$

Furthermore, for any $\alpha \in \mathbb{R}$,

$$P(f(l_{i-1}) \leq \alpha | X_i = 1)$$

$$= \sum_j P(f(l_{i-1}) \leq \alpha, X_{i-1} = j | X_i = 1)$$

$$= \sum_j P(X_{i-1} = j | X_i = 1) P(f(l_{i-1}) \leq \alpha | X_{i-1} = j)$$

$$= \pi_{10} P(f(l_{i-1}) \leq \alpha | X_{i-1} = 0) + (1 - \pi_{10}) P(f(l_{i-1}) \leq \alpha | X_{i-1} = 1),$$

implying with the strict monotonicity of f that

$$\mathcal{L}(l_{i-1} | X_i = 1) = \pi_{10} \mathcal{L}(l_{i-1} | X_{i-1} = 0) + (1 - \pi_{10}) \mathcal{L}(l_{i-1} | X_{i-1} = 1). \tag{35}$$

Similarly, conditioned on the event $X_i = 0$, the two summands on the right-hand side of (16) are independent with

$$(2Z_i - 1) \log\left[\frac{1-\delta}{\delta}\right] = \begin{cases} \log \frac{1-\delta}{\delta} & \text{w.p. } \delta, \\ -\log \frac{1-\delta}{\delta} & \text{w.p. } 1 - \delta, \end{cases} \tag{36}$$

and a computation similar to that leading to (35) gives

$$\mathcal{L}(l_{i-1} | X_i = 0) = \pi_{01} \mathcal{L}(l_{i-1} | X_{i-1} = 1) + (1 - \pi_{01}) \mathcal{L}(l_{i-1} | X_{i-1} = 0). \tag{37}$$

It follows from (16), (34)–(37), and the mentioned conditional independence that any distribution on (Y, U) where $\mathcal{L}(Y) = \mathcal{L}(l_i | X_i = 1)$ and $\mathcal{L}(U) = \mathcal{L}(l_i | X_i = 0)$ gives a marginally stationary process, establishing the first item.

For the second item, let $\{(Y_i, U_i)\}$ be the process initiated by a distribution on (Y, U) with $\mathcal{L}(Y) = \mathcal{L}(l_i | X_i = 1)$ and $\mathcal{L}(U) = \mathcal{L}(l_i | X_i = 0)$. Let $\{(\tilde{Y}_i, \tilde{U}_i)\}$ be the same process started at any other point. Then

$$|\tilde{Y}_i - Y_i| \leq s_i |f(\tilde{U}_{i-1}) - f(U_{i-1})| + (1 - s_i) |f(\tilde{Y}_{i-1}) - f(Y_{i-1})|$$

$$\leq s_i c(\pi_{01}, \pi_{10}) |\tilde{U}_{i-1} - U_{i-1}| + (1 - s_i) c(\pi_{01}, \pi_{10}) |\tilde{Y}_{i-1} - Y_{i-1}|$$

$$\leq c(\pi_{01}, \pi_{10}) \max\{|\tilde{U}_{i-1} - U_{i-1}|, |\tilde{Y}_{i-1} - Y_{i-1}|\}, \tag{38}$$

where the inequality before last follows from (20) (with $c(\pi_{01}, \pi_{10}) < 1$). We similarly obtain

$$|\tilde{U}_i - U_i| \leq c(\pi_{01}, \pi_{10}) \max\{|\tilde{U}_{i-1} - U_{i-1}|, |\tilde{Y}_{i-1} - Y_{i-1}|\}$$

which, combined with (38), yields

$$\max\{|\tilde{U}_i - U_i|, |\tilde{Y}_i - Y_i|\} \le c(\pi_{01}, \pi_{10}) \max\{|\tilde{U}_{i-1} - U_{i-1}|, |\tilde{Y}_{i-1} - Y_{i-1}|\}. \quad (39)$$

Iterating gives

$$\|(\tilde{U}_i, \tilde{Y}_i) - (U_i, Y_i)\|_\infty \le c(\pi_{01}, \pi_{10})^i \|(\tilde{U}_0, \tilde{Y}_0) - (U_0, Y_0)\|_\infty, \quad (40)$$

implying

$$|\tilde{U}_i - U_i| \longrightarrow 0, \quad |\tilde{Y}_i - Y_i| \longrightarrow 0. \quad (41)$$

But $\mathcal{L}(Y_i) = \mathcal{L}(Y)$ and $\mathcal{L}(U_i) = \mathcal{L}(U)$ for all i, thus, if the 'tilded' process is marginally stationary, the only way this could be consolidated with (41) is if $\mathcal{L}(\tilde{Y}_i) = \mathcal{L}(Y)$ and $\mathcal{L}(\tilde{U}_i) = \mathcal{L}(U)$ for all i. □

Henceforth, when referring to the process constructed in Theorem 2.3, it will be implied that it was initiated with a marginally stationary distribution on (U, Y). When combined with (18), Theorem 2.3 implies the following.

Corollary 2.4. *For the process constructed in Theorem 2.3,*

$$\overline{H}(Z) = \frac{\pi_{01}}{\pi_{01} + \pi_{10}} Eh_b \left(\left[\frac{e^{Y_i}}{1 + e^{Y_i}} (1 - \pi_{10}) + \frac{1}{1 + e^{Y_i}} \pi_{01} \right] * \delta \right)$$

$$+ \frac{\pi_{10}}{\pi_{01} + \pi_{10}} Eh_b \left(\left[\frac{e^{U_i}}{1 + e^{U_i}} (1 - \pi_{10}) + \frac{1}{1 + e^{U_i}} \pi_{01} \right] * \delta \right). \quad (42)$$

3 Bounds on the entropy rate for the symmetric chain

In this section we use Corollary 2.2 to bound the entropy rate in the symmetric case, by bounding the expectation on the right side of (29). Assume throughout this section the case of a BSC-corrupted symmetric Markov chain with $0 < \pi_{10} = \pi_{01} = \pi \le 1/2$. The following gives the general form of our bounds, and the condition under which they are valid.

Observation 3.1. Let $\{Y_i\}$ be the stationary Markov process whose evolution is given by (24). Let $\{a_i\}_{i=1}^M, \{b_i\}_{i=1}^M$ be strictly increasing sequences of nonnegative reals such that $b_k \le a_k$ and $b_{k+1} > a_k$ (i.e., the intervals $[b_k, a_k]$ do not intersect). Assume further that $\bigcup_{k=1}^M [b_k, a_k] \cup \bigcup_{k=1}^M [-a_k, -b_k]$ contains the support of Y_i.

Then

$$\sum_{k=1}^{M} P\left(Y_i \in [-a_k, -b_k] \cup [b_k, a_k]\right) h_b \left(\frac{e^{a_k}}{1+e^{a_k}} * \pi * \delta\right)$$

$$\leq \overline{H}(Z) \leq \sum_{k=1}^{M} P\left(Y_i \in [-a_k, -b_k] \cup [b_k, a_k]\right) h_b \left(\frac{e^{b_k}}{1+e^{b_k}} * \pi * \delta\right).$$

Proof. This is immediate from Corollary 2.2 and the decreasing monotonicity of $h_b\left(\frac{e^y}{1+e^y} * \pi * \delta\right)$ in the absolute value of y. □

Evidently, a bound of the type in Observation 3.1 would be applicable only in situations where: (1) the support of Y_i is, in fact, contained in a set of the form $\bigcup_{k=1}^{M}[b_k, a_k] \cup \bigcup_{k=1}^{M}[-a_k, -b_k]$ and (2) the probabilities $P(Y_i \in [-a_k, -b_k] \cup [b_k, a_k])$ can be computed (or bounded from above and below). To get an appreciation for when this can happen, it is instructive to consider first the case $M = 1$, for which Observation 3.1 yields the following.

Corollary 3.2. *Let $\{Y_i\}$ be the process in (24). Let $0 \leq b \leq A$ be such that $[-A, -b] \cup [b, A]$ contains the support of Y_i. Then*

$$h_b \left(\frac{e^A}{1+e^A} * \pi * \delta\right) \leq \overline{H}(Z) \leq h_b \left(\frac{e^b}{1+e^b} * \pi * \delta\right). \tag{43}$$

The lower bound of Corollary 3.2 is clearly optimized when taking A to be the upper end point of the support of Y_i. This point is readily seen, by observation of the dynamics of the process $\{Y_i\}$ in (24), to be the solution to the equation

$$A = f(A) + \log \frac{1-\delta}{\delta}, \tag{44}$$

namely

$$A = \log \frac{\alpha - 1 + (1-\alpha)\pi + \sqrt{4\alpha\pi^2 + (1-\alpha - (1-\alpha)\pi)^2}}{2\pi}, \tag{45}$$

where $\alpha = (1-\delta)/\delta$ (cf. Section 4 of [29]). The obvious symmetry of the support of Y_i around 0 implies that $-A$ is the lower end point of the support of Y_i. In particular, this establishes that the support of Y_i is contained in the interval $[-A, A]$, cf. Figure 1. Similarly, to optimize the upper bound, b should be taken as the lower end point of this support in the positive half of the real line. For the case where δ is small enough so that the first term on the right-hand side

Figure 1 Smallest interval containing the support of Y_i. A is given by (45).

Figure 2 Smallest set of the form $[-A, -b] \cup [b, A]$ containing the support of Y_i. A and b are given, respectively, in (45) and (46).

Figure 3 Smallest set of the form $[-A, -B] \cup [-a, -b] \cup [b, a] \cup [B, A]$ containing the support of Y_i. $A, b, a,$ and B are given, respectively, in (45), (46), (52), and (53). Probabilities of the four intervals are stated in Lemma 3.3.

of (24) uniquely determines the sign of Y_i ("small enough", as will be made explicit below, is any value in the shaded region of Figure 4), the value of this lower end point can be read from the dynamics of the process in (24) (see also the proof of Lemma 3.3 below) to be given by

$$b = -f(A) + \log \frac{1-\delta}{\delta}. \tag{46}$$

Similarly, by symmetry, $-b$ is the upper end point of the support of Y_i in the negative half. This implies then that the support of Y_i is contained in $[-A, -b] \cup [b, A]$ (cf. Figure 2) and that A and b of (45) and (46) are, respectively, the smallest and largest values with this property. Crude as the bounds of Corollary 3.2 may seem, they were shown in [29] (obtained therein directly from the likelihood process (21)) to convey nontrivial information when optimized by substituting the values of A and b from (45) and (46). In particular, this led to varying degrees of precision in characterizing the entropy rate in various asymptotic regimes. Examples include (letting $\overline{H}(\pi, \delta)$ stand for $\overline{H}(Z)$ when π and δ are, respectively, the chain and channel transition probabilities):

- "High SNR": for $0 \le \pi \le 1/2$, as $\delta \to 0$,

$$\overline{H}(\pi, \delta) - h_b(\pi) \asymp \delta. \tag{47}$$

- "Almost memoryless": for $0 \le \delta \le 1/2$, as $\varepsilon \to 0$,

$$\frac{1 - \overline{H}\left(\frac{1}{2} - \varepsilon, \delta\right)}{\varepsilon^2} \sim \frac{2}{\log 2}(1 - 2\delta)^4. \tag{48}$$

- "Low SNR": for $1/4 \le \pi < 1/2$, as $\varepsilon \to 0$,

$$1 - \overline{H}\left(\pi, \frac{1}{2} - \varepsilon\right) \asymp \varepsilon^4. \tag{49}$$

It is instructive to compare with the implications of the bounds of Cover and Thomas [5, Section 4.5] for these regimes. In our setting, a simple calculation shows that $H(Z_0|X_{-1}) = h_b(\pi * \delta)$ and $H(Z_0|Z_{-1}) = h_b(\pi * \delta * \delta)$, so the first-order ($n = 1$) bounds are

$$h_b(\pi * \delta) \le \overline{H}(\pi, \delta) \le h_b(\pi * \delta * \delta), \tag{50}$$

which implies (47) (but no more), recovers the ε^2 behavior in (48) but without the constant, and does not recover the ε^4 behavior in (49). In fact, as was mentioned in [29], there are regimes in which the bounds of [5, Section 4.5], of any order, will not capture the behavior of the entropy rate. For a simple example note that, in our binary symmetric setting, for any n,

$$H(Z_0|Z_{-n+1}^{-1}, X_{-n}) \le H(Z_0|X_{-n}) = h_b(\pi^{*n} * \delta), \tag{51}$$

where π^{*n} denotes binary convolution of π with itself n times. Thus, for example, in the "low SNR" regime where π is fixed and $\delta = 1/2 - \varepsilon$, $H(Z_0|Z_{-n+1}^{-1}, X_{-n}) \le h_b(\pi^{*n} * \delta)) = h_b(1/2 - \varepsilon(1 - 2\pi^{*n}))$ and, in particular, $1 - H(Z_0|Z_{-n+1}^{-1}, X_{-n}) = \Omega(\varepsilon^2)$. In other words, using $H(Z_0|Z_{-n+1}^{-1}, X_{-n})$ to lower bound the entropy rate will give an upper bound on the left-hand side of (49) of order ε^2, failing to provide the true ε^4 order in (49) (and, a fortiori, its refinements we derive below).

Let us now take one step of refinement beyond Corollary 3.2, to study the form of the bounds of Observation 3.1 in the case $M = 2$, and their implications in some asymptotic regimes. Define, in addition to A and b in (44) and (46),

$$a = -f(b) + \log\frac{1 - \delta}{\delta} \tag{52}$$

and

$$B = f(b) + \log\frac{1 - \delta}{\delta}. \tag{53}$$

Lemma 3.3. *Assume that either $\pi \geq 1/4$ and $\delta \leq 1/2$, or $\pi < 1/4$ and $\delta < \frac{1}{2}(1 - \sqrt{1 - 4\pi})$. More compactly, assume that $\delta \leq \frac{1}{2}\left(1 - \sqrt{\max\{1 - 4\pi, 0\}}\right)$ (cf. Figure 4). Then $A, b, a,$ and B (defined in (45), (46), (52), and (53)) satisfy $0 \leq b \leq a < B \leq A$, as well as $P(Y_i \in [B, A]) = (1 - \delta)[\pi * (1 - \delta)]$, $P(Y_i \in [b, a]) = (1 - \delta)[\pi * \delta]$, $P(Y_i \in [-a, -b]) = \delta[\pi * (1 - \delta)]$, and $P(Y_i \in [-A, -B]) = \delta[\pi * \delta]$. In particular, the support of Y_i is contained in $[-A, -B] \cup [-a, -b] \cup [b, a] \cup [B, A]$.*

Proof. That the A solving (44) is the upper end point of the support of Y_i and, by symmetry, $-A$ its lower end point, is evident from (24). It was shown in [29, Corollary 3] that in this region of the $\pi - \delta$ plane $Y_i \geq 0$ if and only if $r_i = 1$, in which case the smallest value Y_i can take is $b = \log \frac{1-\delta}{\delta} - f(A)$. This implies, by symmetry of the support of Y_i, that this support is contained in $[-A, -b] \cup [b, A]$. Furthermore, when $Y_i > 0$ (i.e., $r_i = 1$), there are two possibilities. The first is that the second term on the right-hand side of (24) is negative, in which case the most (least negative) it can be is $-f(b)$, implying that in this case $Y_i \leq \log \frac{1-\delta}{\delta} - f(b) = a$. The second possibility is that this second term is positive, in which case the least it can be is $f(b)$, implying that $Y_i \geq \log \frac{1-\delta}{\delta} + f(b) = B$. It follows that when $Y_i > 0$ either $Y_i \in [b, a]$ or $Y_i \in [B, A]$. Symmetry of the support of Y_i implies that this support is contained in $[-A, -B] \cup [-a, -b] \cup [b, a] \cup [B, A]$. It also follows that Y_i falls in the interval, say $[b, a]$, if and only if both $r_i = 1$ and $s_i f(Y_{i-1}) < 0$, i.e.,

$$P(Y_i \in [b, a]) = P(r_i = 1, s_i f(Y_{i-1}) < 0)$$
$$= P(r_i = 1)P(\{s_i = 1, f(Y_{i-1}) < 0\} \cup \{s_i = -1, f(Y_{i-1}) > 0\})$$
$$= (1 - \delta)[(1 - \pi)\delta + \pi(1 - \delta)]$$
$$= (1 - \delta)[\pi * \delta].$$

Using similar reasoning gives

$$P(Y_i \in [B, A]) = P(r_i = 1, s_i f(Y_{i-1}) > 0) = (1 - \delta)[\pi * (1 - \delta)],$$

$$P(Y_i \in [-a, -b]) = P(r_i = -1, s_i f(Y_{i-1}) > 0) = \delta[\pi * (1 - \delta)],$$

and

$$P(Y_i \in [-A, -B]) = P(r_i = -1, s_i f(Y_{i-1}) < 0) = \delta[\pi * \delta].$$

$\square$

Specializing Observation 3.1 to the case $M = 2$ and combining with Lemma 3.3 gives the following lemma.

Lemma 3.4. *For all* $\delta \leq \frac{1}{2}\left(1 - \sqrt{\max\{1 - 4\pi, 0\}}\right)$,

$$\{(1-\delta)[\pi * (1-\delta)] + \delta[\pi * \delta]\} h_b \left(\frac{e^A}{1+e^A} * \pi * \delta\right)$$

$$+ \{(1-\delta)[\pi * \delta] + \delta[\pi * (1-\delta)]\} h_b \left(\frac{e^a}{1+e^a} * \pi * \delta\right)$$

$$\leq \overline{H}(Z)$$

$$\leq \{(1-\delta)[\pi * (1-\delta)] + \delta[\pi * \delta]\} h_b \left(\frac{e^B}{1+e^B} * \pi * \delta\right)$$

$$+ \{(1-\delta)[\pi * \delta] + \delta[\pi * (1-\delta)]\} h_b \left(\frac{e^b}{1+e^b} * \pi * \delta\right), \tag{54}$$

where $A, B, a,$ and b are as specified in (45), (46), (52), and (53).

As can be expected, the bounds in Lemma 3.4, which are based on the support bound depicted in Figure 3, are considerably tighter, in various asymptotic regimes, than those based on Corollary 3.2, which use the coarser support bound of Figure 2. As a first example, recall that in the "high SNR" regime the analysis in [29, Section 5], which was based on Corollary 3.2, established that $\overline{H}(Z) - h_b(\pi) \asymp \delta$ (recall (47)), while, as we now show, the bounds of Lemma 3.4 recover the constant.

Theorem 3.5. *For $\pi \leq 1/2$ and $\delta \downarrow 0$,*

$$\overline{H}(Z) = h_b(\pi) + \left[2(1 - 2\pi)\log_2 \frac{1-\pi}{\pi}\right] \cdot \delta + o(\delta).$$

The result of Theorem 3.5 was first established in [22], and subsequently derived in [27] and [36]. We give a simple proof of this result in Appendix A, via the bounds of Lemma 3.4.

For the "almost memoryless" regime, the bounds of Lemma 3.4 are tight enough to imply that the term following that characterized in (48) is $o(\varepsilon^3)$. More specifically, by evaluating the bounds of Lemma 3.4 for this regime, the following theorem is proved in Appendix B.

Theorem 3.6. *For $0 \leq \delta \leq 1/2$ and $\pi = 1/2 - \varepsilon$, as $\varepsilon \downarrow 0$,*

$$1 - \overline{H}(Z) = \frac{2}{\log 2}\varepsilon^2(1 - 2\delta)^4 + o(\varepsilon^3).$$

For the "low SNR" regime, the bounds of Lemma 3.4 are shown in Appendix C to imply the following theorem.

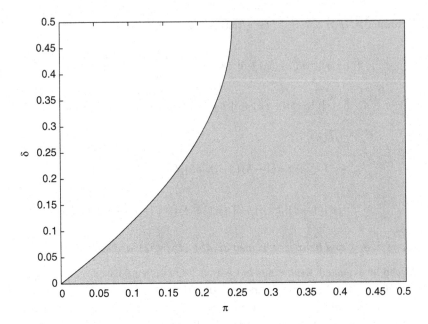

Figure 4 Shaded area below the curve $\frac{1}{2}\left(1-\sqrt{\max\{1-4\pi,0\}}\right)$ is the region in the $\pi-\delta$ plane where the sign of r_i determines the sign of Y_i in (24).

Theorem 3.7. *For $1/4 \le \pi \le 1/2$ and $\delta = \frac{1}{2} - \varepsilon$,*

$$\frac{2(1-2\pi)^2(1-12\pi+48\pi^2-64\pi^3+32\pi^4)}{\pi^2\log 2}$$

$$\le \liminf_{\varepsilon \to 0} \frac{1-\overline{H}(Z)}{\varepsilon^4} \le \limsup_{\varepsilon \to 0} \frac{1-\overline{H}(Z)}{\varepsilon^4}$$

$$\le \frac{2(1-2\pi)^2(1-4\pi+16\pi^2-32\pi^3+32\pi^4)}{\pi^2\log 2}. \tag{55}$$

To see why Theorem 3.7 covers only the range $1/4 \le \pi \le 1/2$, note that Lemma 3.4, on which it relies, applies only in the shaded region of Figure 4. Clearly, for $\pi < 1/4$ and $\delta = 1/2 - \varepsilon$, the point (π, δ) will be outside the shaded region for ε small enough.

Theorem 3.7 should be compared with Corollary 6 of [29], which gives

$$\frac{2}{\log 2}\left[\frac{(4\pi-1)(1-2\pi)}{\pi}\right]^2 \leq \liminf_{\varepsilon\to 0}\frac{1-\overline{H}(Z)}{\varepsilon^4}\leq \limsup_{\varepsilon\to 0}\frac{1-\overline{H}(Z)}{\varepsilon^4}$$

$$\leq \frac{2}{\log 2}\left[\frac{1-2\pi}{\pi}\right]^2 \tag{56}$$

(and on the basis of which (49) was stated). Figure 5 plots the bounds of (56) (the lower and upper curves) and those of Theorem 3.7 (the two internal curves), as a function of π. The bounds become increasingly tight as π approaches $1/2$, all converging to 0. Furthermore, both lower and upper bounds in (56) (and, a fortiori, in (55)) behave as $\sim \frac{8}{\log 2}(1-2\pi)^2$ for $\pi\to 1/2$, implying that $1-\overline{H}(Z)\approx \frac{32}{\log 2}(1/2-\delta)^4(1/2-\pi)^2$ for π and δ close to $1/2$.

Theorems 3.5 through 3.7 were obtained via evaluation of the bounds of Lemma 3.4 in the respective regimes. In turn, Lemma 3.4 is nothing but a specialization of Observation 3.1 to the case $M=2$, optimizing over the choice of constants a_k and b_k. It was seen that these constants define a region of the form depicted in Figure 3, and optimizing their values amounts to finding the smallest region of that form containing the support of Y_i (the alternative Markov process). This optimization was easy to do (in Lemma 3.3), by observation of the dynamics of the process $\{Y_i\}$, as given in Theorem 2.1.

Evidently, moving from the bounds corresponding to $M=1$ (which lead to (47)–(49)) to those of $M=2$ results, for various asymptotic regimes, in characterization of higher-order terms, and refinement of constants. The larger M one takes, the finer will the bounds become, leading, in particular, to finer characterizations in the respective regimes. The development we have detailed for the case $M=2$ scales to any larger value of M. For example, $M=3$ will correspond to an outer bound on the support of Y_i obtained by excluding a subinterval from each of the four intervals of Figure 3. The choice of these subintervals will be optimized analogously as in Lemma 3.3 quite simply, via the dynamics of the process $\{Y_i\}$ in (24). Note that it is only the end points of the new subintervals that need be computed, the remaining end points being identical to those evaluated for $M=2$. More generally, moving from the approximation corresponding to a value of M to the value $M+1$ corresponds to discarding a subinterval from each of the intervals constituting the outer bound of the support obtained at the Mth level. Only the end points of the subintervals that are being discarded need be computed, the remaining ones coinciding with those already obtained in the previous stage.

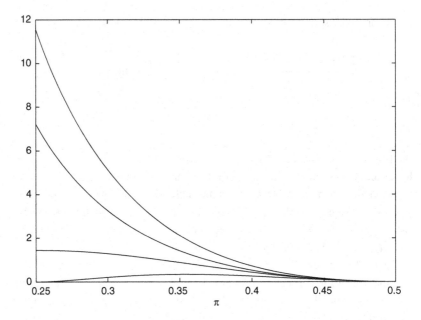

Figure 5 Upper and lower bounds associated with (49) and Theorem 3.7.

4 Non-symmetric case

In this section, we illustrate the use of the process (U_i, V_i) (defined via (31) and (32)) for obtaining bounds on the entropy rate in the nonsymmetric case. More specifically, we use the dynamics of the process (U_i, V_i) detailed in Theorem 2.3 to obtain bounds on the support of (U_i, V_i), which, in turn, we translate to bounds on the expression for the entropy rate given in Corollary 2.4. In particular, paralleling the previous section, we develop bounds corresponding to Observation 3.1 in the case $M = 2$. As a representative example, we use this to obtain the first term in the expansion of the entropy rate in the "high SNR" regime, with the implication that expansions in other regimes, as well as higher-order terms (by using $M > 2$), are obtained similarly.

Bounds for regimes ("high SNR", "almost memoryless", "low SNR") mentioned in the previous section that were obtained via an approximation corresponding to $M = 1$ in [29] were obtained also for the nonsymmetric case. Additional regimes arising in the nonsymmetric setting include the "rare spikes" and "rare bursts" regimes. For example, it was shown (see [29, Theorem 5]) that for $0 \leq \delta \leq 1/2$ and any function $a(\cdot)$ satisfying $0 < a(\varepsilon) \leq \varepsilon$, as $\varepsilon \to 0$,

$$\frac{\overline{H}(a(\varepsilon), 1-\varepsilon, \delta) - h_b(\delta)}{a(\varepsilon)} = \frac{\overline{H}(1-\varepsilon, a(\varepsilon), \delta) - h_b(\delta)}{a(\varepsilon)} \sim (1-2\delta)\log_2 \frac{1-\delta}{\delta}.$$
(57)

The bounds we develop below are applicable, for example, also for refining the characterization in (57).

4.1 The case $\pi_{10} = 1$

The first example we consider is the case where $\pi_{10} = 1$, in the "high SNR" regime. We will establish the following theorem.

Theorem 4.1. *For $\pi_{10} = 1$, $0 \le \pi_{01} < 1$, and δ tending to 0,*

$$\overline{H}(Z) = \overline{H}(X) + \frac{\pi_{01}(2-\pi_{01})}{1+\pi_{01}} \delta \log_2 \frac{1}{\delta} + O(\delta).$$
(58)

Interestingly, the first term in the expansion is of order $\delta \log_2(1/\delta)$, in contrast to that in Theorem 3.5 which is of order δ. As was first shown in [22], and we show in the next subsection in detail, the order of δ behavior in fact reigns for all values of the pair (π_{10}, π_{01}), except when one of the two values equals 1 (in which case Theorem 4.1 asserts that the order is $\delta \log_2(1/\delta)$). This case is left unresolved by the asymptotic expansion of [22], which only hints at the above behavior in that the constant multiplying the order δ term increases to infinity as either π_{01} or π_{10} tends to one. A variation on the (second) proof of Theorem 4.1 appearing below is shown in the next subsection to also recover the expansion of [22] for the case $\pi_{10} < 1, \pi_{01} < 1$.

Note that in the case $\pi_{10} = \pi_{01} = 1$, $\overline{H}(Z) = h_b(\delta)$, while $\overline{H}(X) = 0$, so $\overline{H}(Z) = \overline{H}(X) + \delta \log_2(1/\delta) + O(\delta)$, where the factor multiplying the $\delta \log_2(1/\delta)$ term in (58) is $1/2$ when $\pi_{01} = 1$. The reason for this is that just like there is a transition from order of δ to order of $\delta \log_2(1/\delta)$ when going from $\pi_{10} < 1$ to $\pi_{10} = 1$, a similar term that will be order of δ in our analysis below (where we assume that $\pi_{01} < 1$) becomes order of $\delta \log_2(1/\delta)$ when going from $\pi_{01} < 1$ to $\pi_{01} = 1$. This accounts for the doubling of the said factor from $1/2$ to 1.

As it turns out, Theorem 4.1 is provable via the Cover and Thomas bounds [5, Section 4.5] of order $n = 2$. We detail this proof in Appendix D. In this subsection, we give an alternative proof via our support bounds approach. Throughout the remainder of this subsection, we assume that $\pi_{10} = 1$ and $\pi_{01} < 1$, in which case $f(x)$ simplifies to

$$f(x) = \log \frac{\pi_{01}}{\pi_{01} + e^x},$$

where $\bar{x}$ denotes $1 - x$. Note that $f(x)$ is decreasing in x and is upper bounded by $f(-\infty) = \log \pi_{01}/\overline{\pi_{01}}$. Defining

$$r(x) = \log \frac{1-x}{x}, \tag{59}$$

the upper bound on $f(x)$ just stated is $-r(\pi_{01})$.

In the spirit of the developments in the previous section for bounding the support in the case $M = 2$, considering the alternative process constructed in Theorem 2.3, we will show that the support of $\mathcal{L}(U) = \mathcal{L}(l_i | X_i = 0)$ (and $\mathcal{L}(Y) = \mathcal{L}(l_i | X_i = 1)$, as they have identical supports) is contained in the union of four disjoint intervals on the real line whose boundary points and probabilities (under P_U and P_Y) we characterize explicitly. We will then obtain upper and lower bounds on the entropy rate of $\{Z_i\}$ in terms of the interval boundary points and probabilities, similarly as was done in the derivation of Theorem 3.4. The bounds thus obtained will be shown to lead to the asymptotic behavior of the entropy rate stated in Theorem 4.1.

The following lemma, which follows from elementary calculus, will be used throughout our analysis.

Lemma 4.2. *Suppose that* $p = p_0 + \delta p_1 + O(\delta^2)$. *If* $0 < p_0 \overline{\pi_{10}} + \overline{p_0} \pi_{01} < 1$, *then*

$$h_b([p\overline{\pi_{10}} + \overline{p}\pi_{01}] * \delta)$$

$$= h_b(p_0 \overline{\pi_{10}} + \overline{p_0}\pi_{01})$$

$$- \delta \left[(\overline{p_0}(1 - 2\pi_{01}) + p_0(2\pi_{10} - 1) + p_1(1 - \pi_{01} - \pi_{10})) \log_2 \frac{p_0 \overline{\pi_{10}} + \overline{p_0}\pi_{01}}{\overline{p_0}\pi_{10} + \overline{p_0}\pi_{01}} \right]$$

$$+ O(\delta^2).$$

If $\pi_{10} = p_0 = 1$ *and* $\pi_{01} < 1$, *then*

$$h_b([p\overline{\pi_{10}} + \overline{p}\pi_{01}] * \delta) = (1 - p_1 \pi_{01})\delta \log_2 \frac{1}{\delta} + O(\delta). \tag{60}$$

Define now the following four intervals on the real line, where an interval $[a, b]$ is taken to be empty if $a > b$:

$$I_0 = [-r(\delta) + f(r(\delta) - r(\pi_{01})), -r(\delta) + f(r(\delta) + f(r(\delta) - r(\pi_{01})))],$$

$$I_1 = [-r(\delta) + f(-r(\delta) - r(\pi_{01})), -r(\delta) - r(\pi_{01})],$$

$$I_2 = [r(\delta) + f(r(\delta) - r(\pi_{01})), r(\delta) + f(r(\delta) + f(r(\delta) - r(\pi_{01})))],$$

$$I_3 = [r(\delta) + f(-r(\delta) - r(\pi_{01})), r(\delta) - r(\pi_{01})]. \tag{61}$$

We shall also rely on the following three lemmas. Their proofs, which we defer to Appendix E, are based on ideas similar to those used in the proof of Lemma 3.3.

Lemma 4.3. *The intervals $I_j, j = 0, 1, 2, 3$, are nonempty (i.e. the left end points as specified above are smaller than the right end points).*

Given two intervals I and J, let $I < J$ express the fact that the right end point of I is (strictly) less than the left end point of J.

Lemma 4.4. *For all sufficiently small $\delta > 0$, the intervals $I_j, j = 0, 1, 2, 3$, satisfy $I_0 < I_1 < I_2 < I_3$.*

Lemma 4.5. *The supports of both P_U and P_Y are contained in $I_0 \cup I_1 \cup I_2 \cup I_3$. For all sufficiently small $\delta > 0$, the probabilities of the intervals under P_U and P_Y are given by*

I	$P_Y(I)$	$P_U(I)$
I_0	δ^2	$\overline{\delta}(\pi_{01} * \delta)$
I_1	$\delta\overline{\delta}$	$\overline{\delta}(\pi_{01} * \overline{\delta})$
I_2	$\delta\overline{\delta}$	$\delta(\pi_{01} * \delta)$
I_3	$\overline{\delta}^2$	$\delta(\pi_{01} * \overline{\delta})$

(62)

For any closed interval I on the real line, let $\ell(I)$ denote the smallest value in I (left end point) and $u(I)$ denote the largest value (right end point). Define

$$\beta(x) = \frac{e^x}{1 + e^x},$$

which maps $x = \log Pr(1)/Pr(0)$ to $Pr(1)$ (e.g., log-likelihood ratios to probabilities).

Lemma 4.6. *For all sufficiently small $\delta > 0$, the entropy rate $\overline{H}(Z)$ of the process $\{Z_i\}$ satisfies*

$$\overline{H}(Z) \leq \sum_{j=0}^{3} \left[\frac{1}{1 + \pi_{01}} P_U(I_j) + \frac{\pi_{01}}{1 + \pi_{01}} P_Y(I_j) \right] \max_{x \in I_j} h_b([\overline{\beta(x)}\pi_{01}] * \delta) \quad (63)$$

and

$$\overline{H}(Z) \geq \sum_{j=0}^{3} \left[\frac{1}{1 + \pi_{01}} P_U(I_j) + \frac{\pi_{01}}{1 + \pi_{01}} P_Y(I_j) \right] \min_{x \in I_j} h_b([\overline{\beta(x)}\pi_{01}] * \delta). \quad (64)$$

Proof. The inequality

$$\overline{H}(Z) \geq \sum_{j=0}^{3} \left[\frac{1}{1+\pi_{01}} Pr(l_i \in I_j | X_i = 0) + \frac{\pi_{01}}{1+\pi_{01}} Pr(l_i \in I_j | X_i = 1) \right]$$

$$\times \min_{x \in I_j} h_b([\overline{\beta(x)}\pi_{01}] * \delta)$$

follows from (42) and Lemma 4.5, once δ is sufficiently small for Lemma 4.4 to imply that the I_j are disjoint. The upper bound follows similarly. □

Proof of Theorem 4.1. Lemma 4.5 shows that all but $P_Y(I_3), P_U(I_0)$, and $P_U(I_1)$ are $O(\delta)$. Therefore, since $h_b(\cdot)$ is bounded, the only terms in (63) and (64) that might be greater than $O(\delta)$ are those involving $P_Y(I_3), P_U(I_0)$, and $P_U(I_1)$. First we consider the terms involving $P_U(I_0)$ and $P_U(I_1)$. It follows from elementary calculus that $h_b([\overline{p}\pi_{01}] * \delta)$ is maximized at $\max\{0, (\pi_{01} - 1/2)/(\delta + \pi_{01})\}$. This fact together with the concavity of $h_b([\overline{p}\pi_{01}] * \delta)$ in p, and the fact that both end points of both I_0 and I_1 are tending to $-\infty$, imply that for all sufficiently small $\delta > 0$, $\min_{x \in I_j} h_b([\overline{\beta(x)}\pi_{01}] * \delta)$ and $\max_{x \in I_j} h_b([\overline{\beta(x)}\pi_{01}] * \delta)$ are achieved at either $\ell(I_j)$ or $u(I_j)$ for $j = 0, 1$. It is not difficult to see that for $j = 0, 1$ both $\beta(\ell(I_j))$ and $\beta(u(I_j))$ are ratios of polynomials in δ. In particular, they will be of the form $p_0 + \delta p_1 + O(\delta^2)$ with $p_0 = 0$. Therefore, using Lemmas 4.2 and 4.5,

$$\sum_{j=0}^{1} \frac{1}{1+\pi_{01}} P_U(I_j) \max_{x \in I_j} h_b([\overline{\beta(x)}\pi_{01}] * \delta)$$

$$= \frac{\pi_{01}}{1+\pi_{01}} h_b(\pi_{01}) + \frac{\overline{\pi_{01}}}{1+\pi_{01}} h_b(\pi_{01}) + O(\delta)$$

$$= \overline{H}(X) + O(\delta). \tag{65}$$

Similarly,

$$\sum_{j=0}^{1} \frac{1}{1+\pi_{01}} P_U(I_j) \min_{x \in I_j} h_b([\overline{\beta(x)}\pi_{01}] * \delta) = \overline{H}(X) + O(\delta). \tag{66}$$

Next, we focus on the terms involving $P_Y(I_3)$. In these cases, the above properties (concavity and extremal) of $h_b([\overline{p}\pi_{01}] * \delta)$ viewed as a function of p, and the fact that the left end point of I_3 is greater than the maximizing p, imply that

$$\min_{x \in I_3} h_b([\overline{\beta(x)}\pi_{01}] * \delta) = h_b([\overline{\beta(u(I_3))}\pi_{01}] * \delta)$$

and

$$\max_{x \in I_3} h_b([\overline{\beta(x)}\pi_{01}] * \delta) = h_b([\overline{\beta(\ell(I_3))}\pi_{01}] * \delta).$$

From (61), we see (some algebraic manipulations omitted) that

$$\beta(u(I_3)) = \frac{\bar{\delta}\pi_{01}}{\delta\overline{\pi_{01}} + \bar{\delta}\pi_{01}} \tag{67}$$

$$= 1 - \frac{\delta\overline{\pi_{01}}}{\delta\overline{\pi_{01}} + \bar{\delta}\pi_{01}} \tag{68}$$

$$= 1 - \delta\frac{\overline{\pi_{01}}}{\pi_{01}} + O(\delta^2) \tag{69}$$

and

$$\beta(\ell(I_3)) = \frac{\bar{\delta}^2\pi_{01}\overline{\pi_{01}}}{\pi_{01}^2\delta^2 + \bar{\delta}\delta\overline{\pi_{01}}^2 + \bar{\delta}^2\pi_{01}\overline{\pi_{01}}} \tag{70}$$

$$= 1 - \frac{\pi_{01}^2\delta^2 + \bar{\delta}\delta\overline{\pi_{01}}^2}{\pi_{01}^2\delta^2 + \bar{\delta}\delta\overline{\pi_{01}}^2 + \bar{\delta}^2\pi_{01}\overline{\pi_{01}}} \tag{71}$$

$$= 1 - \delta\frac{\overline{\pi_{01}}}{\pi_{01}} + O(\delta^2). \tag{72}$$

Lemma 4.2, (69) and (72) then imply that

$$\min_{x\in I_3} h_b([\overline{\beta(x)}\pi_{01}] * \delta) = (1+\overline{\pi_{01}})\delta\log_2\frac{1}{\delta} + O(\delta) \tag{73}$$

and

$$\max_{x\in I_3} h_b([\overline{\beta(x)}\pi_{01}] * \delta) = (1+\overline{\pi_{01}})\delta\log_2\frac{1}{\delta} + O(\delta). \tag{74}$$

Equations (65), (66), (73), (74), and the expression for $P_Y(I_3)$ from (62) demonstrate that the combined contribution of the terms involving $P_Y(I_3)$, $P_U(I_0)$, and $P_U(I_1)$ to (63) and (64) is

$$\overline{H}(X) + \frac{\pi_{01}}{1+\pi_{01}}(1+\overline{\pi_{01}})\delta\log_2\frac{1}{\delta} + O(\delta)$$

$$= \overline{H}(X) + \frac{\pi_{01}(2-\pi_{01})}{1+\pi_{01}}\delta\log_2\frac{1}{\delta} + O(\delta)$$

in both cases. The theorem is proved since, as noted above, all the other terms are $O(\delta)$. $\qquad\square$

4.2 The case $0 < \pi_{01}, \pi_{10} < 1$

The analysis of this case is dependent on whether $\pi_{01} + \pi_{10}$ is smaller or greater than 1. If $\pi_{01} + \pi_{10} \leq 1$, then, as shown in Section 2, $f(x)$ (defined in (17)) is nondecreasing and is upper bounded by $f(\infty) = r(\pi_{10})$ and lower bounded by $f(-\infty) = -r(\pi_{01})$, where $r(x)$ is defined in (59). If, on the other hand, $\pi_{01} + \pi_{10} > 1$, then $f(x)$ is decreasing and is upper bounded by $f(-\infty) = -r(\pi_{01})$ and lower bounded by $f(\infty) = r(\pi_{10})$.

We define the four intervals $J_j, j = 0, \ldots, 3$, as

$$J_0 = [-r(\delta) - r(\pi_{01}), -r(\delta) + f(-r(\delta) + r(\pi_{10}))], \tag{75}$$

$$J_1 = [-r(\delta) + f(r(\delta) - r(\pi_{01})), -r(\delta) + r(\pi_{10})], \tag{76}$$

$$J_2 = [r(\delta) - r(\pi_{01}), r(\delta) + f(-r(\delta) + r(\pi_{10}))], \tag{77}$$

$$J_3 = [r(\delta) + f(r(\delta) - r(\pi_{01})), r(\delta) + r(\pi_{10})] \tag{78}$$

and the four intervals K_j as

$$K_0 = [-r(\delta) + r(\pi_{10}), -r(\delta) + f(r(\delta) + r(\pi_{10}))], \tag{79}$$

$$K_1 = [-r(\delta) + f(-r(\delta) - r(\pi_{01})), -r(\delta) - r(\pi_{01})], \tag{80}$$

$$K_2 = [r(\delta) + r(\pi_{10}), r(\delta) + f(r(\delta) + r(\pi_{10}))], \tag{81}$$

$$K_3 = [r(\delta) + f(-r(\delta) - r(\pi_{01})), r(\delta) - r(\pi_{01})]. \tag{82}$$

As before, the intervals $\{J_j\}$ and $\{K_j\}$ are respectively disjoint for sufficiently small δ when $\pi_{01} + \pi_{10} \leq 1$ and $\pi_{01} + \pi_{10} \geq 1$. Additionally, the following analogue of Lemma 4.5 holds.

Lemma 4.7. *If* $\pi_{01} + \pi_{10} \leq 1$, *the supports of both* P_U *and* P_Y *are contained in* $J_0 \cup J_1 \cup J_2 \cup J_3$. *For all sufficiently small* $\delta > 0$, *the probabilities of the intervals* $\{J_j\}$ *under* P_U *and* P_Y *are given by*

I	$P_Y(I)$	$P_U(I)$	
J_0	$\delta(\pi_{10} * \delta)$	$\overline{\delta}(\pi_{01} * \overline{\delta})$	
J_1	$\delta(\pi_{10} * \overline{\delta})$	$\overline{\delta}(\pi_{01} * \delta)$	(83)
J_2	$\overline{\delta}(\pi_{10} * \delta)$	$\delta(\pi_{01} * \overline{\delta})$	
J_3	$\overline{\delta}(\pi_{10} * \overline{\delta})$	$\delta(\pi_{01} * \delta)$	

If $\pi_{01} + \pi_{10} > 1$, *the supports of both* P_U *and* P_Y *are contained in* $K_0 \cup K_1 \cup K_2 \cup K_3$. *For all sufficiently small* $\delta > 0$, *the probabilities of the intervals* $\{K_j\}$

under P_U and P_Y are given by

I	$P_Y(I)$	$P_U(I)$
K_0	$\delta(\pi_{10} * \overline{\delta})$	$\overline{\delta}(\pi_{01} * \delta)$
K_1	$\delta(\pi_{10} * \delta)$	$\overline{\delta}(\pi_{01} * \overline{\delta})$
K_2	$\overline{\delta}(\pi_{10} * \overline{\delta})$	$\delta(\pi_{01} * \delta)$
K_3	$\overline{\delta}(\pi_{10} * \delta)$	$\delta(\pi_{01} * \overline{\delta})$

$$(84)$$

The proof of Lemma 4.7 is similar to that of Lemma 4.5 with (A.64) replaced by

I	Y_i	U_i
J_0	$\{(r_i, s_i, r_{i-1}) = (-1, 0, -1)\} \cup$ $\{(r_i, s_i, q_{i-1}) = (-1, 1, -1)\}$	$\{(q_i, t_i, q_{i-1}) = (-1, 0, -1)\} \cup$ $\{(q_i, t_i, r_{i-1}) = (-1, 1, -1)\}$
J_1	$\{(r_i, s_i, r_{i-1}) = (-1, 0, 1)\} \cup$ $\{(r_i, s_i, q_{i-1}) = (-1, 1, 1)\}$	$\{(q_i, t_i, q_{i-1}) = (-1, 0, 1)\} \cup$ $\{(q_i, t_i, r_{i-1}) = (-1, 1, 1)\}$
J_2	$\{(r_i, s_i, r_{i-1}) = (1, 0, -1)\} \cup$ $\{(r_i, s_i, q_{i-1}) = (1, 1, -1)\}$	$\{(q_i, t_i, q_{i-1}) = (1, 0, -1)\} \cup$ $\{(q_i, t_i, r_{i-1}) = (1, 1, -1)\}$
J_3	$\{(r_i, s_i, r_{i-1}) = (1, 0, 1)\} \cup$ $\{(r_i, s_i, q_{i-1}) = (1, 1, 1)\}$	$\{(q_i, t_i, q_{i-1}) = (1, 0, 1)\} \cup$ $\{(q_i, t_i, r_{i-1}) = (1, 1, 1)\}$

$$(85)$$

for the case that $\pi_{01} + \pi_{10} \leq 1$. For the other case, the events for the intervals K_0, K_1, K_2, and K_3 coincide respectively with those for J_1, J_0, J_3, and J_2, given above.

Additionally, we have the following minor variation on Lemma 4.6.

Lemma 4.8. *For $\pi_{01} + \pi_{10} \leq 1$ and all sufficiently small $\delta > 0$, the entropy rate $\overline{H}(Z)$ of the process $\{Z_i\}$ satisfies*

$$\overline{H}(Z) \leq \sum_{j=0}^{3} \left[\frac{\pi_{10}}{\pi_{01} + \pi_{10}} P_U(J_j) + \frac{\pi_{01}}{\pi_{01} + \pi_{10}} P_Y(J_j) \right]$$
$$\times \max_{x \in J_j} h_b([\overline{\beta(x)}\pi_{01} + \beta(x)\overline{\pi_{10}}] * \delta) \qquad (86)$$

and

$$\overline{H}(Z) \geq \sum_{j=0}^{3} \left[\frac{\pi_{10}}{\pi_{01} + \pi_{10}} P_U(J_j) + \frac{\pi_{01}}{\pi_{01} + \pi_{10}} P_Y(J_j) \right]$$
$$\times \min_{x \in J_j} h_b([\overline{\beta(x)}\pi_{01} + \beta(x)\overline{\pi_{10}}] * \delta). \qquad (87)$$

For $\pi_{01} + \pi_{10} > 1$ and all sufficiently small $\delta > 0$, the entropy rate satisfies (86) and (87) with J_j replaced by K_j.

Theorem 4.9. For $0 < \pi_{01}, \pi_{10} < 1$ and δ tending to 0,

$$\overline{H}(Z) = \overline{H}(X) + \delta \left(\frac{1}{\pi_{01} + \pi_{10}} \left[(\pi_{01} + \pi_{10} - 4\pi_{01}\pi_{10}) \log_2 \frac{\overline{\pi_{01}}\, \overline{\pi_{10}}}{\pi_{01}\, \pi_{10}} \right. \right.$$

$$\left. \left. + (\pi_{10} - \pi_{01}) \log_2 \frac{\overline{\pi_{01}}}{\overline{\pi_{10}}} \right] \right) + O(\delta^2). \tag{88}$$

Note that (88) applies regardless of the value of $\pi_{01} + \pi_{10}$ even though our proof treats the cases when this sum is smaller or greater than one differently. The expression (88) was first obtained in [22] using a different technique. The factor multiplying δ can be shown to equal $D(P_{X_0,X_1,X_2} \| P_{X_0,\overline{X}_1,X_2})$ (where $D(P \| Q)$ denotes the relative entropy or Kullback–Leibler divergence between distributions P and Q), which is the form given in [22].

Proof of Theorem 4.9. When $0 < \pi_{01}, \pi_{10} < 1$, straightforward calculus shows that the maximum of $h_b([\overline{p}\pi_{01} + p\overline{\pi_{10}}] * \delta)$ over p is bounded away from 0 and 1. Note also that both end points of J_0, J_1, K_0, and K_1 tend to $-\infty$ while the end points of J_2, J_3, K_2, and K_3 tend to ∞. Therefore, for δ sufficiently small, the concavity of $h_b([\overline{p}\pi_{01} + p\overline{\pi_{10}}] * \delta)$ in p implies that $\max_{x \in I} h_b([\overline{\beta(x)}\pi_{01} + \beta(x)\overline{\pi_{10}}] * \delta)$ is achieved at $x = u(I)$ for $I \in \{J_0, J_1, K_0, K_1\}$ and at $x = \ell(I)$ for $I \in \{J_2, J_3, K_2, K_3\}$ and that $\min_{x \in I} h_b([\overline{\beta(x)}\pi_{01} + \beta(x)\overline{\pi_{10}}] * \delta)$ is achieved at $x = \ell(I)$ for $I \in \{J_0, J_1, K_0, K_1\}$ and at $x = u(I)$ for $I \in \{J_2, J_3, K_2, K_3\}$. Thus, by Lemma 4.8, for $\pi_{01} + \pi_{10} \le 1$,

$$\overline{H}(Z) \le \sum_{j=0}^{1} \left[\frac{\pi_{10}}{\pi_{01} + \pi_{10}} P_U(J_j) + \frac{\pi_{01}}{\pi_{01} + \pi_{10}} P_Y(J_j) \right]$$

$$\times h_b([\overline{\beta(u(J_j))}\pi_{01} + \beta(u(J_j))\overline{\pi_{10}}] * \delta) \tag{89}$$

$$+ \sum_{j=2}^{3} \left[\frac{\pi_{10}}{\pi_{01} + \pi_{10}} P_U(J_j) + \frac{\pi_{01}}{\pi_{01} + \pi_{10}} P_Y(J_j) \right]$$

$$\times h_b([\overline{\beta(\ell(J_j))}\pi_{01} + \beta(\ell(J_j))\overline{\pi_{10}}] * \delta) \tag{90}$$

and

$$\overline{H}(Z) \ge \sum_{j=0}^{1} \left[\frac{\pi_{10}}{\pi_{01} + \pi_{10}} P_U(J_j) + \frac{\pi_{01}}{\pi_{01} + \pi_{10}} P_Y(J_j) \right]$$

$$\times h_b([\overline{\beta(\ell(J_j))}\pi_{01} + \beta(\ell(J_j))\overline{\pi_{10}}] * \delta) \tag{91}$$

$$+\sum_{j=2}^{3}\left[\frac{\pi_{10}}{\pi_{01}+\pi_{10}}P_U(J_j)+\frac{\pi_{01}}{\pi_{01}+\pi_{10}}P_Y(J_j)\right]$$

$$\times h_b([\overline{\beta(u(J_j))}\pi_{01}+\beta(u(J_j))\overline{\pi_{10}}]*\delta). \tag{92}$$

Corresponding expressions hold for $\pi_{01}+\pi_{10}>1$, with $\{K_j\}$ replacing $\{J_j\}$.

The next step is to show that $\beta(u(I))-\beta(\ell(I))=O(\delta^2)$ for I equal to each of $\{J_j\}$ and $\{K_j\}$ and to express $\beta(u(I))$ (and hence $\beta(\ell(I))$) using the asymptotic approximation $p_0+\delta p_1+O(\delta^2)$, where p_0 and p_1 depend on I. We give the details for $I=J_3$ with the other cases following similarly. For $I=J_3$, we have

$$\beta(u(J_3))=\frac{\delta\overline{\pi_{10}}}{\delta\pi_{10}+\overline{\delta\pi_{10}}} \tag{93}$$

$$=1-\frac{\delta\pi_{10}}{\delta\pi_{10}+\overline{\delta\pi_{10}}} \tag{94}$$

$$=1-\delta\frac{\pi_{10}}{\overline{\pi_{10}}}+O(\delta^2) \tag{95}$$

and

$$\beta(\ell(J_3))=\frac{\overline{\delta}\delta\pi_{01}+\overline{\delta}^2(\overline{\pi_{10}}\pi_{01}/\overline{\pi_{01}})}{\delta^2\overline{\pi}_{01}+\delta\overline{\delta}(\pi_{10}\pi_{01}/\overline{\pi_{01}})+\overline{\delta}\delta\pi_{01}+\overline{\delta}^2(\overline{\pi_{10}}\pi_{01}/\overline{\pi_{01}})} \tag{96}$$

$$=1-\frac{\delta^2\overline{\pi}_{01}+\delta\overline{\delta}(\pi_{10}\pi_{01}/\overline{\pi_{01}})}{\delta^2\overline{\pi}_{01}+\delta\overline{\delta}(\pi_{10}\pi_{01}/\overline{\pi_{01}})+\overline{\delta}\delta\pi_{01}+\overline{\delta}^2(\overline{\pi_{10}}\pi_{01}/\overline{\pi_{01}})} \tag{97}$$

$$=1-\delta\frac{(\pi_{10}\pi_{01}/\overline{\pi_{01}})}{(\overline{\pi_{10}}\pi_{01}/\overline{\pi_{01}})}+O(\delta^2) \tag{98}$$

$$=1-\delta\frac{\pi_{10}}{\overline{\pi_{10}}}+O(\delta^2). \tag{99}$$

The asymptotic expressions for the intervals $\{J_j\}$ are given by (100). The expressions for K_0, K_1, K_2, and K_3 coincide respectively with those for J_1, J_0, J_3, and J_2.

I	p_0	p_1
J_0	0	$\pi_{01}/\overline{\pi_{01}}$
J_1	0	$\overline{\pi_{10}}/\pi_{10}$
J_2	1	$-\overline{\pi_{01}}/\pi_{01}$
J_3	1	$-\pi_{10}/\overline{\pi_{10}}$

$$\tag{100}$$

The asymptotic expression (88) for the entropy rate is then obtained by substituting the expressions (100) into (90) and (91), invoking Lemma 4.2, substituting the interval probabilities (83) and (84), and combining terms. It is easy to see that the two cases $\pi_{01} + \pi_{10} \leq 1$ and $\pi_{01} + \pi_{10} > 1$ should result in the same expression, since the asymptotic expressions for the interval end points are permuted in the same manner as the interval probabilities. □

5 A deterministic approximation algorithm

In this section, we present and analyze an entropy rate approximation scheme, which is based on approximating the stationary distribution of the alternative Markov processes constructed in Section 2. We remark that a somewhat similar scheme has previously been described, though not analyzed, in [32]. Throughout, "operations" refers to arithmetic operations.

5.1 The symmetric case

Assume without loss of generality that $\pi < 1/2$. Since $|f| \leq \log(1-\pi)/\pi$, from (24) it is clear that the support of Y_i is contained in the interval

$$I_{\pi,\delta} \overset{\Delta}{=} \left[-\log \frac{(1-\pi)(1-\delta)}{\pi\delta}, \log \frac{(1-\pi)(1-\delta)}{\pi\delta} \right].$$

Let Q be an M-level quantizer of $I_{\pi,\delta}$ (i.e., a mapping from $I_{\pi,\delta}$ to M real numbers) with the property that

$$\max_{x \in I_{\pi,\delta}} |Q(x) - x| \leq \varepsilon. \tag{101}$$

For example, a uniform quantizer of $I_{\pi,\delta}$ with

$$M \geq \frac{1}{\varepsilon} \log \frac{(1-\pi)(1-\delta)}{\pi\delta}$$

levels has this property. Consider now the finite-state Markov process (with M states) evolving with the process in (24) according to

$$\tilde{Y}_i = Q\left(r_i \log \frac{1-\delta}{\delta} + s_i f(\tilde{Y}_{i-1}) \right) \tag{102}$$

and initiated (at time $i = 0$) with its stationary distribution (say independently of Y_0). Then

$$|\tilde{Y}_i - Y_i| \le \varepsilon + |f(\tilde{Y}_{i-1}) - f(Y_{i-1})| \tag{103}$$

$$\le \varepsilon + (1 - 2\pi)|\tilde{Y}_{i-1} - Y_{i-1}| \tag{104}$$

$$\le \varepsilon + (1 - 2\pi)[\varepsilon + |f(\tilde{Y}_{i-2}) - f(Y_{i-2})|] \tag{105}$$

$$\vdots \tag{106}$$

$$\le \varepsilon \sum_{j=0}^{i} (1 - 2\pi)^j |\tilde{Y}_0 - Y_0| \tag{107}$$

$$\le \frac{\varepsilon}{2\pi} |\tilde{Y}_0 - Y_0| \tag{108}$$

$$\le \varepsilon \frac{1}{\pi} \log \frac{(1 - \pi)(1 - \delta)}{\pi \delta} \tag{109}$$

$$\le \varepsilon \frac{1}{\pi} \log \frac{1}{\pi \delta}, \tag{110}$$

where (104) follows from (23), and (108) from the fact that both $\tilde{Y}_0$ and Y_0 belong to the interval $I_{\pi,\delta}$. Let

$$\lambda_{\pi,\delta} = \max_{y \in I_{\pi,\delta}} \left| \frac{\partial h_b \left(\frac{e^y}{1+e^y} * \pi * \delta \right)}{\partial y} \right|.$$

Note that $\lambda_{\pi,\delta} < \infty$ when π or δ are bounded away from 0 and 1. Combining the fact that

$$\left| h_b \left(\frac{e^y}{1+e^y} * \pi * \delta \right) - h_b \left(\frac{e^{y'}}{1+e^{y'}} * \pi * \delta \right) \right| \le \lambda_{\pi,\delta} |y - y'|$$

for $y, y' \in I_{\pi,\delta}$ with Corollary 2.2, we obtain

$$\left| \overline{H}(Z) - E h_b \left(\frac{e^{\tilde{Y}_i}}{1+e^{\tilde{Y}_i}} * \pi * \delta \right) \right| \le \lambda_{\pi,\delta} \varepsilon \frac{1}{\pi} \log \frac{1}{\pi \delta}. \tag{111}$$

In particular, for the M-level uniform quantizer mentioned above,

$$\varepsilon \le \frac{1}{M} \log \frac{(1 - \pi)(1 - \delta)}{\pi \delta},$$

so (111) implies that

$$\left| \overline{H}(Z) - Eh_b\left(\frac{e^{\tilde{Y}_i}}{1+e^{\tilde{Y}_i}} * \pi * \delta\right) \right| \leq \frac{1}{M}\frac{\lambda_{\pi,\delta}}{\pi}\left[\log\frac{1}{\pi\delta}\right]^2. \qquad (112)$$

Thus, for a given precision ε we would need to take M such that

$$\frac{1}{M}\frac{\lambda_{\pi,\delta}}{\pi}\left[\log\frac{1}{\pi\delta}\right]^2 \leq \varepsilon,$$

find the M-dimensional stationary distribution vector, and use it to compute $Eh_b\left(\frac{e^{\tilde{Y}_i}}{1+e^{\tilde{Y}_i}} * \pi * \delta\right)$. More specifically, we have the following approximation algorithm.

Algorithm 5.1.

Input: M, π, δ

(1) Let Q denote the M-level uniform quantizer of the interval $I_{\pi,\delta}$ and $q_1,\ldots,q_M$ denote the quantization levels. Let P_M be the $M \times M$ stochastic matrix given by

$$P_M(i,j) = \begin{array}{l}[(1-\delta)(1-\pi)]\,1\left(q_j = Q\left(\log\frac{1-\delta}{\delta}+f(q_i)\right)\right)+ \\ [\delta(1-\pi)]\,1\left(q_j = Q\left(-\log\frac{1-\delta}{\delta}+f(q_i)\right)\right)+ \\ [(1-\delta)\pi]\,1\left(q_j = Q\left(\log\frac{1-\delta}{\delta}-f(q_i)\right)\right)+ \\ [\delta\pi]\,1\left(q_j = Q\left(-\log\frac{1-\delta}{\delta}-f(q_i)\right)\right), \end{array} \qquad (113)$$

where $1(\cdot)$ is the indicator function of the condition in the argument (i.e., $1(\cdot) = 1$ if the condition in the argument is true and $1(\cdot) = 0$ otherwise).
(2) Compute the stationary distribution of P_M, i.e., the M-dimensional row vector $\mathbf{a}_M$ solving $\mathbf{a}_M \cdot P_M = \mathbf{a}_M$.
(3) Compute the entropy estimate

$$\hat{H} = \sum_{i=1}^{M}\mathbf{a}_M(i) \cdot h_b\left(\frac{e^{q_i}}{1+e^{q_i}} * \pi * \delta\right). \qquad (114)$$

Output: $\hat{H}$.

Note that $\hat{H}$ in (114) is nothing but the expression $Eh_b\left(\frac{e^{\tilde{Y}_i}}{1+e^{\tilde{Y}_i}} * \pi * \delta\right)$ that appears in (112), where $\{\tilde{Y}_i\}$ is the quantized process defined in (102) (initiated at its stationary distribution). One possible brute force method for finding the stationary distribution of an $M \times M$ stochastic matrix is via Gaussian elimination ($2M^3/3$ operations), back substitution (M^2 operations), and normalization (M operations) [12]. Since the remaining steps in the algorithm require $O(M)$ operations, the overall number of operations required is $O(M^3)$. From (112), it follows that the resulting precision is $O(1/M)$. In summary, we have established the following theorem.

Theorem 5.2. *For fixed π, δ, Algorithm 5.1 requires $O(M^3)$ operations and guarantees precision of $O(1/M)$. In other words, N operations buy precision $O\left(N^{-1/3}\right)$.*

Theorem 5.2 was derived via a rather rough analysis. Two ingredients that are likely to significantly improve the bound on the approximation–precision tradeoff are as follows.

(1) Using a nonuniform quantizer, with finer resolution near 0 (where f is least contractive) and coarser resolution towards the end points of the quantized interval (where f is highly contractive).

(2) The main part of the computational burden is finding the stationary distribution of the stochastic matrix P_M given in (113). The upper bound of $O(M^3)$ that was used on the number of operations that this requires holds for any $M \times M$ stochastic matrix. This does not use the particular structure of P_M, a very sparse matrix with the same four nonzero entries in each row.

Thus, a nonuniform quantization followed by an efficient procedure for finding the stationary distribution of P_M should result in an implementation of Algorithm 5.1 with higher precision than that guaranteed in the theorem. Theorem 5.2, however, suffices to make our main point, which is the independence of the bound on the precision order on the process parameters (in this case π and δ). This should be contrasted with the hitherto best known precision–complexity tradeoff among deterministic approximation schemes obtained via (11). Specifically, the difference between the upper and lower bounds in (11), conveniently expressed as the mutual information $I(Z_0; X_{-n-1} | Z_{-n}^{-1})$, is known since [3] to decay exponentially with n. The best known bounds have been obtained in [19] and are of the form

$$I(Z_0; X_{-n-1} | Z_{-n}^{-1}) \le C(\pi, \delta) \rho(\pi, \delta)^n, \tag{115}$$

where $C(\pi, \delta), \rho(\pi, \delta)$ are positive constants and $\rho(\pi, \delta) < 1$. On the other hand, the number of operations required to compute the Cover and Thomas bounds (11) is exponential in n (the exponential rate depending on the size of the alphabet). When combined, these bounds imply precision $O(N^{-\eta})$, for

$\eta = \eta(\pi, \delta) > 0$. However, $\eta(\pi, \delta)$ is arbitrarily small for appropriate values of the parameters, since in the known bound (115) $\rho(\pi, \delta)$ is arbitrarily close to 1 for appropriate values of the parameters.

5.2 The nonsymmetric case

Let us now derive a similar algorithm for the nonsymmetric chain. Paralleling the development of the previous section, the idea is to couple the process pair $\{(Y_i, U_i)\}$ with a quantized process pair $\{(\tilde{Y}_i, \tilde{U}_i)\}$ evolving as

$$\tilde{Y}_i = Q\left(r_i \log \frac{1-\delta}{\delta} + s_i f(\tilde{U}_{i-1}) + (1-s_i) f(\tilde{Y}_{i-1})\right) \tag{116}$$

and

$$\tilde{U}_i = Q\left(q_i \log \frac{1-\delta}{\delta} + (1-t_i) f(\tilde{U}_{i-1}) + t_i f(\tilde{Y}_{i-1})\right), \tag{117}$$

where $\{q_i\}$, $\{r_i\}$, $\{s_i\}$, and $\{t_i\}$ are as defined in Theorem 2.3. We have

$$|\tilde{Y}_i - Y_i| \leq \varepsilon + s_i |f(\tilde{U}_{i-1}) - f(U_{i-1})| + (1-s_i)|f(\tilde{Y}_{i-1}) - f(Y_{i-1})| \tag{118}$$

$$\leq \varepsilon + \max\{|f(\tilde{U}_{i-1}) - f(U_{i-1})|, |f(\tilde{Y}_{i-1}) - f(Y_{i-1})|\} \tag{119}$$

$$\leq \varepsilon + c(\pi_{01}, \pi_{10}) \max\{|\tilde{U}_{i-1} - U_{i-1}|, |\tilde{Y}_{i-1} - Y_{i-1}|\}, \tag{120}$$

with $c(\pi_{01}, \pi_{10})$ defined in (20). Since the right-hand side will similarly also bound $|\tilde{U}_i - U_i|$, we have

$$\max\{|\tilde{U}_i - U_i|, |\tilde{Y}_i - Y_i|\} \leq \varepsilon + c(\pi_{01}, \pi_{10}) \max\{|\tilde{U}_{i-1} - U_{i-1}|, |\tilde{Y}_{i-1} - Y_{i-1}|\}. \tag{121}$$

Iterating gives, similarly as in (110),

$$\max\{|\tilde{U}_i - U_i|, |\tilde{Y}_i - Y_i|\}$$

$$\leq \frac{\varepsilon}{1 - c(\pi_{01}, \pi_{10})} \max\{|\tilde{U}_0 - U_0|, |\tilde{Y}_0 - Y_0|\} \leq \varepsilon \tilde{c}(\pi_{01}, \pi_{10}, \delta), \tag{122}$$

where

$$\tilde{c}(\pi_{01}, \pi_{10}, \delta) = \frac{1}{1 - c(\pi_{01}, \pi_{10})} \max_{x \in \text{support}(U_i) \cup \text{support}(Y_i)} |Q(x) - x|.$$

It follows similarly as in (111) that by letting $\hat{H}(\pi_{01}, \pi_{10}, \delta)$ denote the expectation on the right-hand side of (42), with $\tilde{Y}_i$ replacing Y and $\tilde{U}_i$ replacing U, if the quantizer used has resolution ε, then

$$\left|\overline{H}(Z) - \hat{H}(\pi_{01}, \pi_{10}, \delta)\right| \leq O(\varepsilon). \tag{123}$$

Similarly as in the symmetric case, most of the burden in computing $\hat{H}(\pi_{01}, \pi_{10}, \delta)$ is in computing the stationary distribution for $(\tilde{Y}_i, \tilde{U}_i)$, which is a Markov chain with state space of size M^2 and transition kernel with the same 16 nonzero entries per line, regardless of (sufficiently large) M. Done brute force (without exploiting the structure of the transition matrix), this requires no more than $(M^2)^3 = M^6$ operations. Since $\varepsilon = O(1/M)$, we get by (123) precision $O(1/M)$ for $O(M^6)$ operations, or similarly as in Theorem 5.2, precision $O\left(1/N^{1/6}\right)$ for N operations. As in the previous subsection, the analysis can be refined to improve the order of this polynomial dependence. Beyond possible refinements that were already mentioned for the symmetric case, further simplification is possible for the nonsymmetric case by noting that rather than the joint distribution of $(\tilde{Y}_i, \tilde{U}_i)$, it is only its two marginals that are needed. The analysis given, however, suffices to make the main point, which is the independence of the order of the polynomial on the process parameters.

5.3 Larger alphabet sizes

Extending the above approximation algorithm and its analysis to larger state and observation alphabet sizes is nontrivial and we leave it for future work. Briefly, however, such an extension would first require an extension of the alternative Markov process, the components of which would be $(|\mathcal{X}| - 1)$-dimensional vectors, corresponding to the conditional probability distribution of the underlying state variables conditioned on current and past observations. The components of the alternative Markov process would then capture the evolution of this conditional probability distribution conditioned on the possible values of the underlying state variable. An approximation algorithm could then be obtained by applying vector quantization to the vector-valued components of the process, in analogy to the scalar quantization step above. As noted for a similar algorithm in [32], the complexity of such an algorithm for a given approximation accuracy would scale rather adversely with the state alphabet size $|\mathcal{X}|$ since, by vector quantization theory (also rate distortion theory), the number of quantization points required for a given level of quantization error per dimension of the Markov process components increases exponentially with the dimension, which is $|\mathcal{X}| - 1$. The complexity scaling, however, is much more benign in the observation alphabet size $|\mathcal{Z}|$.

Interestingly, the reverse seems to be true for the Cover and Thomas approximation bounds. The complexity of computing these bounds for a given conditioning history n, assuming that observation sequence probabilities are computed efficiently using the well-known "forward" recursion, is easily seen to be proportional to $|\mathcal{Z}|^n(n|\mathcal{X}|^2)$. Thus, the required computation increases considerably faster with the observation alphabet size than with the state alphabet size, for even moderate n.

The above considerations suggest that, among deterministic algorithms, a quantization-based approach, along the lines we have presented, may be the best choice for approximating the entropy rate when $|\mathcal{X}|$ is small, while the Cover and Thomas bounds may be the better choice when $|\mathcal{X}|$ is large but $|\mathcal{Z}|$ is small. A rigorous comparison of the precision–complexity tradeoff of these two approaches for larger alphabet sizes is left for future work.

6 Conclusions and discussion

We have presented an approach to approximating the entropy rate of a hidden Markov process via approximations of the stationary distribution of a related Markov process. It was illustrated how the approach leads to characterization of the entropy rate in various asymptotic regimes. It was seen that a refinement of the bounding technique in [29], whereby the support is partitioned into a small number of nonoverlapping regions with easily computed probabilities, can lead to significantly tighter bounds and finer characterizations of the asymptotics. It was argued that the bounds derived can be further tightened by further refining this partition, leading, in various asymptotic regimes, to characterization of higher-order terms. Finally, a deterministic algorithm for approximating the entropy rate of the HMP was derived. This scheme, based on approximating the stationary distribution of the related Markov process, was shown to achieve the best known precision–complexity tradeoff.

Though focus was on the binary case, the approach developed for bounding the entropy rate, for asymptotic characterizations, and for approximation are applicable in the general finite-alphabet case.

Results for some of the asymptotic regimes (e.g., the "high SNR" regime when $\pi_{10} = 1$) were shown to be derivable also via the Cover and Thomas bounds [5, Section 4.5]. On the other hand, for other regimes (e.g., the "low SNR" one), our bounds were shown to yield precise characterizations of the asymptotics, while the Cover and Thomas bounds of arbitrarily high order were shown to fall short of implying such characterizations.

A key ingredient in the bounds developed is bounding the support of the belief process. As such, the asymptotic regimes characterized via these bounds are ones that exhibit a "concentration of the support", meaning that the conditional distribution of the state given the past and present HMP components lies, with probability one, in a very small subset of the simplex of possible distributions. For example, in the "high SNR" regime, this belief was seen to fall, with probability one, in a region of the simplex corresponding to very high certainty (that the value is either 0 or 1, depending primarily on the present observation and very weakly on the remaining ones from the past). In the "low SNR" regime, the belief falls, with probability one, in a small region of the simplex corresponding to very low certainty. In the "almost memoryless" regime, as a final example, the belief falls in a small region, concentrated near the belief of a "singlet filter" [6] (which in the binary case consists of two point masses).

Asymptotics of the entropy rate can be obtained also in regimes that lack this concentration property via a more delicate study of the dynamics of the alternative Markov processes of Section 2.3. One such example is the "rare transitions" regime[1] considered in [26], and recently conclusively characterized in [31]. As is argued in [26], this regime is another example of one whose asymptotics are not captured by the Cover and Thomas bounds of arbitrarily high order.

Appendices

A Proof of Theorem 3.5

We first argue that the bounds of Lemma 3.4 continue to hold (and are relaxed) when A, B, a, b are replaced, respectively, by $\tilde{A}, \tilde{B}, \tilde{a}, \tilde{b}$ that have the following simpler expressions:

$$\tilde{A} = \log \frac{1-\pi}{\pi} + \log \frac{1-\delta}{\delta}, \tag{A.1}$$

$$\tilde{b} = -f(\tilde{A}) + \log \frac{1-\delta}{\delta}, \tag{A.2}$$

$$\tilde{a} = -f(\tilde{b}) + \log \frac{1-\delta}{\delta}, \tag{A.3}$$

and

$$\tilde{B} = f(\tilde{b}) + \log \frac{1-\delta}{\delta}. \tag{A.4}$$

---

[1] In this regime, "most" of the time, between the transitions, there is high certainty regarding the value of the underlying state yet, every once in a while, around the occurrences of state transitions, an observer of the HMP will be uncertain regarding the exact location of the transition and, hence, the state values in the neighborhood of these transitions. Consequently, the support of the belief process is a large part of the simplex, which includes regions corresponding to varying degrees of certainty.

To see this, note that $\tilde{A} = f(\infty) + \log \frac{1-\delta}{\delta} \geq f(A) + \log \frac{1-\delta}{\delta} = A$, implying by the monotonicity of f that $\tilde{b} \leq b$, in turn implying that both $\tilde{a} \geq a$ and $\tilde{B} \leq B$. Combined with the decreasing monotonicity of $h_b\left(\frac{e^x}{1+e^x} * \pi * \delta\right)$ for $x > 0$, this implies that indeed substituting the tilded quantities in (54) increases the upper bound and decreases the lower bound. Noting now that

$$(1-\delta)[\pi * (1-\delta)] + \delta[\pi * \delta] = 1 - \pi - \delta(2 - 4\pi) + \delta^2 2(1 - 2\pi), \quad \text{(A.5)}$$

$$(1-\delta)[\pi * \delta] + \delta[\pi * (1-\delta)] = \pi + \delta(2 - 4\pi) - \delta^2 2(1 - 2\pi), \quad \text{(A.6)}$$

and

$$\frac{1}{1 + e^{\tilde{A}}} \sim \frac{\pi}{1 - \pi}\delta,$$

we obtain

$$
\begin{aligned}
h_b\left(\frac{e^{\tilde{A}}}{1 + e^{\tilde{A}}} * \pi * \delta\right) &= h_b\left(\frac{1}{1 + e^{\tilde{A}}} * \pi * \delta\right) \\
&= h_b\left(\left[\frac{\pi}{1 - \pi}\delta(1 + o(1))\right] * \pi * \delta\right) \\
&= h_b\left(\left[\left(\frac{\pi}{1 - \pi} + 1\right)\delta(1 + o(1))\right] * \pi\right) \\
&= h_b\left(\pi + \frac{1 - 2\pi}{1 - \pi}\delta + o(\delta)\right) \\
&= h_b(\pi) + h_b'(\pi)\frac{1 - 2\pi}{1 - \pi}\delta + o(\delta). \quad \text{(A.7)}
\end{aligned}
$$

Similarly,

$$\frac{1}{1 + e^{\tilde{b}}} \sim \frac{1}{e^{\tilde{b}}} \sim \frac{e^{\tilde{A}}(1 - \pi) + \pi}{e^{\tilde{A}}\pi + (1 - \pi)}\frac{\delta}{1 - \delta} \sim \frac{(1 - \pi)}{\pi}\delta,$$

$$\frac{1}{1 + e^{\tilde{a}}} \sim \frac{1}{e^{\tilde{a}}} \sim \frac{e^{\tilde{b}}(1 - \pi) + \pi}{e^{\tilde{b}}\pi + (1 - \pi)}\frac{\delta}{1 - \delta} \sim \frac{(1 - \pi)}{\pi}\delta,$$

and

$$\frac{1}{1 + e^{\tilde{B}}} \sim \frac{1}{e^{\tilde{B}}} \sim \frac{e^{\tilde{b}}\pi + (1 - \pi)}{e^{\tilde{b}}(1 - \pi) + \pi}\frac{\delta}{1 - \delta} \sim \frac{\pi}{1 - \pi}\delta.$$

Thus, we get also

$$h_b\left(\frac{e^{\tilde{a}}}{1+e^{\tilde{a}}}*\pi*\delta\right)=h_b\left(\frac{1}{1+e^{\tilde{a}}}*\pi*\delta\right)$$

$$=h_b\left(\left[\frac{1-\pi}{\pi}\delta(1+o(1))\right]*\pi*\delta\right)$$

$$=h_b\left(\left[\frac{1}{\pi}\delta(1+o(1))\right]*\pi\right)$$

$$=h_b\left(\pi+\frac{1-2\pi}{\pi}\delta+o(\delta)\right)$$

$$=h_b(\pi)+h'_b(\pi)\frac{1-2\pi}{\pi}\delta+o(\delta) \qquad (A.8)$$

and similarly obtain

$$h_b\left(\frac{e^{\tilde{b}}}{1+e^{\tilde{b}}}*\pi*\delta\right)=h_b(\pi)+h'_b(\pi)\frac{1-2\pi}{\pi}\delta+o(\delta) \qquad (A.9)$$

and

$$h_b\left(\frac{e^{\tilde{B}}}{1+e^{\tilde{B}}}*\pi*\delta\right)=h_b(\pi)+h'_b(\pi)\frac{1-2\pi}{1-\pi}\delta+o(\delta). \qquad (A.10)$$

Combining Lemma 3.4 (with $\tilde{A},\tilde{B},\tilde{a},\tilde{b}$ replacing A,B,a,b) with (A.5), (A.6), (A.7), (A.8), (A.9), and (A.10) gives

$$\overline{H}(Z)=[(1-\pi)-\delta(2-4\pi)+o(\delta)]\left[h_b(\pi)+h'_b(\pi)\frac{1-2\pi}{1-\pi}\delta+o(\delta)\right]$$

$$+[\pi+\delta(2-4\pi)+o(\delta)]\left[h_b(\pi)+h'_b(\pi)\frac{1-2\pi}{\pi}\delta+o(\delta)\right]$$

$$=h_b(\pi)+\delta\left[h'_b(\pi)(1-2\pi)-(2-4\pi)h_b(\pi)+h'_b(\pi)(1-2\pi)\right.$$

$$\left.+(2-4\pi)h_b(\pi)\right]+o(\delta)$$

$$=h_b(\pi)+\delta 2h'_b(\pi)(1-2\pi)+o(\delta)$$

$$=h_b(\pi)+\delta 2(1-2\pi)\log_2\frac{1-\pi}{\pi}+o(\delta). \qquad \square$$

B Proof of Theorem 3.6

It is easily checked that in this regime

$$f\left(\log\frac{1}{\delta}\frac{\delta}{\delta}+o(1)\right)=4\varepsilon(1-2\delta)+o(\varepsilon),$$

implying that

$$A=f(A)+\log\frac{1-\delta}{\delta}=\log\frac{1-\delta}{\delta}+4\varepsilon(1-2\delta)+o(\varepsilon), \tag{A.11}$$

$$b=-f(A)+\log\frac{1-\delta}{\delta}=\log\frac{1-\delta}{\delta}-4\varepsilon(1-2\delta)+o(\varepsilon), \tag{A.12}$$

$$a=-f(b)+\log\frac{1-\delta}{\delta}=\log\frac{1-\delta}{\delta}-4\varepsilon(1-2\delta)+o(\varepsilon), \tag{A.13}$$

and

$$B=f(b)+\log\frac{1-\delta}{\delta}=\log\frac{1-\delta}{\delta}+4\varepsilon(1-2\delta)+o(\varepsilon). \tag{A.14}$$

It follows that

$$\frac{1}{1+e^A}=\delta+4\delta(1-\delta)(1-2\delta)\varepsilon+o(\varepsilon), \tag{A.15}$$

$$\frac{1}{1+e^B}=\delta+4\delta(1-\delta)(1-2\delta)\varepsilon+o(\varepsilon), \tag{A.16}$$

$$\frac{1}{1+e^a}=\delta-4\delta(1-\delta)(1-2\delta)\varepsilon+o(\varepsilon), \tag{A.17}$$

and

$$\frac{1}{1+e^b}=\delta-4\delta(1-\delta)(1-2\delta)\varepsilon+o(\varepsilon). \tag{A.18}$$

Now

$$(1-\delta)[\pi*(1-\delta)]+\delta[\pi*\delta]=(1-\delta)[(1/2-\varepsilon)*(1-\delta)]+\delta[(1/2-\varepsilon)*\delta]$$
$$=\frac{1}{2}+\varepsilon(1-2\delta)^2 \tag{A.19}$$

and

$$(1-\delta)[\pi*\delta]+\delta[\pi*(1-\delta)]=\frac{1}{2}-\varepsilon(1-2\delta)^2. \tag{A.20}$$

Now

$$\frac{1}{1+e^A}*\pi*\delta=[\delta+4\delta(1-\delta)(1-2\delta)\varepsilon+o(\varepsilon)]*\left(\frac{1}{2}-\varepsilon\right)*\delta$$
$$=\frac{1}{2}-\varepsilon(1-2\delta)^2+\varepsilon^28\delta(1-\delta)(1-2\delta)^2+o(\varepsilon^2) \tag{A.21}$$

and, similarly,

$$\frac{1}{1+e^B} * \pi * \delta = [\delta + 4\delta(1-\delta)(1-2\delta)\varepsilon + o(\varepsilon)] * \left(\frac{1}{2} - \varepsilon\right) * \delta$$

$$= \frac{1}{2} - \varepsilon(1-2\delta)^2 + \varepsilon^2 8\delta(1-\delta)(1-2\delta)^2 + o(\varepsilon^2), \quad \text{(A.22)}$$

$$\frac{1}{1+e^a} * \pi * \delta = [\delta + 4\delta(1-\delta)(1-2\delta)\varepsilon + o(\varepsilon)] * \left(\frac{1}{2} - \varepsilon\right) * \delta$$

$$= \frac{1}{2} - \varepsilon(1-2\delta)^2 - \varepsilon^2 8\delta(1-\delta)(1-2\delta)^2 + o(\varepsilon^2), \quad \text{(A.23)}$$

and

$$\frac{1}{1+e^b} * \pi * \delta = [\delta + 4\delta(1-\delta)(1-2\delta)\varepsilon + o(\varepsilon)] * \left(\frac{1}{2} - \varepsilon\right) * \delta$$

$$= \frac{1}{2} - \varepsilon(1-2\delta)^2 - \varepsilon^2 8\delta(1-\delta)(1-2\delta)^2 + o(\varepsilon^2). \quad \text{(A.24)}$$

It follows now from Lemma 3.4, (A.49), the above displays and

$$h_b\left(\frac{1}{2} - \varepsilon\right) = 1 + \frac{1}{2}h_b''(1/2)\varepsilon^2 + \frac{1}{4!}h_b^{(4)}(1/2)\varepsilon^4 + O(\varepsilon^6) = 1 - \frac{2}{\log 2}\varepsilon^2 + O(\varepsilon^4)$$

$$\text{(A.25)}$$

(note that h_b is symmetric around $1/2$, so odd-ordered terms annihilate) that

$$\overline{H}(Z) = \left[\frac{1}{2} + \varepsilon(1-2\delta)^2\right]$$

$$\times h_b\left(\frac{1}{2} - \varepsilon(1-2\delta)^2 + \varepsilon^2 8\delta(1-\delta)(1-2\delta)^2 + o(\varepsilon^2)\right) \quad \text{(A.26)}$$

$$+ \left[\frac{1}{2} - \varepsilon(1-2\delta)^2\right]$$

$$\times h_b\left(\frac{1}{2} - \varepsilon(1-2\delta)^2 - \varepsilon^2 8\delta(1-\delta)(1-2\delta)^2 + o(\varepsilon^2)\right) \quad \text{(A.27)}$$

$$= \left[\frac{1}{2} + \varepsilon(1-2\delta)^2 \right]$$

$$\times \left\{ 1 - \frac{2}{\log 2} \left[\varepsilon(1-2\delta)^2 - \varepsilon^2 8\delta(1-\delta)(1-2\delta)^2 \right]^2 + o(\varepsilon^3) \right\} \tag{A.28}$$

$$+ \left[\frac{1}{2} - \varepsilon(1-2\delta)^2 \right]$$

$$\times \left\{ 1 - \frac{2}{\log 2} \left[\varepsilon(1-2\delta)^2 + \varepsilon^2 8\delta(1-\delta)(1-2\delta)^2 \right]^2 + o(\varepsilon^3) \right\}. \tag{A.29}$$

So,

$$1 - \overline{H}(Z) = \frac{2}{\log 2} \left\{ \left[\frac{1}{2} + \varepsilon(1-2\delta)^2 \right] \right.$$

$$\times \left[\varepsilon(1-2\delta)^2 - \varepsilon^2 8\delta(1-\delta)(1-2\delta)^2 \right]^2 \tag{A.30}$$

$$+ \left[\frac{1}{2} - \varepsilon(1-2\delta)^2 \right] \left[\varepsilon(1-2\delta)^2 + \varepsilon^2 8\delta(1-\delta)(1-2\delta)^2 \right]^2 \right\}$$

$$+ o(\varepsilon^3) \tag{A.31}$$

$$= \frac{2}{\log 2} \varepsilon^2 (1-2\delta)^4 + o(\varepsilon^3). \tag{A.32}$$

C Proof of Theorem 3.7

Proof. By (A.5) and (A.6),

$$(1-\delta)[\pi * (1-\delta)] + \delta[\pi * \delta] = 1 - \pi - \left(\frac{1}{2} - \varepsilon \right)(2 - 4\pi) + \left(\frac{1}{2} - \varepsilon \right)^2$$

$$2(1-2\pi) = \frac{1}{2} + \varepsilon^2 2(1-2\pi) \tag{A.33}$$

and

$$(1-\delta)[\pi * \delta] + \delta[\pi * (1-\delta)] = \pi + \left(\frac{1}{2} - \varepsilon \right)(2 - 4\pi) - \left(\frac{1}{2} - \varepsilon \right)^2$$

$$2(1-2\pi) = \frac{1}{2} - \varepsilon^2 2(1-2\pi). \tag{A.34}$$

Now, recalling (45),

$$A = \log\left[(\alpha - 1)\frac{1 - \pi}{2\pi} + \sqrt{\alpha + \left[(\alpha - 1)\frac{1 - \pi}{2\pi}\right]^2} \right],$$

where $\alpha \overset{\triangle}{=} \dfrac{1 - \delta}{\delta} = \dfrac{1 + 2\varepsilon}{1 - 2\varepsilon} = 1 + 4\varepsilon + o(\varepsilon)$, so

$$A = \log\left[(4\varepsilon + o(\varepsilon))\frac{1 - \pi}{2\pi} + \sqrt{1 + 4\varepsilon + o(\varepsilon) + \left[(4\varepsilon + o(\varepsilon))\frac{1 - \pi}{2\pi}\right]^2} \right]$$

$$\text{(A.35)}$$

$$= \log\left[(4\varepsilon + o(\varepsilon))\frac{1 - \pi}{2\pi} + \sqrt{1 + 4\varepsilon + o(\varepsilon)} \right] \tag{A.36}$$

$$= \log\left[(4\varepsilon + o(\varepsilon))\frac{1 - \pi}{2\pi} + 1 + 2\varepsilon + o(\varepsilon) \right] \tag{A.37}$$

$$= \log\left[1 + \varepsilon\left(4 \cdot \frac{1 - \pi}{2\pi} + 2\right) + o(\varepsilon) \right] \tag{A.38}$$

$$= \log\left[1 + \varepsilon \cdot \frac{2}{\pi} + o(\varepsilon) \right] \tag{A.39}$$

$$= \varepsilon \cdot \frac{2}{\pi} + o(\varepsilon), \tag{A.40}$$

implying that

$$\frac{1}{1 + e^A} = \frac{1}{2 + \varepsilon \cdot \frac{2}{\pi} + o(\varepsilon)} = \frac{1}{2} - \frac{1}{2\pi}\varepsilon + o(\varepsilon). \tag{A.41}$$

Now, recalling that $f(0) = 0$ and using (19),

$$b = -f(A) + \log\frac{1 - \delta}{\delta} = -f'(0)A + o(A) + \log[1 + 4\varepsilon + o(\varepsilon)]$$

$$= -(1 - 2\pi)\varepsilon \cdot \frac{2}{\pi} + 4\varepsilon + o(\varepsilon) = (8 - 2/\pi)\varepsilon + o(\varepsilon), \tag{A.42}$$

implying that

$$\frac{1}{1 + e^b} = \frac{1}{2 + \varepsilon \cdot (8 - 2/\pi) + o(\varepsilon)} = \frac{1}{2} - \left(2 - \frac{1}{2\pi}\right)\varepsilon + o(\varepsilon). \tag{A.43}$$

Moving to a, we have

$$a = -f(b) + \log \frac{1-\delta}{\delta} = -f'(0)h + o(h) + \log[1 + 4\varepsilon + o(\varepsilon)]$$
$$= -(1-2\pi)(8-2/\pi)\varepsilon + 4\varepsilon + o(\varepsilon) = (16\pi + 2/\pi - 8)\varepsilon + o(\varepsilon), \quad \text{(A.44)}$$

implying that

$$\frac{1}{1+e^a} = \frac{1}{2 + \varepsilon \cdot (16\pi + 2/\pi - 8) + o(\varepsilon)} = \frac{1}{2} - \left(4\pi + \frac{1}{2\pi} - 2\right)\varepsilon + o(\varepsilon).$$
$$\text{(A.45)}$$

Finally,

$$B = f(b) + \log \frac{1-\delta}{\delta} = f'(0)b + o(b) + \log[1 + 4\varepsilon + o(\varepsilon)]$$
$$= (1-2\pi)(8-2/\pi)\varepsilon + 4\varepsilon + o(\varepsilon) = (-16\pi - 2/\pi + 16)\varepsilon + o(\varepsilon), \quad$$
$$\text{(A.46)}$$

implying that

$$\frac{1}{1+e^B} = \frac{1}{2 + \varepsilon \cdot (-16\pi - 2/\pi + 16) + o(\varepsilon)} = \frac{1}{2} - \left(-4\pi - \frac{1}{2\pi} + 4\right)\varepsilon + o(\varepsilon).$$
$$\text{(A.47)}$$

Using the easily verified identity

$$\left(\frac{1}{2} - c\varepsilon\right) * \pi * \left(\frac{1}{2} - \varepsilon\right) = \frac{1}{2} - 2c(1-2\pi)\varepsilon^2, \quad \text{(A.48)}$$

the Taylor expansion

$$h_b\left(\frac{1}{2} - \varepsilon\right) = 1 + \frac{1}{2}h_b''(1/2)\varepsilon^2 + o(\varepsilon^2) = 1 - \frac{2}{\log 2}\varepsilon^2 + o(\varepsilon^2), \quad \text{(A.49)}$$

and combining with (A.33), (A.34), (A.41), and (A.45) gives

$$\{(1-\delta)[\pi*(1-\delta)]+\delta[\pi*\delta]\}\,h_b\left(\frac{e^A}{1+e^A}*\pi*\delta\right)$$

$$+\{(1-\delta)[\pi*\delta]+\delta[\pi*(1-\delta)]\}\,h_b\left(\frac{e^a}{1+e^a}*\pi*\delta\right)$$

$$=\left[\frac{1}{2}+\varepsilon^2 2(1-2\pi)\right]h_b\left(\frac{1}{2}-\frac{1-2\pi}{\pi}\varepsilon^2+o(\varepsilon^2)\right) \tag{A.50}$$

$$+\left[\frac{1}{2}-\varepsilon^2 2(1-2\pi)\right]$$

$$\times h_b\left(\frac{1}{2}-2\left(4\pi+\frac{1}{2\pi}-2\right)(1-2\pi)\varepsilon^2+o(\varepsilon^2)\right) \tag{A.51}$$

$$=\left[\frac{1}{2}+\varepsilon^2 2(1-2\pi)\right]\left\{1-\frac{2}{\log 2}\left[\frac{1-2\pi}{\pi}\varepsilon^2\right]^2+o(\varepsilon^4)\right\}$$

$$+\left[\frac{1}{2}-\varepsilon^2 2(1-2\pi)\right] \tag{A.52}$$

$$\times\left\{1-\frac{2}{\log 2}\left[2\left(4\pi+\frac{1}{2\pi}-2\right)(1-2\pi)\varepsilon^2\right]^2+o(\varepsilon^4)\right\} \tag{A.53}$$

$$=1-\frac{1}{\log 2}\left\{\left[\frac{1-2\pi}{\pi}\right]^2+\left[2\left(4\pi+\frac{1}{2\pi}-2\right)(1-2\pi)\right]^2\right\}\varepsilon^4+o(\varepsilon^4) \tag{A.54}$$

$$=1-\frac{2(1-2\pi)^2(1-4\pi+16\pi^2-32\pi^3+32\pi^4)}{\pi^2\log 2}\varepsilon^4+o(\varepsilon^4). \tag{A.55}$$

Similarly, using (A.43) and (A.47) in lieu of (A.41) and (A.45), we obtain

$$\{(1-\delta)[\pi*(1-\delta)]+\delta[\pi*\delta]\}\,h_b\left(\frac{e^B}{1+e^B}*\pi*\delta\right)$$

$$+\{(1-\delta)[\pi*\delta]+\delta[\pi*(1-\delta)]\}\,h_b\left(\frac{e^b}{1+e^b}*\pi*\delta\right)$$

$$=\left[\frac{1}{2}+\varepsilon^2 2(1-2\pi)\right]$$

$$\times\left\{1-\frac{2}{\log 2}\left[2\left(-4\pi-\frac{1}{2\pi}+4\right)(1-2\pi)\varepsilon^2\right]^2+o(\varepsilon^4)\right\} \tag{A.56}$$

$$+\left[\frac{1}{2}-\varepsilon^2 2(1-2\pi)\right]\left\{1-\frac{2}{\log 2}\left[2\left(2-\frac{1}{2\pi}\right)(1-2\pi)\varepsilon^2\right]^2+o(\varepsilon^4)\right\}$$

$$=1-\frac{1}{\log 2}\left\{\left[2\left(-4\pi-\frac{1}{2\pi}+4\right)(1-2\pi)\right]^2\right. \tag{A.57}$$

$$\left.+\left[2\left(2-\frac{1}{2\pi}\right)(1-2\pi)\right]^2\right\}\varepsilon^4+o(\varepsilon^4) \tag{A.58}$$

$$=1-\frac{2(1-2\pi)^2(1-12\pi+48\pi^2-64\pi^3+32\pi^4)}{\pi^2\log 2}\varepsilon^4+o(\varepsilon^4). \tag{A.59}$$

Combining (A.55) and (A.59) with Lemma 3.4 completes the proof.　□

D Proof of Theorem 4.1 via Cover and Thomas bounds

In this section, we show that the result of Section 4.1 can be obtained using the upper and lower bounds on the entropy rate of an HMM given in the book of Cover and Thomas. These upper and lower bounds are $H(Z_k|Z^k)$ and $H(Z_k|Z_2^k,X_1)$, respectively, and are valid for any k. We analyze the asymptotic behavior of these bounds for $k=3$ for the setting of Section 4.1, and show that the factor multiplying the $\delta\log_2(1/\delta)$ term in both the lower and upper bounds agrees with that given in Theorem 4.1.

We first treat the upper bound $H(Z_3|Z_2,Z_1)$ and expand it as $H(Z_3,Z_2,Z_1)-H(Z_2,Z_1)$. In each resulting joint entropy the $-p(\cdot)\log_2 p(\cdot)$ terms contributing to the $\delta\log_2(1/\delta)$ factor are those for which $p(z_3,z_2,z_1)$ and $p(z_2,z_1)$ tend to zero no faster than δ. The remaining $-p(\cdot)\log_2 p(\cdot)$ terms contribute to higher-order asymptotics and we ignore these. The $\Omega(\delta)$ probabilities arise from those sequences (z_3,z_2,z_1) and (z_2,z_1) that have zero probability under the Markov chain distribution $P_{X_3X_2X_1}$, and differ from at least one nonzero probability sequence, again under the Markov chain distribution, in precisely one position. The factor contributed to the $\delta\log_2(1/\delta)$ term by any such zero probability sequence is then the probability, under the Markov chain distribution, of the set of sequences at Hamming distance 1. The sequences at Hamming distance 2 or greater contribute terms of $\delta^2\log_2(1/\delta)$ or smaller.

Let $m_0=Pr(X_i=0)=\pi_{10}/(\pi_{01}+\pi_{10})$ and $m_1=Pr(X_i=1)=1-m_0$. For the $H(Z_2,Z_1)$ term, $(1,1)$ is the only zero probability sequence and the probability of the Hamming distance 1 sequences $(1,0)$ and $(0,1)$ is $m_0\pi_{01}+m_1$. For the $H(Z_3,Z_2,Z_1)$ term, there are three zero probability sequences $(1,1,1)$, $(1,1,0)$, and $(0,1,1)$. For $(1,1,1)$, the only Hamming distance 1 sequence with nonzero probability is $(1,0,1)$ and its probability is $m_1 \pi_{01}$. For $(1,1,0)$, the nonzero probability Hamming distance 1 sequences are $(0,1,0)$ and $(1,0,0)$ with a combined probability of $m_0 \pi_{01} + m_0 \overline{\pi_{01}} \pi_{01}$. Similarly, for $(0,1,1)$, the contributing sequences are $(0,1,0)$ and $(0,0,1)$ with a combined probability of $m_0 \pi_{01} + m_1 \overline{\pi_{01}}$. Thus, the overall factor multiplying the $\delta \log_2(1/\delta)$ term is

$$m_1 \pi_{01} + m_0 \pi_{01} + m_0 \overline{\pi_{01}} \pi_{01} + m_0 \pi_{01} + m_1 \overline{\pi_{01}} - m_0 \pi_{01} - m_1$$

$$= m_0 \overline{\pi_{01}} \pi_{01} + m_0 \pi_{01} \tag{A.60}$$

$$= \frac{\pi_{01}}{1 + \pi_{01}} (2 - \pi_{01}). \tag{A.61}$$

The lower bound $H(Z_3 | Z_2, X_1)$ is similarly expanded to $H(Z_3, Z_2, X_1) - H(Z_2, X_1)$ and the two joint entropies are analyzed as above. In this case, the Hamming distance 1 sequences must differ from the zero probability sequences (z_3, z_2, x_1) and (z_2, x_1) only in the z_i positions. For the $H(Z_2, X_1)$ term, again $(1,1)$ is the only zero probability sequence, and the only allowed nonzero probability Hamming distance 1 sequence under the new restriction is $(0,1)$, the probability of which is m_1. For the $H(Z_3, Z_2, X_1)$ term, the three zero probability sequences are again $(1,1,1)$, $(1,1,0)$, and $(0,1,1)$. The contributions from $(1,1,1)$ and $(1,1,0)$ are as above. For $(0,1,1)$, the only contributing Hamming distance 1 sequence is $(0,0,1)$ (since the sequence $(0,1,0)$ differs in the x position), and its probability is $m_1 \overline{\pi_{01}}$. The overall factor multiplying $\delta \log_2(1/\delta)$ for the lower bounds is

$$m_1 \pi_{01} + m_0 \pi_{01} + m_0 \overline{\pi_{01}} \pi_{01} + m_1 \overline{\pi_{01}} - m_1 = m_0 \overline{\pi_{01}} \pi_{01} + m_0 \pi_{01} \tag{A.62}$$

$$= \frac{\pi_{01}}{1 + \pi_{01}} (2 - \pi_{01}). \tag{A.63}$$

The claim of Theorem 4.1 then follows from the agreement of the upper and lower bound factors. □

E Proofs of lemmas used in proving Theorem 4.1

Proof of Lemma 4.3. As noted, $-r(\pi_{01}) = f(-\infty) \geq f(r(\delta) - r(\pi_{01}))$. This fact together with the fact that $f(x)$ is decreasing shows that I_0 and I_2 are nonempty. The other two intervals are handled by similarly noting that $-r(\pi_{01}) \geq f(-r(\delta) - r(\pi_{01}))$. □

Proof of Lemma 4.4. For x large and positive the difference between $f(x)$ and $-x + \log \pi_{01}$ converges to 0. Therefore, δ tends to 0, the right end point of

I_0 behaves like $-r(\delta) + f(r(\pi_{01}) + \log \pi_{01}) = -r(\delta) + f(\log \overline{\pi_{01}}) = -r(\delta) - r(\pi_{01}) - \log 2$, while the left end point of I_1 behaves like $-r(\delta) - f(-\infty) = -r(\delta) - r(\pi_{01})$. It thus follows that $I_0 < I_1$ for all sufficiently small $\delta > 0$. The right end point of I_1 tends to $-\infty$, while the left end point of I_2, based on the preceding observations, tends to $r(\pi_{01}) + \log \pi_{01} = \log \overline{\pi_{01}}$. Consequently, $I_1 < I_2$ for all sufficiently small $\delta > 0$. Finally, the right end point of I_2 (similarly to the right end point of I_0) behaves like $r(\delta) - r(\pi_{01}) - \log 2$, while the left end point of I_3 behaves like $r(\delta) - r(\pi_{01})$, implying here as well that $I_2 < I_3$ for all sufficiently small $\delta > 0$. $\qquad\square$

Proof of Lemma 4.5. As a consequence of the above upper bound on $f(x)$, it follows from the transitions of $\{(U_i, Y_i)\}$ that the supports of P_U and P_Y are bounded from above by

$$u = \log \frac{\overline{\delta}}{\delta} + \log \frac{\pi_{01}}{\overline{\pi_{01}}} = r(\delta) - r(\pi_{01}).$$

Furthermore, since $f(x) \geq f(u)$ for $x \leq u$, the U_i and Y_i are also bounded from below by

$$\ell = -r(\delta) + f(u).$$

Next, we argue that the values of q_i, t_i, r_i, q_{i-1}, and r_{i-1}, appearing in the Markov chain transitions, determine intervals among $I_j, j = 0, 1, 2, 3$, into which U_i and Y_i must fall and do so according to the table below.

I	Y_i	U_i
I_0	$\{(r_i, q_{i-1}) = (-1, 1)\}$	$\{(q_i, t_i, q_{i-1}) = (-1, 0, 1)\} \cup \{(q_i, t_i, r_{i-1}) = (-1, 1, 1)\}$
I_1	$\{(r_i, q_{i-1}) = (-1, -1)\}$	$\{(q_i, t_i, q_{i-1}) = (-1, 0, -1)\} \cup \{(q_i, t_i, r_{i-1}) = (-1, 1, -1)\}$
I_2	$\{(r_i, q_{i-1}) = (1, 1)\}$	$\{(q_i, t_i, q_{i-1}) = (1, 0, 1)\} \cup \{(q_i, t_i, r_{i-1}) = (1, 1, 1)\}$
I_3	$\{(r_i, q_{i-1}) = (1, -1)\}$	$\{(q_i, t_i, q_{i-1}) = (1, 0, -1)\} \cup \{(q_i, t_i, r_{i-1}) = (1, 1, -1)\}$

$$(A.64)$$

The entries in the row corresponding to interval I_j and columns corresponding to Y_i and U_i specify the values of the above variables that force Y_i and U_i, respectively, to fall in I_j. The table is derived by inspecting the transitions of $\{(U_i, Y_i)\}$. As a representative example of how the table is filled, we consider the entry in row I_2 and column U_i and show how $\{(q_i, t_i, q_{i-1}) = (1, 0, 1)\} \cup \{(q_i, t_i, r_{i-1}) = (1, 1, 1)\}$ implies that $U_i \in I_2$. In the case that $\{(q_i, t_i, q_{i-1}) = (1, 0, 1)\}$, the transitions of $\{(U_i, Y_i)\}$ imply that

$$U_i \in \left[\min_{x \in [\ell, u]} r(\delta) + f(r(\delta) + f(x)), \max_{x \in [\ell, u]} r(\delta) + f(r(\delta) + f(x)) \right] \quad \text{(A.65)}$$

$$\overset{(a)}{\subset} [r(\delta) + f(r(\delta) + f(-\infty)), r(\delta) + f(r(\delta) + f(u))] \quad \text{(A.66)}$$

$$= [r(\delta) + f(r(\delta) - r(\pi_{01})), r(\delta) + f(r(\delta) + f(r(\delta) - r(\pi_{01})))], \text{(A.67)}$$

where (a) follows from the fact noted above that $f(x)$ is decreasing in x. The case of $\{(q_i, t_i, r_{i-1}) = (1, 1, 1)\}$ and the other entries of the table are obtained in a similar fashion.

Lemma 4.4 guarantees that for all sufficiently small $\delta > 0$, the intervals I_j, $j = 0, 1, 2, 3$, are disjoint. From this and the fact that the events in each column of the above table are exhaustive, we can conclude that for all sufficiently small $\delta > 0$, the probabilities of the interval I_j under P_{Y_i} and P_{U_i}, for all $i > 2$, and hence under P_Y and P_U, coincide with the probabilities of the corresponding events in the respective columns of (A.64). The entries of (62) are simply the probabilities of the events in (A.64) as specified by the Markov chain transitions. □

References

[1] L. Arnold, L. Demetrius, and M. Gundlach. *Evolutionary formalism for products of positive random matrices.* Ann. Appl. Prob. **4** (1994) 859–901

[2] T. Berger and J. D. Gibson. *Lossy source coding.* IEEE Trans. Inf. Theory **44**(6) (1998) 2693–2723

[3] J. J. Birch. *Approximations for the entropy for functions of Markov chains.* Ann. Math. Statist., **33** (1962) 930–938

[4] D. Blackwell. *The entropy of functions of finite-state Markov chains.* In Trans. First Prague Conf. Information Theory, Statistical Decision Functions, Random Processes, 1957, pp. 13–20

[5] T. M. Cover and J. A. Thomas. *Elements of Information Theory*, 2nd edn. Wiley, Hoboken, NJ, 2006

[6] J. L. Devore. *A note on the observation of a Markov source through a noisy channel.* IEEE Trans. Inf. Theory **20** (1974) 762–764

[7] S. Egner, V. B. Balakirsky, L. M. G. M. Tolhuizen, S. P. M. J. Baggen, and H. D. L. Hollmann. *On the entropy rate of a hidden Markov model.* In Proc. Int. Symp. Information Theory, Chicago, IL, June 2004, p. 12

[8] E. O. Elliott. *Estimates of error rates for codes on burst-noise channels.* Bell Syst. Tech. J. **42** (1963) 1977–1997

[9] Y. Ephraim and N. Merhav. *Hidden Markov processes.* IEEE Trans. Inf. Theory **48**(6) (2002) 1518–1569

[10] E. N. Gilbert. *Capacity of a burst-noise channel.* Bell Syst. Tech. J. **39** (1960) 1253–1265

[11] F. Le Gland and L. Mevel. *Exponential forgetting and geometric ergodicity in hidden Markov models.* Math. Control Signals Syst. **13**(1) (2000) 63–93

[12] G. H. Golub and C. F. Van Loan. *Matrix Computations*, 3rd edn. Johns Hopkins, Baltimore, MD, 1996

[13] R. M. Gray. *Information rates of autoregressive processes.* IEEE Trans. Inf. Theory **16**(2) (1970) 412–421

[14] R. M. Gray. *Rate distortion functions for finite-state finite-alphabet Markov sources.* IEEE Trans. Inf. Theory **17**(2) (1971) 127–134

[15] G. Han and B. Marcus. *Analyticity of entropy rate of hidden Markov chains.* IEEE Trans. Inf. Theory **52**(12) (2006) 5251–5266

[16] G. Han and B. Marcus. *Derivatives of entropy rate in special families of hidden Markov chains.* IEEE Trans. Inf. Theory **53**(7) (2007) 2642–2652

[17] G. Han and B. Marcus. *Asymptotics of noisy constrained channel capacity.* Ann. Appl. Prob. **19**(3) (2009) 1063–1091

[18] G. Han, B. Marcus, and Y. Peres. *A note on a complex Hilbert metric with application to domain of analyticity for entropy rate of hidden Markov processes.* This volume, Chapter 3, 2011

[19] B. M. Hochwald and P. R. Jelenković. *State learning and mixing in entropy of hidden Markov processes and the Gilbert–Elliott channel.* IEEE Trans. Inf. Theory **45**(1) (1999) 128–138

[20] T. Holliday, P. Glynn, and A. Goldsmith. *Capacity of finite state Markov channels with general inputs.* In Int. Symp. Information Theory, Yokohama, Japan, June–July 2003, p. 289

[21] T. Holliday, P. Glynn, and A. Goldsmith. *Capacity of finite state channels based on Lyapunov exponents of random matrices.* IEEE Trans. Inf. Theory **52**(8) (2006) 3509–3532

[22] P. Jacquet, G. Seroussi, and W. Szpankowski. *On the entropy of a hidden Markov process.* In Proc. Data Compression Conf., Snowbird, UT, June 2004, pp. 362–371

[23] J. Luo and D. Guo. *On the entropy rate of hidden Markov processes observed through arbitrary memoryless channels.* IEEE Trans. Inf. Theory **55**(4) (2009) 1460–1467

[24] D. J. C. MacKay. *Equivalence of Boltzmann chains and hidden Markov models.* Neural Comput. **8** (1996) 178–181

[25] M. Mushkin and I. Bar-David. *Capacity and coding for the Gilbert–Elliott channel.* IEEE Trans. Inf. Theory **35** (1989) 1277–1290

[26] C. Nair, E. Ordentlich, and T. Weissman. *On asymptotic filtering and entropy rate for a hidden Markov process in the rare transitions regime.* In Int. Symp. Information Theory, Adelaide, Australia, September 2005, pp. 1838–1842

[27] E. Ordentlich and T. Weissman. *New bounds on the entropy of hidden Markov processes.* In Proc. IEEE Information Theory Workshop, San Antonio, TX, October 2004, pp. 117–122

[28] E. Ordentlich and T. Weissman. *Approximations for the entropy rate of a hidden Markov process.* In Int. Symp. Information Theory, Adelaide, Australia, September 2005, pp. 2198–2202

[29] E. Ordentlich and T. Weissman. *On the optimality of symbol by symbol filtering and denoising.* IEEE Trans. Inf. Theory **52**(1) (2006) 19–40

[30] Y. Peres. Analytic dependence of Lyapunov exponents on transition probabilities. In *Lyapunov Exponents. Proc. Oberwolfach Conf.*, Lecture Notes in Mathematics **1486**. Springer, Berlin, 1991, pp. 64–80

[31] Y. Peres and A. Quas. *Entropy rate for hidden Markov chains with rare transitions.* This volume, Chapter 5, 2011

[32] H. Pfister. *On the Capacity of Finite-State Channels and the Analysis of Convolutional Accumulate-m Codes.* PhD thesis, University of California, San Diego, CA, 2003

[33] M. Talagrand. *The Sherrington–Kirkpatrick model: a challenge to mathematicians.* Prob. Theory Relat. Fields **110** (1998) 109–176

[34] T. Weissman and N. Merhav. *On competitive predictability and its relation to rate-distortion theory.* IEEE Trans. Inf. Theory **49** (2003) 3185–3193

[35] T. Weissman and E. Ordentlich. *The empirical distribution of rate-constrained source codes.* IEEE Trans. Inf. Theory **51**(11) (2005) 3718–3733

[36] O. Zuk, I. Kanter, and E. Domany. *Asymptotics of the entropy rate for a hidden Markov process.* In Proc. Data Compression Conf., Snowbird, UT, 2005 pp. 173–182

5

Entropy rate for hidden Markov chains
with rare transitions

YUVAL PERES

Microsoft Research, One Microsoft Way, Redmond, WA 98052, USA
E-mail address: peres@microsoft.com

ANTHONY QUAS

Department of Mathematics and Statistics, University of Victoria, Victoria,
BC V8W 3R4, Canada
E-mail address: aquas@uvic.ca

Abstract. We consider hidden Markov chains obtained by passing a Markov chain with rare transitions through a noisy memoryless channel. We obtain asymptotic estimates for the entropy of the resulting hidden Markov chain as the transition rate is reduced to zero.

Let (X_n) be a Markov chain with finite state space S and transition matrix $P(p)$ and let (Y_n) be the hidden Markov chain observed by passing (X_n) through a homogeneous noisy memoryless channel (i.e., Y takes values in a set T, and there exists a matrix Q such that $\mathbb{P}(Y_n = j | X_n = i, X_{-\infty}^{n-1}, X_{n+1}^{\infty}, Y_{-\infty}^{n-1}, Y_{n+1}^{\infty}) = Q_{ij})$. We make the additional assumption on the channel that the rows of Q are distinct. In this case we call the channel *statistically distinguishing*.

We assume that $P(p)$ is of the form $I + pA$, where A is a matrix with negative entries on the diagonal, nonnegative entries in the off-diagonal terms, and zero row sums. We further assume that for small positive p, the Markov chain with transition matrix $P(p)$ is irreducible. Notice that for Markov chains of this form, the invariant distribution $(\pi_i)_{i \in S}$ does not depend on p. In this case, we say that for small positive values of p, the Markov chain is in a *rare transition regime*.

We will adopt the convention that H is used to denote the entropy of a finite partition, whereas h is used to denote the entropy of a process (the *entropy rate* in information theory terminology). Given an irreducible Markov chain with transition matrix P, we let $h(P)$ be the entropy of the Markov chain (i.e., $h(P) = -\sum_{i,j} \pi_i P_{ij} \log P_{ij}$, where π_i is the (unique) invariant distribution of the Markov chain and as usual we adopt the convention that $0 \log 0 = 0$). We also

Entropy of Hidden Markov Processes and Connections to Dynamical Systems: Papers from the Banff International Research Station Workshop, ed. B. Marcus, K. Petersen, and T. Weissman. Published by Cambridge University Press. © Cambridge University Press 2011.

let $H_{\text{chan}}(i)$ be the entropy of the output of the channel when the input symbol is i (i.e., $H_{\text{chan}}(i) = -\sum_{j \in T} Q_{ij} \log Q_{ij}$). Let $h(Y)$ denote the entropy of Y (i.e., $h(Y) = -\lim_{N \to \infty} (1/N) \sum_{w \in T^N} \mathbb{P}(Y_1^N = w) \log \mathbb{P}(Y_1^N = w))$.

Theorem 1. *Consider the hidden Markov chain (Y_n) obtained by observing a Markov chain with irreducible transition matrix $P(p) = I + Ap$ through a statistically distinguishing channel with transition matrix Q. Then there exists a constant $C > 0$ such that, for all small $p > 0$,*

$$h(P(p)) + \sum_i \pi_i H_{\text{chan}}(i) - Cp \leq h(Y) \leq h(P(p)) + \sum_i \pi_i H_{\text{chan}}(i), \quad (1)$$

where $(\pi_i)_{i \in S}$ is the invariant distribution of $P(p)$.

If in addition the channel has the property that there exist i, i', and j such that $P_{ii'} > 0$, $Q_{ij} > 0$, and $Q_{i'j} > 0$, then there exists a constant $c > 0$ such that

$$h(Y) \leq h(P(p)) + \sum_i \pi_i H_{\text{chan}}(i) - cp. \quad (2)$$

The entropy rate in the rare transition regime was considered previously in the special case of a 0–1 valued Markov chain with transition matrix

$$P(p) = \begin{pmatrix} 1-p & p \\ p & 1-p \end{pmatrix}$$

and where the channel was the binary symmetric channel with cross-over probability ϵ, i.e.,

$$Q = \begin{pmatrix} 1-\epsilon & \epsilon \\ \epsilon & 1-\epsilon \end{pmatrix}.$$

It is convenient to introduce the notation $g(p) = -p \log p - (1-p) \log(1-p)$. In [4], Nair *et al.* proved that $g(\epsilon) - (1-2\epsilon)^2 p \log p/(1-\epsilon) \leq h(Y) \leq g(p) + g(\epsilon)$. For comparison, with our result, this is essentially of the form $g(\epsilon) + a(\epsilon)g(p) \leq h(Y) \leq g(p) + g(\epsilon)$, where $a(\epsilon) < 1$ but $a(\epsilon) \to 1$ as $\epsilon \to 0$ (i.e., $h(Y) = g(p) + g(\epsilon) - O(p \log p)$). A second paper due to Chigansky [1] shows that $g(\epsilon) + b(\epsilon)g(p) \leq h(Y)$ for a function $b(\epsilon) < 1$ satisfying $b(\epsilon) \to 1$ as $\epsilon \to 1/2$ (again giving an $O(p \log p)$ error). Our result states in this case that there exist $C > c > 0$ such that $g(p) + g(\epsilon) - Cp \leq h(Y) \leq g(p) + g(\epsilon) - cp$ (i.e., $h(Y) = g(p) + g(\epsilon) - \Theta(p)$).

We note that as part of the proof we attempt a reconstruction of (X_n) from the observed data (Y_n). In our case, the reconstruction of the nth symbol of X_n depended on past and future values of Y_m. A related but harder problem of filtering is to try to reconstruct X_n given only Y_1^n. This problem was addressed in essentially the same scenario by Khasminskii and Zeitouni [3], where they gave a lower bound for the asymptotic reconstruction error of the form $Cp|\log p|$

for an explicit constant C (i.e., for an arbitrary reconstruction scheme, the probability of wrongly guessing X_n is bounded below in the limit as $n \to \infty$ by $Cp|\log p|$). Our scheme shows that if one is allowed to use future as well as past observations then the asymptotic reconstruction error is $O(p)$. This was previously observed by Shue *et al.* [5], who used a similar scheme to ours.

Before giving the proof of the theorem, we discuss the strategy. We start from the equality

$$h(X) + h(Y|X) = h(X,Y) = h(Y) + h(X|Y). \tag{3}$$

Since $h(X)$ and $h(Y|X)$ are known to be $h(P(p))$ and $\sum_i \pi_i H_{\text{chan}}(i)$, the estimates for the entropy of Y are obtained by estimating $h(X|Y)$. The inequality (1) is equivalent to showing that $0 \le h(X|Y) \le Cp$ for some $C > 0$. The lower bound here is trivial, whereas the main part of the proof is the upper bound for $h(X|Y)$ (giving a lower bound for $h(Y)$). The second part of the proof, showing (2) lowering the upper bound for $h(Y)$ under additional conditions, is proved by showing that $h(X|Y) \ge cp$ for some $c > 0$.

We explain briefly the underlying idea of the upper bound $h(X|Y) = O(p)$. Since the transitions in the (X_n) sequence are rare, given a realization of (Y_n), the Y_n values allow one to guess (using the statistical distinguishing property) the X_n values from which the Y_n values are obtained. This provides for an accurate reconstruction except that where there is a transition in the X_n there is some uncertainty as to its location as estimated using the Y_n. It turns out that by using maximum likelihood estimation, the transition locations may be pinpointed up to an error with exponentially small tail. Since the transitions occur with rate p, there is an $O(p)$ entropy error in reconstructing (X_n) from (Y_n).

We make use of a number of notational conventions, some standard and others less so. Firstly, we shall denote events by set notation so that $\{X_0 = X_2\}$ denotes the event that the random variables X_0 and X_2 agree. We make extensive use of relative entropy. For two partitions $\mathcal{P}$ and $\mathcal{Q}$, the relative entropy is defined by $H(\mathcal{Q}|\mathcal{P}) = H(\mathcal{P} \vee \mathcal{Q}) - H(\mathcal{P})$. When conditioning, we shall not distinguish between random variables and the partitions and σ-algebras that they induce (so that for example $H(X_0^{N-1})$ is $-\sum_{w \in S^N} \mathbb{P}(X_0^{N-1} = w) \log \mathbb{P}(X_0^{N-1} = w)$ and $H(X_0|Y)$ is the conditional entropy of X_0 relative to the σ-algebra generated by $\{Y_n : n \in \mathbb{Z}\}$). On the other hand, if A is an event, we use $H(\mathcal{P}|A)$ to mean the entropy of the partition with respect to the conditional measure $\mathbb{P}_A(B) = \mathbb{P}(A \cap B)/\mathbb{P}(A)$. For jointly stationary processes $(X_n)_{n \in \mathbb{Z}}$ and $(Y_n)_{n \in \mathbb{Z}}$, the relative entropy of the processes is given by $h(Y|X) = h((X_n, Y_n)_{n \in \mathbb{Z}}) - h((X_n)_{n \in \mathbb{Z}}) = \lim_{N \to \infty} (1/N)(H(X_0^{N-1} \vee Y_0^{N-1}) - H(X_0^{N-1})) = \lim_{N \to \infty} (1/N) H(Y_0^{N-1}|X_{-\infty}^\infty) = H(Y_0|X_{-\infty}^\infty, Y_{-\infty}^{-1})$.

Given a measurable partition $\mathcal{Q}$ of the space, an event A, and a σ-algebra $\mathcal{F}$ we will write $H(\mathcal{Q}|\mathcal{F}|A)$ for the entropy of $\mathcal{Q}$ relative to $\mathcal{F}$ with respect to $\mathbb{P}_A$. In the case where A is $\mathcal{F}$-measurable (as will always be the case in what follows), we have

$$H(\mathcal{Q}|\mathcal{F}|A) = \int \left(-\sum_{B \in \mathcal{Q}} \mathbb{P}(B|\mathcal{F}) \log \mathbb{P}(B|\mathcal{F}) \right) d\mathbb{P}_A.$$

If $A_1, \ldots, A_k$ form an $\mathcal{F}$-measurable partition of the space, then we have the following equality:

$$H(\mathcal{Q}|\mathcal{F}) = \sum_{j=1}^{k} \mathbb{P}(A_j) H(\mathcal{Q}|\mathcal{F}|A_j). \tag{4}$$

Proof of Theorem 1. Note that $((X_n, Y_n))_{n \in \mathbb{Z}}$ forms a Markov chain with transition matrix $\bar{P}$ given by $\bar{P}_{(i,j),(i',j')} = P_{ii'} Q_{i'j'}$ and invariant distribution $\bar{\pi}_{(i,j)} = \pi_i Q_{ij}$. The standard formula for the entropy of a Markov chain then gives $h(X, Y) = h(P(p)) + \sum_i \pi_i H_{\text{chan}}(i)$. Since $h(X, Y) = h(Y) + h(X|Y)$, one obtains

$$h(Y) = h(X, Y) - h(X|Y) = h(P(p)) + \sum_i \pi_i H_{\text{chan}}(i) - h(X|Y). \tag{5}$$

This establishes the upper bound in the first part of the theorem.

We now establish the lower bound. We are aiming to show that $h(X|Y) = O(p)$ (for which it suffices to show that $H(X_0^{L-1}|Y) = O(Lp)$ for some L). Setting $L = |\log p|^4$ and letting $\mathcal{P}$ be a suitable partition, we estimate $H(X_0^{L-1}|Y, \mathcal{P})$ and use the inequality

$$H(X_0^{L-1}|Y) \leq H(X_0^{L-1}|Y, \mathcal{P}) + H(\mathcal{P}). \tag{6}$$

We define the partition $\mathcal{P}$ as follows: set $K = |\log p|^2$ and let $\mathcal{P} = \{E_m, E_b, E_{g1}, E_{g2}\}$. Here E_m (for many) is the event that there are at least two transitions in X_0^{L-1}, E_b (for boundary) is the event that there is exactly one transition and that it takes place within a distance K of the boundary of the block, and finally E_g (for good) is the event that there is at most one transition and if it takes place, then it occurs at a distance at least K from the boundary of the block. This will later be subdivided into E_{g1} and E_{g2}.

If E_m holds, then we bound the entropy contribution by the entropy of the equidistributed case, whereas, if E_b holds, there are $2K|S|(|S| - 1) = O(K)$

possible values of X_0^{L-1}. This yields the following estimates:

$$\mathbb{P}(E_m) = O(p^2 L^2) = o(p), \tag{7}$$

$$H(X_0^{L-1}|E_m) \leq L\log|\mathring{S}|, \tag{8}$$

$$\mathbb{P}(E_b) = O(pK), \tag{9}$$

$$H(X_0^{L-1}|E_b) = O(\log K). \tag{10}$$

It follows that $\mathbb{P}(E_g) = 1 - O(pK)$. Given that the event E_g holds, the sequence X_0^{L-1} belongs to $B = \{a^L : a \in S\} \cup \{a^i b^{L-i} : a, b \in S, K \leq i \leq L - K\}$.

Given a sequence $u \in B$, the log-likelihood of u being the input sequence yielding the output Y_0^{L-1} is $L_u(Y_0^{L-1}) = \sum_{i=0}^{L-1} \log Q_{u_i Y_i}$. We define Z_0^{L-1} to be the sequence in B for which $L_Z(Y_0^{L-1})$ is maximized (breaking ties lexicographically if necessary). We will then show using large deviation methods that when E_g holds, Z_0^{L-1} is a good reconstruction of X_0^{L-1} with small error.

We calculate, for $u, v \in B$,

$$\mathbb{P}\left(L_v(Y_0^{L-1}) \geq L_u(Y_0^{L-1})|X_0^{L-1} = u\right)$$

$$= \mathbb{P}\left(\sum_{i=0}^{L-1} \log(Q_{v_i Y_i}/Q_{u_i Y_i}) \geq 0 \Big| X_0^{L-1} = u\right)$$

$$= \mathbb{P}\left(\sum_{i \in \Delta} \log(Q_{v_i Y_i}/Q_{u_i Y_i}) \geq 0 \Big| X_0^{L-1} = u\right),$$

where $\Delta = \{i : u_i \neq v_i\}$. For each $i \in \Delta$, given that $X_0^{L-1} = u$, we have that $\log(Q_{v_i Y_i}/Q_{u_i Y_i})$ is an independent random variable taking the value $\log(Q_{v_{ij}}/Q_{u_{ij}})$ with probability $Q_{u_{ij}}$.

It is well known (and easy to verify using elementary calculus) that for a given probability distribution π on a set T, the probability distribution σ maximizing $\sum_{j \in T} \pi_j \log(\sigma_j/\pi_j)$ is $\sigma = \pi$ (for which the maximum is 0). Accordingly, we see that given that $X_0^{L-1} = u$, $L_v(Y_0^{L-1}) - L_u(Y_0^{L-1})$ is the sum of $|\Delta|$ random variables, each having one of $|S|(|S| - 1)$ distributions, each with negative expectation. It follows from Hoeffding's inequality [2] that there exist $C > 0$ and $\eta < 1$ independent of p such that $\mathbb{P}(L_v(Y_0^{L-1}) \geq L_u(Y_0^{L-1})|X_0^{L-1} = u) \leq C\eta^{|\Delta|}$.

We deduce that, for $u, v \in B$,

$$\mathbb{P}(Z_0^{L-1} = v|X_0^{L-1} = u) \leq C\eta^{\delta(u,v)}, \tag{11}$$

where $\delta(u, v)$ is the number of places in which u and v differ.

We split E_g into two subsets:

$$E_{g1} = E_g \cap \{\delta(X_0^{L-1}, Z_0^{L-1}) < K\} \text{ and}$$
$$E_{g2} = E_g \cap \{\delta(X_0^{L-1}, Z_0^{L-1}) \geq K\}.$$

Since there are fewer than $|S|^2 L$ elements in B, we see using (11) and recalling that $K = |\log p|^2$ that

$$\mathbb{P}(E_{g2}) \leq |S|^2 LC\eta^K = o(p), \tag{12}$$
$$H(X_0^{L-1}|E_{g2}) \leq \log(|S|^2 L). \tag{13}$$

Combining (12) with (9) and (7), we see that $\mathbb{P}(E_{g1}) = 1 - O(pK)$. We then obtain

$$H(\mathcal{P}) = O(pK \log(pK)) = o(pL). \tag{14}$$

Conditioned on being in E_{g1}, if $Z_0 = Z_{L-1}$, then $X_0^{L-1} = Z_0^{L-1}$, so we have

$$H(X_0^{L-1}|Z_0^{L-1} \vee \mathcal{P}|E_{g1} \cap \{Z_0 = Z_{L-1}\}) = 0. \tag{15}$$

Given that E_{g1} holds, if $X_0^{L-1} = a^i b^{L-i}$, then Z_0^{L-1} must be of the form $a^j b^{L-j}$ for some j satisfying $-K < j - i < K$. Denote this difference $j - i$ by the random variable N. We have

$$H(X_0^{L-1}|Y_0^{L-1} \vee \mathcal{P}|E_{g1} \cap \{Z_0 \neq Z_{L-1}\})$$
$$\leq H(X_0^{L-1}|Z_0^{L-1} \vee \mathcal{P}|E_{g1} \cap \{Z_0 \neq Z_{L-1}\})$$
$$= H(N|Z_0^{L-1} \vee \mathcal{P}|E_{g1} \cap \{Z_0 \neq Z_{L-1}\})$$
$$\leq H(N|E_{g1} \cap \{Z_0 \neq Z_{L-1}\}),$$

where the first inequality follows because Z_0^{L-1} is determined by Y_0^{L-1}, so the partition generated by Y_0^{L-1} is finer than that generated by Z_0^{L-1}; and the equality follows because given Z_0^{L-1} and conditioned on being in E_{g1}, knowing N is sufficient to reconstruct X_0^{L-1}, so the partition generated by N is the same as the partition generated by X_0^{L-1}.

Since $E_{g1} \cap \{Z_0 \neq Z_{L-1}\} = E_{g1} \cap \{X_0 \neq X_{L-1}\}$, we have for $|k| < K$, $\mathbb{P}(N = k|E_{g1} \cap \{Z_0 \neq Z_{L-1}\}) = \mathbb{P}(N = k|E_{g1} \cap \{X_0 \neq X_{L-1}\})$. From (11), this is bounded above by $C\eta^{|k|}$. Since a distribution with these bounds has entropy bounded above independently of p, it follows from this that $H(N|E_{g1} \cap \{Z_0 \neq Z_{L-1}\}) = O(1)$ and hence that

$$H(X_0^{L-1}|Y_0^{L-1} \vee \mathcal{P}|E_{g1} \cap \{Z_0 \neq Z_{L-1}\}) = O(1). \tag{16}$$

Finally, we have $\mathbb{P}(E_{g1} \cap \{Z_0 \neq Z_{L-1}\}) = O(pL)$.

We now have $H(X_0^{L-1}|Y_0^{L-1}) \leq H(X_0^{L-1}|Y_0^{L-1} \vee \mathcal{P}) + H(\mathcal{P})$. We estimate the right-hand side using (4), splitting the space up into the sets E_b, E_m, E_{g2}, $E_{g1} \cap \{Z_0 = Z_{L-1}\}$, and $E_{g1} \cap \{Z_0 \neq Z_{L-1}\}$. All of these sets are $Y_0^{L-1} \vee \mathcal{P}$ measurable. Calculating the contribution to the entropy from each of the sets, each part contributes at most $O(pL)$, yielding the estimate $H(X_0^{L-1}|Y_0^{L-1}) = O(pL)$, so that $h(X|Y) = O(p)$, as required. This completes the first part of the proof.

For the second part of the proof, suppose that the additional properties are satisfied (the existence of i, i', and j such that $P_{ii'} > 0$, $Q_{ij} > 0$, and $Q_{i'j} > 0$). We need to show that $h(X|Y) \geq cp$ for some $c > 0$ or equivalently that $H(X_0|Y, X_{-\infty}^{-1}) \geq cp$. In fact, we show the stronger statement: $H(X_0|Y, (X_n)_{n \neq 0}) \geq cp$. Let A be the event that $X_{-1} = i$ and $X_1 = i'$ and $Y_0 = j$. We now estimate $H(X_0|Y, (X_n)_{n \neq 0}|A)$.

For $x \in A$, we have

$$\mathbb{P}(X_0 = i|Y, (X_n)_{n \neq 0})(x) = \frac{P_{ii}P_{ii'}Q_{ij}}{P_{ii}P_{ii'}Q_{ij} + P_{ii'}P_{i'i'}Q_{i'j} + \sum_{k \notin \{i,i'\}} P_{ik}P_{ki'}Q_{kj}},$$

$$\mathbb{P}(X_0 = i'|Y, (X_n)_{n \neq 0})(x) = \frac{P_{ii'}P_{i'i'}Q_{i'j}}{P_{ii}P_{ii'}Q_{ij} + P_{ii'}P_{i'i'}Q_{i'j} + \sum_{k \notin \{i,i'\}} P_{ik}P_{ki'}Q_{kj}}.$$

As $p \to 0$, we have $\mathbb{P}(X_0 = i|Y, (X_n)_{n \neq 0})(x) \to Q_{ij}/(Q_{ij} + Q_{i'j})$ and $\mathbb{P}(X_0 = i'|Y, (X_n)_{n \neq 0})(x) \to Q_{i'j}/(Q_{ij} + Q_{i'j})$. From this, we see that $H(X_0|Y, (X_n)_{n \neq 0}|A)$ converges to a nonzero constant as $p \to 0$. Since A has probability $\Omega(p)$, applying (4) we obtain the lower bound $h(X|Y) \geq cp$. From this, we deduce the claimed upper bound for $h(Y)$:

$$h(Y) \leq h(X) + \sum_i \pi_i H_{\text{chan}}(i) - cp.$$

In this case we therefore have $h(Y) = h(X) + \sum_i \pi_i H_{\text{chan}}(i) + \Theta(p)$. This completes the proof of the theorem. $\square$

References

[1] P. Chigansky. *The entropy rate of a binary channel with slowly switching input.* arXiv: cs/0602074v1, 2006
[2] W. Hoeffding. *Probability inequalities for sums of bounded random variables.* J. Amer. Statist. Assoc. **58** (1963) 13–30
[3] R. Z. Khasminskii and O. Zeitouni. *Asymptotic filtering for finite state Markov chains.* Stoch. Process. Appl. **63** (1996) 1–10
[4] C. Nair, E. Ordentlich, and T. Weissman. *Asymptotic filtering and entropy rate of a hidden Markov process in the rare transitions regime.* In Int. Symp. Information Theory, 2005, pp. 1838–1842
[5] L. Shue, B. Anderson, and F. DeBruyne. *Asymptotic smoothing errors for hidden Markov models.* IEEE Trans. Signal Process. **48** (2000) 3289–3302

6
The capacity of finite-state channels in the high-noise regime

HENRY D. PFISTER

Department of Electrical and Computer Engineering, Texas A&M University,
3128 TAMU, College Station, TX 77843, USA
E-mail address: hpfister@tamu.edu

Abstract. This article considers the derivative of the entropy rate of a hidden Markov process with respect to the observation probabilities. The main result is a compact formula for the derivative that can be evaluated easily using Monte Carlo methods. It is applied to the problem of computing the capacity of a finite-state channel (FSC) and, in the high-noise regime, the formula has a simple closed-form expression that enables series expansion of the capacity of an FSC. This expansion is evaluated for a binary-symmetric channel under a (0, 1) run-length-limited constraint and an intersymbol-interference channel with Gaussian noise.

1 Introduction

1.1 The hidden Markov process

A hidden Markov process (HMP) is a discrete-time finite-state Markov chain (FSMC) observed through a memoryless channel. The HMP has become ubiquitous in statistics, computer science, and electrical engineering because it approximates many processes well using a dependency structure that leads to many efficient algorithms. While the roots of the HMP lie in the "grouped Markov chains" of Harris [20] and the "functions of a finite-state Markov chain" of Blackwell [8], the HMP first appears (in full generality) as the output process of a finite-state channel (FSC) [9]. The statistical inference algorithm of Baum and Petrie [5], however, cemented the HMP's place in history and is responsible for great advances in fields such as speech recognition and biological sequence analysis [22, 24]. An exceptional survey of HMPs, by Ephraim and Merhav, gives a nice summary of what is known in this area [12].

Entropy of Hidden Markov Processes and Connections to Dynamical Systems: Papers from the Banff International Research Station Workshop, ed. B. Marcus, K. Petersen, and T. Weissman. Published by Cambridge University Press. © Cambridge University Press 2011.

Definition 1.1. Let $\mathcal{Q}$ be the state set of an irreducible aperiodic FSMC $\{Q_t\}_{t\in\mathbb{Z}}$ with state transition matrix P and define

$$p_{ij} \triangleq [P]_{i,j} = \Pr(Q_{t+1} = j \mid Q_t = i)$$

for $i,j \in \mathcal{Q}$. Let $\mathcal{Y}$ be a finite set of possible observations and $\{Y_t\}_{t\in\mathbb{Z}}$ be the stochastic process where $Y_t \in \mathcal{Y}$ is generated by the transition from Q_t to Q_{t+1}. The distribution of the observation conditioned on the FSMC transition[1] is given by

$$h_{ij}(y) \triangleq \begin{cases} \Pr(Y_t = y \mid Q_t = i, Q_{t+1} = j) & \text{if } (i,j) \in \mathcal{V}, \\ 0 & \text{otherwise,} \end{cases}$$

for $i,j \in \mathcal{Q}$, where $\mathcal{V} = \{(i,j) \in \mathcal{Q} \times \mathcal{Q} \mid p_{ij} > 0\}$ is the set of valid transitions. The ergodic process $\{Y_t\}_{t\in\mathbb{Z}}$ is called a *hidden Markov process*. With proper initialization, the process is also stationary.

Although the notation of this article assumes that $\mathcal{Y}$ is a finite set, many results remain correct when $\mathcal{Y} = \mathbb{R}$ if $h_{ij}(y)$ is assumed to be a continuous probability density function and sums over $\mathcal{Y}$ are converted to integrals over $\mathbb{R}$.

1.2 The entropy rate

The entropy rate of a stationary stochastic process $\{Y_t\}_{t\in\mathbb{Z}}$ is defined to be

$$H(\mathcal{Y}) \triangleq \lim_{n\to\infty} \frac{1}{n} H(Y_1, \ldots, Y_n),$$

where $H(Y_1) \triangleq -E[\ln \Pr(Y_1)]$ is the entropy of the random variable Y_1 and the limit exists and is finite if $H(Y_1) < \infty$ [11]. Computing the exact entropy rate of an HMP in closed form appears to be difficult, however. In [8], Blackwell states that

"In this paper we study the entropy of the $\{y_n\}$ [hidden Markov] process; our result suggests that this entropy is intrinsically a complicated function of [the parameters of the hidden Markov process] M and Φ."

---

[1] In general, HMPs are defined by noisy observations of the FSMC states (rather than the transitions). This paper uses the "transition observation" model instead because of its natural connection with finite-state channels. Moreover, any random process that can be represented by the "transition observation" HMP model with M states can also be represented by the "state observation" model with M^2 states.

On the other hand, the Shannon–McMillan–Breiman theorem shows that the empirical entropy rate $-(1/n)\ln\Pr(y_1^n)$ converges almost surely to the entropy rate $H(\mathcal{Y})$ (in nats) as $n \to \infty$ [11]. Therefore, simulation-based (i.e., Monte Carlo) approaches work well in many cases [29, 16, 1, 37, 36, 3, 2].

Other early work related to the entropy rate of HMPs can be found in [7, 35, 38, 34]. Recently, interest in HMPs has surged and there have been a large number of papers discussing the entropy rate of HMPs. These range from bounds [36, 30, 31] to establishing the analyticity of the entropy rate [17] to computing series expansions of the entropy rate [44, 43, 19].

1.3 The finite-state channel

The work in this article is largely motivated by the analysis of a class of time-varying channels known as FSCs. An FSC is a discrete-time channel where the distribution of the channel output depends on both the channel input and the underlying channel state [15]. This allows the channel output to depend implicitly on previous inputs and outputs via the channel state. In practice, there are three types of channel variation which FSCs are typically used to model. A *flat fading* channel is a time-varying channel whose state is independent of the channel inputs. An *intersymbol-interference* (ISI) channel is a time-varying channel whose state is a deterministic function of the previous channel inputs. Channels which exhibit both fading and ISI can also be modeled, and their state is a stochastic function of the previous channel inputs. An *indecomposable FSC* is, roughly speaking, an FSC where the effect of the initial state decays with time. The output process of an indecomposable FSC with an ergodic Markov input is an HMP.

Consider an indecomposable FSC with state set $\mathcal{S}$, finite input alphabet $\mathcal{X}$, and output alphabet $\mathcal{Y}$. The channel is defined by its input–output state-transition probability $W(y, s'|x, s)$, which is defined for $x \in \mathcal{X}$, $y \in \mathcal{Y}$, and $s, s' \in \mathcal{S}$. Using this notation, $W(y, s'|x, s)$ is the conditional probability that the channel output is y and the new channel state is s' given that the channel input was x and the initial state was s. The n-step transition probability for a sequence of n channel uses (with input x_1^n and output y_1^n) is given by

$$\Pr\left(Y_1^n = y_1^n \mid X_1^n = x_1^n\right) = \sum_{s_1^{n+1} \in \mathcal{S}^{n+1}} \Pr\left(S_1 = s_1\right) \prod_{t=1}^{n} W\left(y_t, s_{t+1} \mid x_t, s_t\right).$$

When $\mathcal{Y} = \mathbb{R}$, we will also use $W(y, s'|x, s)$ to represent a conditional probability density function for the channel outputs.

The achievable information rate of an FSC with Markov inputs is intimately related to the entropy rate of an HMP [1, 37, 23, 2, 41, 21]. Computing this entropy rate exactly is usually quite difficult, and often the main obstacle in the computation of achievable rates.

1.4 Main results

The main result of this article, given in Theorem 3.2, is a compact formula for the derivative, with respect to the observation probability $h_{ij}(y)$, of the entropy rate of a general HMP. A Monte Carlo estimator for this derivative follows easily because the formula is an expectation over distributions that are relatively easy to sample. The formula is also amenable to analysis in some asymptotic regimes. In particular, Theorem 3.6 derives a simple formula for the first two nontrivial terms in the expansion of the entropy rate in the high-noise regime.

In Section 4, this derivative formula also allows one to consider the derivative of achievable information rates for FSCs. For example, a closed-form expression for the capacity of a binary symmetric channel (BSC) under a $(0, 1)$ run-length-limited (RLL) constraint is derived in the high-noise limit. Section 2 provides the mathematical background necessary for the later sections.

2 Mathematical background

2.1 Notation

Calligraphic letters are used to denote sets (e.g., $\mathcal{Q}, \mathcal{Y}, \mathcal{V}$) and $1_{\mathcal{Y}}(\cdot)$ is the indicator function of the set $\mathcal{Y}$. Capital letters are used to denote random variables (e.g., Q_t, Y_t) and matrices (e.g., M, P). Lower-case letters are used to represent realizations of random variables (e.g., q_t, y_t), column vectors (e.g., π, α, β, u, v), and indices (e.g., i, j, k, l). The ith element of the vector π is denoted $\pi(i)$.

The following sets will also be used: $\mathbb{R}_+ = \{a \in \mathbb{R} \,|\, a > 0\}$, $\mathcal{A} = \mathbb{R}^{|\mathcal{Q}|}$, $\mathcal{A}_\delta = \{u \in \mathcal{A} \,|\, u(q) > \delta, q \in \mathcal{Q}\}$, $\mathcal{P} = \{u \in \mathcal{A} \,|\, \sum_q u(q) = 1\}$, and $\mathcal{P}_\delta = \mathcal{A}_\delta \cap \mathcal{P}$. We note that the symbols $\pi, \alpha_t \in \mathcal{P}_0$ are used interchangeably to denote distributions over $\mathcal{Q}$ and $|\mathcal{Q}|$-dimensional column vectors (e.g., $\pi^{\mathsf{T}} P = \pi^{\mathsf{T}}$). The standard p-norm of the vector u is denoted by $\|u\|_p \triangleq \left(\sum_i |u(i)|^p\right)^{1/p}$ and the induced matrix norm is $\|M\|_p \triangleq \sup_{\|u\|_p = 1} \|Mu\|_p$.

2.2 The forward–backward algorithm

One of the primary reasons for the popularity of HMPs is that the forward and backward state estimation problems have a simple recursive structure. Let us assume that the Markov chain $\{Q_t\}_{t \in \mathbb{Z}}$ is stationary and that $\pi \in \mathcal{P}_0$ is the unique stationary distribution that satisfies $\pi^{\mathsf{T}} P = \pi^{\mathsf{T}}$. For a length-$n$ block, let the forward state probability $\alpha_t \in \mathcal{P}$ and the backward state probability $\beta_t \in \mathcal{A}$ be defined by

$$\alpha_t(i) \triangleq \Pr\left(Q_t = i \mid Y_1^{t-1} = y_1^{t-1}\right),$$

$$\beta_t(j) \triangleq \frac{1}{\pi(j)} \Pr\left(Q_t = j \mid Y_t^n = y_t^n\right)$$

for $i, j \in \mathcal{Q}$. These definitions lead naturally to the recursions

$$\alpha_{t+1}(j) = \frac{1}{\psi_{t+1}} \sum_{i \in \mathcal{Q}} \alpha_t(i) p_{ij} h_{ij}(y_t),$$

$$\beta_{t-1}(i) = \frac{1}{\phi_{t-1}} \sum_{j \in \mathcal{Q}} \beta_t(j) p_{ij} h_{ij}(y_{t-1})$$

for $i, j \in \mathcal{Q}$, where ψ_{t+1} is chosen so that $\sum_{i \in \mathcal{Q}} \alpha_{t+1}(i) = 1$ and ϕ_{t-1} is chosen[2] so that $\sum_{j \in \mathcal{Q}} \pi(j) \beta_{t-1}(j) = 1$. It is worth noting that $\psi_{t+1} = \Pr(Y_t = y_t \mid Y_1^{t-1} = y_1^{t-1})$ and therefore we find that

$$-\frac{1}{n} \sum_{t=1}^{n} \ln \psi_{t+1} = -\frac{1}{n} \ln \Pr\left(Y_1^n = y_1^n\right) \overset{\text{a.s.}}{\to} H(\mathcal{Y}) \text{ nats.}$$

This simple connection between the forward recursion and the entropy rate implies a simple Monte Carlo approach to estimating the achievable information rates of FSCs [1, 37, 36, 3, 2].

2.3 The matrix perspective

2.3.1 The forward–backward algorithm revisited

In this section, we review a natural connection between the product of random matrices and the forward–backward recursions. This connection is interesting in its own right, but will also be very helpful in understanding the results of later sections.

Definition 2.1. For any $y \in \mathcal{Y}$, the *transition observation probability matrix*, $M(y)$, is a $|\mathcal{Q}| \times |\mathcal{Q}|$ matrix defined by

$$[M(y)]_{ij} \triangleq \Pr(Y_t = y, Q_{t+1} = j \mid Q_t = i) = p_{ij} h_{ij}(y). \tag{1}$$

---

[2] We believe that this normalization for $\beta_{t-1}(q)$ is new and it appears to be the natural choice for the problem considered in this article (and perhaps in general).

These matrices behave similarly to transition probability matrices because their sequential products compute the n-step transition observation probabilities of the form

$$[M(y_t)M(y_{t+1})\cdots M(y_{t+k})]_{ij} = \Pr\left(Y_t^{t+k} = y_t^{t+k}, Q_{t+k+1} = j \mid Q_t = i\right).$$

This means that we can write $\Pr(Y_1^n = y_1^n)$ as the matrix product[3]

$$\Pr\left(Y_1^n = y_1^n\right) = \pi^{\mathrm{T}}\left(\prod_{t=1}^{n} M(y_t)\right)\mathbf{1}, \tag{2}$$

where $\mathbf{1}$ is a $|\mathcal{Q}|$-dimensional column vector of ones. When $\mathcal{Y} = \mathbb{R}$, the above expressions are understood to be probability density functions with respect to the observations and the joint probability becomes the joint density.

Likewise, the forward/backward recursions can be written in matrix form as

$$\alpha_{t+1}^{\mathrm{T}} = \frac{\alpha_t^{\mathrm{T}} M(y_t)}{\alpha_t^{\mathrm{T}} M(y_t)\mathbf{1}}, \qquad \beta_{t-1} = \frac{M(y_{t-1})\beta_t}{\pi^{\mathrm{T}} M(y_{t-1})\beta_t},$$

where $\pi^{\mathrm{T}}\mathbf{1} = 1$, $\alpha_{t+1}^{\mathrm{T}}\mathbf{1} = 1$, and $\pi^{\mathrm{T}}\beta_{t-1} = 1$. We will also make use of the shorthand notation

$$M(y_k^l) \triangleq \prod_{t=k}^{l} M(y_t).$$

2.3.2 Contraction coefficients

This section summarizes some standard results on the contractive properties of positive matrices and their connections to HMPs. More details can be found in [39, 26, 25].

Definition 2.2. For any two vectors $u, v \in \mathcal{A}_0$, the *Hilbert projective metric* is

$$d(u, v) \triangleq \ln \frac{\max_i (u(i)/v(i))}{\min_j (u(j)/v(j))} = \ln \max_{i,j} \frac{u(i)v(j)}{v(i)u(j)} = -\ln \min_{i,j} \frac{u(i)v(j)}{v(i)u(j)}.$$

It is a metric on $\mathcal{A}_0 \backslash \sim$, where $\sim$ is the equivalence relation with $u \sim v$ if $au = v$ for some $a \in \mathbb{R}_+$.

---

[3] Since matrix multiplication is not commutative, we use the convention that $\prod_{t=1}^{n} M(y_t) = M(y_1)M(y_2)\cdots M(y_n)$.

Proposition 2.3. *For $u, v, w \in \mathcal{A}_0$ such that $w^{\mathrm{T}} u = w^{\mathrm{T}} v$, the Hilbert projective metric characterizes the element-wise relative distance between two vectors in the sense that, for any $i \in \mathcal{Q}$,*

$$d_M(u(i), v(i)) \triangleq \frac{|u(i) - v(i)|}{\max(u(i), v(i))} \leq 1 - e^{-d(u,v)} \leq d(u, v),$$

$$d_m(u(i), v(i)) \triangleq \frac{|u(i) - v(i)|}{\min(u(i), v(i))} \leq e^{d(u,v)} - 1 \overset{\text{"}d(u,v) \leq 1\text{"}}{\leq} 2d(u, v),$$

where d_M is a metric on $\mathbb{R}_+$ and d_m is a semimetric on $\mathbb{R}_+$ (i.e., the triangle inequality does not hold).

Proof. If $u(k) \geq v(k)$, then we have

$$u(k)e^{-d(u,v)} = u(k) \min_j \frac{v(j)}{u(j)} \min_i \frac{u(i)}{v(i)} \leq v(k) \min_i \frac{u(i)}{v(i)} \leq v(k),$$

where $\min_i(u(i)/v(i)) \leq 1$ because $w^{\mathrm{T}} u = w^{\mathrm{T}} v$. The stated results follow from $u(k) - v(k) \leq e^{d(u,v)} v(k) - v(k)$, $u(k) - v(k) \leq u(k) - u(k)e^{-d(u,v)}$, and simple bounds on e^x. Both distances are clearly symmetric and positive definite. The triangle inequality and other properties of d_M are discussed in [42]. □

Lemma 2.4. *For any vectors $u, v, w \in \mathcal{A}_0$ such that $w^{\mathrm{T}} u = w^{\mathrm{T}} v$, we have*

$$\|u - v\|_1 \leq \left(1 - e^{-d(u,v)}\right) \sum_{i \in \mathcal{Q}} \max(u(i), v(i)) \leq (\|u\|_1 + \|v\|_1) d(u, v),$$

$$\|u - v\|_1 \leq \left(e^{d(u,v)} - 1\right) \sum_{i \in \mathcal{Q}} \min(u(i), v(i)) \leq \left(e^{d(u,v)} - 1\right) \min(\|u\|_1, \|v\|_1).$$

Proof. The expressions follow from direct calculation of $\|u - v\|_1$ using the bounds in Proposition 2.3. □

The following theorem of Birkhoff plays an important role in the remainder of this article.

Theorem 2.5. [39, Chapter 3] *Consider any nonnegative matrix M with at least one positive entry in every row and column. Then, for all $u, v \in \mathcal{A}_0$, we have*

$$d(Mu, Mv) \leq \tau(M) d(u, v),$$

where

$$\tau(M) \triangleq \frac{1-\phi(M)^{1/2}}{1+\phi(M)^{1/2}} = \tau\left(M^{\mathsf{T}}\right) \leq 1$$

is the Birkhoff contraction coefficient *and*

$$\phi(M) = \min_{i,j,k,l} \frac{[M]_{ik}[M]_{jl}}{[M]_{jk}[M]_{il}} \geq \left(\frac{\min_{i,j}[M]_{ij}}{\max_{i,j}[M]_{ij}}\right)^2. \tag{3}$$

The following results connect our HMP definition with Birkhoff's contraction coefficients. An FSMC that is irreducible and aperiodic is called *primitive*. Since the underlying Markov chain is primitive, the matrix P must have at least one nonzero entry in each row and column.

Condition 2.6. For some $\delta \geq 0$, the joint probability of every valid transition and output is greater than δ. In other words, this means that $p_{ij}h_{ij}(y) > \delta \geq 0$ for all $(i,j) \in \mathcal{V}$ and $y \in \mathcal{Y}$.

Under Condition 2.6, the matrix $M(y)$ has exactly the same pattern of zero/nonzero entries as P for all $y \in \mathcal{Y}$. Since P is a transition matrix for an ergodic Markov chain, one finds that $M(y)$ must also have at least one nonzero entry in each row and column for all $y \in \mathcal{Y}$. Therefore, $\tau(M(y)) \leq 1$ for all $y \in \mathcal{Y}$.

Definition 2.7. An HMP is said to be (ϵ,k)-*primitive* if $\min_{i,j}\left[M(y_1^k)\right]_{ij} > k\epsilon$ for all $y_1^k \in \mathcal{Y}$. This gives a uniform lower bound on the probability that a k-step transition of the HMP simultaneously moves between any two states and generates any output sequence y_1^k. An HMP is said to be ϵ-*primitive* if there exists a $k < \infty$ such it is (ϵ,k)-*primitive*.

Lemma 2.8. *An HMP is* (ϵ,k)-*primitive if it satisfies Condition 2.6 with* $\delta \geq k^{1/k}\epsilon^{1/k}$ *and* P^k *is a positive matrix. Moreover, this implies that* $\pi(i) \geq k\epsilon$ *(i.e., strictly positive) for all* $i \in \mathcal{Q}$.

Proof. First, we note that P^k positive implies that there is a length-k path between any two states. Next, we write

$$\left[M(y_1^k)\right]_{q_1,q_{k+1}} = \sum_{q_2,\dots,q_k \in \mathcal{Q}^{k-1}} \prod_{t=1}^{k} p_{q_t,q_{t+1}} h_{q_t,q_{t+1}}(y_t)$$

$$> \sum_{q_2,\dots,q_k \in \mathcal{Q}^{k-1}} \prod_{t=1}^{k} 1_{\mathcal{V}}((q_t,q_{t+1}))\delta$$

$$\overset{(a)}{\geq} \delta^k,$$

where the last step follows from the fact that there is a length-k path between any two states. Since $\delta^k > k\epsilon$, we see that the HMP is (ϵ, k)-primitive according to Definition 2.7. Note that, for any $u \in \mathcal{A}_0$, we have

$$\sum_{i \in \mathcal{Q}} u(i) \left[M\left(y_1^k\right) \right]_{ij} \geq \left(\sum_{i \in \mathcal{Q}} u(i) \right) k\epsilon \geq \|u\|_1 k\epsilon, \tag{4}$$

which implies that $\pi(i) \geq k\epsilon$ for all $i \in \mathcal{Q}$. □

Lemma 2.9. *For any ϵ-primitive HMP, there exists a $k_0 < \infty$ such that, for all $y_1^k \in \mathcal{Y}^k$ and all $k \geq k_0$,*

$$\tau\left(M\left(y_1^k\right) \right) \leq e^{-2k_0 \lfloor k/k_0 \rfloor \epsilon}.$$

Proof. From Definition 2.7, we can assume that the HMP is (ϵ, k_0)-primitive. Using the bound (3), we see that

$$\phi\left(M\left(y_1^{k_0}\right) \right) \geq \left(\frac{\min_{i,j} [M]_{ij}}{\max_{i,j} [M]_{ij}} \right)^2 \geq \left(\frac{k_0 \epsilon}{1} \right)^2$$

and

$$\tau\left(M\left(y_1^{k_0}\right) \right) \leq \frac{1 - k_0 \epsilon}{1 + k_0 \epsilon} \leq e^{-2k_0 \epsilon}.$$

Since we can break any length-k sequence into at least $\lfloor k/k_0 \rfloor$ length-k_0 pieces and $\tau(M(y)) \leq 1$ for the remaining pieces, we have $\tau\left(M\left(y_1^k\right) \right) \leq \left(e^{-2k_0 \epsilon} \right)^{\lfloor k/k_0 \rfloor}$. □

2.3.3 Lyapunov exponents

Consider any stationary stochastic process, $\{Y_i\}_{i \in \mathbb{Z}}$, equipped with a function, $M(y)$, that maps each $y \in \mathcal{Y}$ to a matrix. Now, consider the limit

$$\lim_{n \to \infty} \frac{1}{n} \log \left\| u^{\mathsf{T}} \prod_{i=1}^n M(Y_i) \right\|,$$

where u is any nonzero vector and $\|\cdot\|$ is any vector norm. Oseledec's multiplicative ergodic theorem says that this limit is deterministic for almost all realizations [33]. An earlier ergodic theorem of Furstenberg and Kesten [13] gives a nice proof that

$$\lim_{n \to \infty} \frac{1}{n} \log \left\| \prod_{i=1}^n M(Y_i) \right\| \overset{\text{a.s.}}{=} \gamma_1,$$

where $\|\cdot\|$ is any matrix norm and γ_1 is known as the top Lyapunov exponent. The connection with entropy rate is given by the fact that, for an HMP, choosing $M(y)$ according to (1) implies that $H(\mathcal{Y}) = -\gamma_1$ [36, 21].

2.4 Stationary measures

The forward and backward state probability vectors play a very important role in the analysis of HMPs. These vectors, $\alpha_i, \beta_i \in \mathcal{A}_0$, are themselves random variables which often have well-defined stationary distributions. To illustrate the mixing properties, we exploit the stationarity of the HMP and focus on time zero by defining the random variables

$$U_n(i) \triangleq \Pr\left(Q_0 = i \mid Y_{-n}^{-1}\right),$$

$$V_n(i) \triangleq \frac{1}{\pi(i)} \Pr\left(Q_0 = i \mid Y_0^{n-1}\right).$$

It is worth noting that $U_n(i)$ is a deterministic function of y_{-n}^{-1} and $V_n(i)$ is a deterministic function of y_0^{n-1}. The following sufficient condition characterizes some of the HMPs that have stationary distributions.

Definition 2.10. An HMP is called *almost-surely mixing* if there exist $C < \infty$, $\gamma < 1$, and $k < \infty$ such that

$$\Pr\left(d\left(U_m, U_n\right) > C\gamma^n\right) \leq C\gamma^n,$$

$$\Pr\left(d\left(V_m, V_n\right) > C\gamma^n\right) \leq C\gamma^n$$

for all $m \geq n + k \geq k + 1$. This implies that the forward and backward recursions both forget their initial conditions at an exponential rate that is uniform over all but an exponentially small set of received sequences.

Definition 2.11. An HMP is called *sample-path mixing* if there exist $C < \infty$, $\gamma < 1$, and $k < \infty$ such that

$$d\left(U_m, U_n\right) \leq C\gamma^n,$$

$$d\left(V_m, V_n\right) \leq C\gamma^n$$

for all $m \geq n + k \geq k + 1$ and all received sequences $y_{-m}^{m-1} \in \mathcal{Y}^{2m}$. This implies that the forward and backward recursions both forget their initial conditions at an exponential rate that is uniform over all received sequences. It is easy to see that sample path mixing implies almost-surely mixing.

Lemma 2.12. *An (ϵ, k)-primitive HMP is sample-path mixing with $\gamma = e^{-2\epsilon}$ and $C = -2\ln(k\epsilon)\gamma^{-k}$.*

Proof. For each y_{-n}^{-1}, the realization of $U_n(i)$ is given by

$$u_n(i) = \Pr\left(Q_0 = i \mid Y_{-n}^{-1} = y_{-n}^{-1}\right) = \left[\frac{\pi^{\mathrm{T}} M\left(y_{-n}^{-1}\right)}{\pi^{\mathrm{T}} M\left(y_{-n}^{-1}\right) \mathbf{1}}\right]_i.$$

First, we let $w^{\mathrm{T}} = \pi^{\mathrm{T}} M\left(y_{-m}^{-n-1}\right)$ and note that (4) implies that

$$d\left(w^{\mathrm{T}}, \pi^{\mathrm{T}}\right) = \ln\max_{i,j} \frac{w(i)\pi(j)}{\pi(i)w(j)} \leq \ln\max_{i,j} \frac{1}{\pi(i)w(j)} \leq \ln\left(\left(\frac{1}{k\epsilon}\right)^2\right)$$

when $m \geq n + k$. Next, we use Theorem 2.5 and Lemma 2.9 to see that

$$d\left(u_m, u_n\right) = d\left(w^{\mathrm{T}} M\left(y_{-n}^{-1}\right), \pi^{\mathrm{T}} M\left(y_{-n}^{-1}\right)\right) d\left(w^{\mathrm{T}}, \pi^{\mathrm{T}}\right)$$

$$\leq \tau\left(M\left(y_{-n}^{-1}\right)\right) \ln\left((k\epsilon)^{-2}\right)$$

$$\leq -2\ln(k\epsilon) e^{-2\lfloor n/k \rfloor k\epsilon}.$$

This gives an exponential rate of $\gamma = e^{-2\epsilon}$ and $C = -2\ln(k\epsilon)\gamma^{-k}$ is chosen to handle the floor function and constant. For the backward recursion, the proof is identical except that the constant C is smaller by a factor of 2 because

$$d\left(M(y_n^{m-1})\mathbf{1}, \mathbf{1}\right) = \ln\max_{i,j} \frac{w(i)}{w(j)} \leq \ln\max_{i,j} \frac{\mathbf{1}^{\mathrm{T}} \mathbf{1}}{\mathbf{1}^{\mathrm{T}} \mathbf{1} k\epsilon} \leq \ln\left(\frac{1}{k\epsilon}\right).$$

$\square$

Lemma 2.13. *A $(0, k)$-primitive HMP is almost-surely mixing for some $\gamma < 1$ and $C < \infty$ if*

$$\max_{q \in \mathcal{Q}} E\left[\frac{\max_{i,j}\left[M\left(Y_1^k\right)\right]_{ij}}{\min_{i,j}\left[M\left(Y_1^k\right)\right]_{ij}} \,\middle|\, Q = q\right] < \infty.$$

In particular, this can be applied to HMPs with continuous observations.

Proof. This lemma follows, with slight modifications, from the arguments in [26]. Its proof is beyond the scope of this work. $\square$

Proposition 2.14. *The joint process $\{Q_t, \alpha_t\}_{t \in \mathbb{Z}}$ forms a Markov chain. If the HMP is almost-surely mixing, then the marginal distribution converges weakly to a unique stationary measure $\mu_q(A)$.*

Proof. One can see this is a Markov chain by considering the following method of generating the sequence. Starting from state q_t at time t, we first choose q_{t+1} according to $p_{q_t,q_{t+1}}$, then choose y_t according to $h_{q_t,q_{t+1}}(y_t)$, and finally compute $\alpha_{t+1}(\ )$ from $\alpha_t(\cdot)$ and y_t. In most cases, this Markov chain will not have a finite state space because $\alpha_t(\cdot)$ may take uncountably many values. Of course, this process depends on the initialization of the first α_t but this dependence decays with time if the HMP is almost-surely mixing. For simplicity, one may assume that the initialization $\alpha_1 = \pi$ is used.

To show that $\mu_q^{(t)}(A) \triangleq \Pr(Q_0 = q, U_t \in A)$ converges weakly to the probability measure $\mu_q(A)$ for all Borel subsets $A \subseteq \mathcal{P}_0$, we observe that $\mu_q^{(t)}(A)$ is a Cauchy sequence with respect to the Prohorov metric. This is sufficient because the Prohorov metric metrizes weak convergence on separable spaces and $\mathcal{P}_0$ is separable [6, page 72]. Let $d(u, A) \triangleq \inf_{v \in A} d(u, v)$ and $A^\delta \triangleq \{u \in \mathcal{P}_0 | d(u, A) < \delta\}$ so that the Prohorov metric is given by

$$d_P\left(\mu, \mu'\right) = \inf\left\{\delta \in \mathbb{R}_+ \mid \mu'(A) \le \mu\left(A^\delta\right) + \delta \ \forall \text{Borel } A \subseteq \mathcal{P}_0\right\}.$$

Since the HMP is almost-surely mixing, we can use the fact that $\Pr\left(d(U_{t+k}, U_t) > C\gamma^t\right) \le C\gamma^t$, for all $k \ge 0$, to see that

$$\mu_q^{(t+k)}(A) = \Pr(Q_0 = q, U_{t+k} \in A) \le \Pr(Q_0 = q, U_t \in A^{C\gamma^t}) + C\gamma^t$$
$$= \mu_q^{(t)}(A^{C\gamma^t}) + C\gamma^t.$$

This implies that $d_P(\mu_q^{(t)}, \mu_q^{(t+k)}) \le C\gamma^t$ for all $k \ge 0$. Therefore, $\mu_q^{(t)}(A)$ is a Cauchy sequence with respect to d_P and it converges weakly to some probability measure. Therefore, we can define $\mu_q(A)$ to be the weak limit of $\mu_q^{(t)}(A)$. □

Definition 2.15. The *(forward) Furstenberg measure* is the unique stationary measure (when it exists) of the joint process $\{Q_t, \alpha_t\}_{t \in \mathbb{Z}}$ and is given by the weak limit

$$\Pr(Q_t = q, \alpha_t \in A) \xrightarrow{w} \mu_q(A)$$

for any Borel measurable set $A \subseteq \mathcal{P}_0$. While this does not depend on the initialization of α_t, one may assume the initialization $\alpha_1 = \pi$ for simplicity.

Remark 2.16. This name is chosen because the measure first appears in the work of Furstenberg and Kifer [14] and is closely related to the work that was started by Furstenberg and Kesten [13].

2.4.1 Consistency of the a posteriori probability (APP)

The following lemma will be used to make connections between the measures defined in this section.

Lemma 2.17. *Let X, Y be discrete random variables and let the APP function be $E_y(x) \triangleq \Pr(X = x | Y = y)$. Then $E_Y(x) = \Pr(X = x | Y)$ is a random function (due to Y) and we have*

$$\Pr(X = x, E_Y(\cdot) = e(\cdot)) = \Pr(E_Y(\cdot) = e(\cdot)) e(x).$$

Proof. Applying the chain rule and the definition of $E_Y(\cdot)$ gives

$$\Pr(X = x, E_Y(\cdot) = e(\cdot)) = \Pr(E_Y(\cdot) = e(\cdot)) \Pr(X = x | E_Y(\cdot) = e(\cdot))$$
$$= \Pr(E_Y(\cdot) = e(\cdot)) \Pr(X = x | Y)$$
$$= \Pr(E_Y(\cdot) = e(\cdot)) e(x),$$

where the second step follows from the fact that $E_Y(\cdot)$ is a sufficient statistic for X (e.g., X can be faithfully generated from Y using the Markov chain $Y \to E_Y(\cdot) \to X$). $\quad\square$

Proposition 2.18. *The process $\{\alpha_t\}_{t \in \mathbb{Z}}$ forms a Markov chain. If the HMP is almost-surely mixing, then it converges weakly to a unique stationary measure $\mu(A)$.*

Proof. One can see that $\{\alpha_t\}_{t \in \mathbb{Z}}$ is Markov by considering another method of generating the sequence. At time t, we choose q_t according to $\alpha_t(\cdot)$, then choose q_{t+1} according to $p_{q_t, q_{t+1}}$, then choose y_t according to $h_{q_t, q_{t+1}}(y_t)$, and finally compute $\alpha_{t+1}(\cdot)$ from $\alpha_t(\cdot)$ and y_t. Of course, this process depends on the initialization of the first α_t but this dependence decays with time if the HMP is almost-surely mixing. For simplicity, one may assume that the initialization $\alpha_1 = \pi$ is used.

Comparing this to Proposition 2.14, one sees that we are now using $\alpha_t(\cdot)$ as a proxy distribution for Q_t. This works because Lemma 2.17 shows that

$$\Pr(\alpha_t \in A) \inf_{\tilde{\alpha} \in A} \tilde{\alpha}(q) \leq \Pr(Q_t = q, \alpha_t \in A) \leq \Pr(\alpha_t \in A) \sup_{\tilde{\alpha} \in A} \tilde{\alpha}(q)$$

for any open set $A \subseteq \mathcal{P}_0$. By making A arbitrarily small, one can force the left-hand side and right-hand side to be arbitrarily close. The proof of weak convergence to a unique stationary distribution as $t \to \infty$ is essentially identical to the corresponding proof for Proposition 2.14. $\quad\square$

Definition 2.19. The *(forward) Blackwell measure* is the unique stationary measure (when it exists) of the process $\{\alpha_t\}_{t \in \mathbb{Z}}$ and is given by the weak limit

$$\Pr(\alpha_t \in A) \overset{w}{\to} \mu(A)$$

for any Borel measurable set $A \subseteq \mathcal{P}_0$. From the definition of μ_q, we see also that $\mu(A) = \sum_{q \in \mathcal{Q}} \mu_q(A)$.

Remark 2.20. This name is chosen because this measure first appears in the work of Blackwell [8] and is now commonly called the Blackwell measure [17].

Lemma 2.21. *The Radon–Nikodym derivative* $\mathrm{d}\mu_q/\mathrm{d}\mu$ *of the (forward) Furstenberg measure* μ_q *with respect to the (forward) Blackwell measure* μ *exists and satisfies*

$$\frac{\mathrm{d}\mu_q}{\mathrm{d}\mu}(\alpha) = \Pr(Q_t = q | \alpha_t = \alpha)$$

μ-*almost everywhere (assuming stationary* $\{Q_t, \alpha_t\}_{t \in \mathbb{Z}}$). *This implies that*

$$\mu_q(\mathrm{d}\alpha) = \alpha(q)\mu(\mathrm{d}\alpha).$$

Proof. First, we note that $\mu(A) = \sum_{q \in \mathcal{Q}} \mu_q(A)$ implies that μ_q is absolutely continuous with respect to μ. Therefore, the Radon–Nikodym derivative $\mathrm{d}\mu_q/\mathrm{d}\mu$ exists. Since

$$\frac{\mu_q(A)}{\mu(A)} = \frac{\Pr(Q_t = q, \alpha_t \in A)}{\Pr(\alpha_t \in A)} = \Pr(Q_t = q | \alpha_t \in A),$$

the first result can be seen by choosing A to be arbitrarily small. The second result holds because $\alpha_t(\cdot)$ is the APP estimate of Q_t given $Y_{-\infty}^{t-1}$ and this (e.g., see Lemma 2.17) implies that

$$\Pr(Q_t = q | \alpha_t = \alpha) = \alpha(q).$$

$\square$

Theorem 2.22. [8] *In terms of the Blackwell measure, the entropy rate (in nats) of an HMP is*

$$H(\mathcal{Y}) = -\int_{\mathcal{P}_0} \mu(\mathrm{d}\alpha) \sum_{y \in \mathcal{Y}} \alpha^{\mathrm{T}} M(y) \mathbf{1} \ln\left(\alpha^{\mathrm{T}} M(y) \mathbf{1}\right). \tag{5}$$

Proof. Consider the sequence $H(Y_t | Y_1^{t-1})$ for any stationary process. This sequence is nonnegative and nonincreasing and therefore must have a limit. Moreover, the entropy rate

$$H(\mathcal{Y}) \triangleq \lim_{n \to \infty} \frac{1}{n} H(Y_1, \ldots, Y_n) = \lim_{n \to \infty} \frac{1}{n} \sum_{t=1}^{n} H(Y_t | Y_1^{t-1})$$

is the Cesàro mean of this sequence and must have the same limit. Next, we note that

$$\alpha_t^T M(y)\mathbf{1} = \sum_{i,j \in \mathcal{Q}} \alpha_t(i) p_{i,j} h_{i,j}(y) = \Pr\left(Y_t = y \mid Y_1^{t-1}\right).$$

Therefore, (5) is simply the expression for $\lim_{t \to \infty} H(Y_t \mid Y_1^{t-1})$. □

2.4.2 Once again, this time in reverse ...

One can also reverse time for these Markov processes so that $\{Q_t, \beta_t\}_{t \in \mathbb{Z}}$ forms a backward Markov chain. Starting from q_t and working backwards, one first chooses q_{t-1} according to $\Pr(Q_{t-1} = q_{t-1} \mid Q_t = q_t) = p_{q_{t-1}, q_t} \pi_{q_{t-1}} / \pi_{q_t}$. Then, one generates y_{t-1} according to $h_{q_{t-1}, q_t}(y_{t-1})$ and computes β_{t-1} from β_t and y_{t-1}.

This process also depends on the initialization of the first β_t but this dependence decays with time if the HMP is almost-surely mixing. For simplicity, one may assume that the initialization $\beta_1 = \mathbf{1}$ is used. If the HMP is almost-surely mixing, then the joint distribution of Q_t, β_t converges weakly to a unique stationary distribution as $t \to -\infty$; the proof is very similar to the corresponding part of the proof of Proposition 2.14. This allows us to define the stationary distribution of the backwards state probability vector.

As with the forward process, we can reduce the state space to $\{\beta_t\}_{t \in \mathbb{Z}}$. At each step, one chooses q_t according to $\Pr(Q_t = q_t) = \beta_t(q_t) \pi_{q_t}$, then continues as described above to generate with q_{t-1}, y_{t-1}, and β_{t-1}. Let $B \subseteq \{u \in \mathcal{A}_0 \mid \pi^T u = 1\}$ be any open measurable set. Then, using $\beta_t(q)\pi_q$ as a proxy distribution for Q_t works because Lemma 2.17 shows that

$$\Pr(\beta_t \in B)\pi(q) \inf_{\widetilde{\beta} \in B} \widetilde{\beta}(q) \leq \Pr(Q_t = q, \beta_t \in B) \leq \Pr(\beta_t \in B)\pi(q) \sup_{\widetilde{\beta} \in B} \widetilde{\beta}(q),$$

and choosing B arbitrarily small allows the left-hand side and right-hand side to be made arbitrarily close. This process also depends on the initialization of β_t, but if the HMP is almost-surely mixing, then it converges weakly to a unique stationary distribution.

Definition 2.23. The *backward Furstenberg measure* is the unique stationary measure (when it exists) of the backwards process $\{Q_t, \beta_t\}_{t \in \mathbb{Z}}$ and is given by the weak limit

$$\Pr(Q_t = q, \beta_t \in B) \xrightarrow{w} \nu_q(B)$$

for any Borel measurable set $B \subseteq \{u \in \mathcal{A}_0 \mid \pi^T u = 1\}$.

Definition 2.24. The *backward Blackwell measure* is the unique stationary measure (when it exists) of the backwards process $\{\beta_t\}_{t \in \mathbb{Z}}$ and is given by the weak limit

$$\Pr(\beta_t \in B) \xrightarrow{w} \nu(B)$$

for any Borel measurable set $B \subseteq \{u \in \mathcal{A}_0 \mid \pi^{\mathsf{T}} u = 1\}$. From the definition of ν_q, we see also that $\nu(B) = \sum_{q \in Q} \nu_q(B)$.

Lemma 2.25. *The Radon–Nikodym derivative* $\mathrm{d}\nu_q/\mathrm{d}\nu$ *of the backwards Furstenberg measure* ν_q *with respect to the backwards Blackwell measure* ν *exists and satisfies*

$$\frac{\mathrm{d}\nu_q}{\mathrm{d}\nu}(\beta) = \Pr(Q_t = q \mid \beta_t = \beta)$$

ν-*almost everywhere (assuming stationary* $\{Q_t, \beta_t\}_{t \in \mathbb{Z}}$*). This implies that*

$$\nu_q(\mathrm{d}\beta) = \pi(q)\beta(q)\nu(\mathrm{d}\beta).$$

Proof. First, we note that $\nu(B) = \sum_{q \in Q} \nu_q(B)$ implies that ν_q is absolutely continuous with respect to ν. Therefore, the Radon–Nikodym derivative $\mathrm{d}\nu_q/\mathrm{d}\nu$ exists. Since

$$\frac{\nu_q(B)}{\nu(B)} = \frac{\Pr(Q_t = q, \beta_t \in B)}{\Pr(\beta_t \in B)} = \Pr(Q_t = q \mid \beta_t \in B),$$

the first result can be seen by choosing B to be arbitrarily small. The second result holds because $\beta_t(\cdot)$ is the APP estimate of Q_t given Y_t^∞ and this (e.g., see Lemma 2.17) implies that

$$\Pr(Q_t = q \mid \beta_t = \beta) = \pi(q)\beta(q). \qquad \square$$

3 Taking the derivative

3.1 The derivative shortcut

In this section, we introduce a shortcut often used in the statistical physics community. It was introduced to the author by Méasson *et al.* in [27, 28]. It has also been applied to the problem under consideration by Zuk *et al.* in [44, 43].

Let $D \subset \mathbb{R}$ be a compact set and $g_n : D^n \to \mathbb{R}$ be a sequence of functions which essentially depend on a single parameter $\theta \in D$ in n different ways. Abusing notation, we also let $g_n : D \to \mathbb{R}$ be the same function where this dependence is combined so that $g_n(\theta) = g_n(\theta, \ldots, \theta)$. The total derivative of g_n can be written as

$$\frac{\mathrm{d}}{\mathrm{d}\theta} g_n(\theta) = \sum_{i=1}^{n} \frac{\partial}{\partial\theta_i} g_n(\theta_1, \ldots, \theta_n) \bigg|_{(\theta_1, \ldots, \theta_n) = (\theta, \ldots, \theta)}.$$

This motivates us to define

$$g_n'(\theta_1,\ldots,\theta_n) \triangleq \sum_{i=1}^{n} \frac{\partial}{\partial \theta_i} g_n(\theta_1,\ldots,\theta_n).$$

Since the abuse of notation is habit forming, we will also define $g_n'(\theta) \triangleq g_n'(\theta,\ldots,\theta)$.

The focus in this article is the limit of these functions as n goes to infinity, so a few technical details are required. If $g_n(\theta) \to f(\theta)$ uniformly over $\theta \in D$ and $\lim_{n\to\infty} g_n'(\theta)$ converges uniformly over $\theta \in D$, then it follows that $f'(\theta) = \lim_{n\to\infty} g_n'(\theta)$ [4]. One might assume that it is necessary to prove uniform convergence for both of these sequences, but the following standard problem in analysis shows that it suffices to consider only the sequence of derivatives.

Lemma 3.1. *Let $g_n : D \to \mathbb{R}$ be a sequence of functions that are continuously differentiable on a compact set $D \subset \mathbb{R}$. If $g_n(\theta_0)$ converges for some $\theta_0 \in D$ and $g_n'(\theta)$ converges uniformly on D, then the limits*

$$f(\theta) \triangleq \lim_{n\to\infty} g_n(\theta),$$

$$f'(\theta) \triangleq \lim_{n\to\infty} g_n'(\theta)$$

both exist and are uniformly continuous on D.

Proof. First, we note that each $g_n'(\theta)$ is uniformly continuous because D is compact. Since $g_n'(\theta)$ converges uniformly, we find that $f'(\theta)$ exists and is uniformly continuous (and hence bounded) on D. Interchanging the limit and integral, based on uniform convergence, implies that

$$\lim_{n\to\infty} [g_n(\theta) - g_n(\theta_0)] = \lim_{n\to\infty} \int_{\theta_0}^{\theta} g_n'(x)\,dx = \int_{\theta_0}^{\theta} \lim_{n\to\infty} g_n'(x)\,dx$$

$$= \int_{\theta_0}^{\theta} f'(x)\,dx = f(\theta) - f(\theta_0).$$

This implies that $g_n(\theta)$ converges to $f(\theta)$. Finally, we note that $f(\theta)$ is uniformly continuous on D because $f'(\theta)$ exists and is bounded on D. $\qquad\square$

3.2 Warmup example: the derivative of the log spectral radius

The spectral radius of a real $|\mathcal{Q}| \times |\mathcal{Q}|$ matrix M is defined to be

$$\rho(M) \triangleq \lim_{n\to\infty} \left\| M^n \right\|^{1/n}$$

for any matrix norm. Likewise, the log spectral radius (LSR) of a real matrix M is given by

$$\ln \rho(M) = \lim_{n \to \infty} \frac{1}{n} \log \|M^n\|$$

for any matrix norm. Moreover, if M has nonnegative entries, then

$$\ln \rho(M) = \lim_{n \to \infty} \frac{1}{n} \log \left(u^T M^n v \right)$$

for any vectors $u, v \in \mathcal{A}_0$.

Let M_θ be a mapping from a compact set $D \subset \mathbb{R}$ to the set of nonnegative real matrices. Assume further that M_θ has a unique real eigenvalue λ_1 of maximum modulus (i.e., the second largest eigenvalue λ_2 satisfies $|\lambda_2/\lambda_1| < \gamma < 1$) for all $\theta \in D$. Using the shorthand notation $M \triangleq M_{\theta^*}$ for $\theta^* \in D$, we let $a, b \in \mathcal{A}$ be left/right (column) eigenvectors of M with eigenvalue $\rho(M)$; they satisfy $a^T M = \rho(M) a^T$ and $Mb = \rho(M) b$. In this case, it is known that the derivative of the LSR is given by

$$\left. \frac{d}{d\theta} \ln \rho(M_\theta) \right|_{\theta=\theta^*} = \frac{a^T M_{\theta^*}' b}{a^T M_{\theta^*} b},$$

where $M' \triangleq M_{\theta^*}'$ is the element-wise derivative defined by $\left[M_\theta' \right]_{ij} \triangleq \frac{d}{d\theta} [M_\theta]_{ij}$. Of course, one must assume that M' exists and satisfies $\|M'\| < \infty$.

One can prove this by applying the derivative shortcut to $f(\theta) = \log \rho(M_\theta)$ using

$$g_n(\theta_1, \ldots, \theta_n) = \frac{1}{n} \ln \left(u^T \left(\prod_{t=1}^{n} M_{\theta_t} \right) v \right)$$

for any vectors $u, v \in \mathcal{A}_0$. Based on Lemma 3.1, we focus on $g_n'(\theta)$ by writing

$$g_n'(\theta^*) = \sum_{i=1}^{n} \frac{\partial}{\partial \theta_i} \frac{1}{n} \ln \left(u^T \left(\prod_{t=1}^{n} M_{\theta_t} \right) v \right) \Bigg|_{(\theta_1, \ldots, \theta_n) = (\theta^*, \ldots, \theta^*)}$$

$$= \frac{1}{n} \sum_{i=1}^{n} \frac{\partial}{\partial \theta_i} \ln \left(u^T \left(\prod_{t=1}^{i-1} M_{\theta_t} \right) M(\theta_i) \left(\prod_{t=i+1}^{n} M_{\theta_t} \right) v \right) \Bigg|_{\theta_1^n = (\theta^*, \ldots, \theta^*)}$$

$$= \frac{1}{n} \sum_{i=1}^{n} \frac{u^T \left(\prod_{t=1}^{i-1} M_{\theta_t} \right) M_{\theta_i}' \left(\prod_{t=i+1}^{n} M_{\theta_t} \right) v}{u^T \left(\prod_{t=1}^{i-1} M_{\theta_t} \right) M_{\theta_i} \left(\prod_{t=i+1}^{n} M_{\theta_t} \right) v} \Bigg|_{\theta_1^n = (\theta^*, \ldots, \theta^*)}$$

$$= \frac{1}{n} \sum_{i=1}^{n} \frac{u^T M^{i-1} M' M^{n-i} v}{u^T M^{i-1} M M^{n-i} v},$$

where we have used

$$\frac{d}{d\theta} x^T M_\theta y = \sum_{k,l} x_k \frac{d}{d\theta} [M_\theta]_{k,l} y_l = x^T M'_\theta y.$$

Since M_θ satisfies $|\lambda_2/\lambda_1| < \gamma$ for all $\theta \in D$, it follows that

$$\frac{u^T M^{i-1}}{\|u^T M^{i-1}\|} = \frac{a^T}{\|a\|} + O\left(\gamma^{i-1}\right),$$

$$\frac{M^{n-i} v}{\|M^{n-i} v\|} = \frac{b}{\|b\|} b + O\left(\gamma^{n-i}\right).$$

Treating the boundary and interior terms, in the sum, separately gives

$$g'_n(\theta^*) = O\left(\frac{\lfloor (\ln n)^2 \rfloor}{n} \frac{\|M'\| (a^T b)}{\rho(M) \frac{u^T b}{\|u\|} \frac{a^T v}{\|v\|}}\right)$$

$$+ \frac{1}{n} \sum_{i=\lfloor (\ln n)^2 \rfloor + 1}^{n - \lfloor (\ln n)^2 \rfloor} \frac{a^T M' b + O\left(\gamma^{(\ln n)^2}\right) \|M'\|}{a^T M b + O\left(\gamma^{(\ln n)^2}\right) \|M\|}.$$

Therefore, $g_n(\theta)$ and $g'_n(\theta)$ converge uniformly for all $\theta \in D$ and we find that

$$f'(\theta^*) = \frac{a^T M' b}{a^T M b}.$$

3.3 The derivative of the entropy rate

Let $M_\theta(y)$ be the transition observation probability matrix of an HMP, which depends on the real parameter θ, and let π be the stationary distribution of the underlying Markov chain. To compute the derivative of the entropy rate, we define

$$g_n(\theta_1, \ldots, \theta_n) = -\frac{1}{n} \sum_{y_1^n \in \mathcal{Y}^n} \Pr\left(Y_1^n = y_1^n; \theta_1^n\right) \ln \Pr\left(Y_1^n = y_1^n; \theta_1^n\right)$$

$$= -\frac{1}{n} \sum_{y_1^n \in \mathcal{Y}^n} \pi^T \left(\prod_{i=1}^{n} M_{\theta_i}(y_i)\right) \mathbf{1}$$

$$\cdot \ln \left[\pi^T \left(\prod_{i=1}^{n} M_{\theta_i}(y_i)\right) \mathbf{1}\right]\Bigg|_{(\theta_1, \ldots, \theta_n) = (\theta^*, \ldots, \theta^*)}.$$

This implies that $f(\theta) = \lim_{n \to \infty} g_n(\theta) = H(\mathcal{Y}; \theta)$ in nats.

Theorem 3.2. *Let $D \subset \mathbb{R}$ be a compact set and assume that $\frac{d}{d\theta}\pi = 0$ and $M'_\theta(y) \triangleq \frac{d}{d\theta}M_\theta(y)$ exists for all $\theta \in D$. Then, if the HMP is well defined and ϵ-primitive for all $\theta \in D$,*

$$f'(\theta^*) = \frac{d}{d\theta}H(\mathcal{Y}; \theta)\Big|_{\theta=\theta^*}$$

$$= -\int_{\mathcal{A}_0} \mu(d\alpha) \int_{\mathcal{A}_0} \nu(d\beta) \sum_{y \in \mathcal{Y}} \alpha^{\mathsf{T}} M'_{\theta^*}(y)\beta \ln\left(\alpha^{\mathsf{T}} M_{\theta^*}(y)\beta\right), \quad (6)$$

where μ and ν are the forward/backward Blackwell measures of the HMP at $\theta = \theta^*$. Moreover, $f(\theta)$ and $f'(\theta)$ are uniformly continuous on D.

Proof. The following shorthand is used throughout: $\pi_t(q) \triangleq \mathrm{Pr}\,(Q_t = q)$, $M(y) \triangleq M_{\theta^*}(y)$, $M'(y) \triangleq M'_{\theta^*}(y)$, and $M(y_j^k) \triangleq \prod_{t=j}^{k} M_{\theta^*}(y_t)$. For the HMP to be well defined, the transition matrices must satisfy $\sum_{y \in \mathcal{Y}} M_\theta(y)\mathbf{1} = \mathbf{1}$ and $\sum_{y \in \mathcal{Y}} M'_\theta(y)\mathbf{1} = \mathbf{0}$ for all $\theta \in D$. It follows that, for any $u \in \mathcal{P}_0$, one has

$$\sum_{y_1^n \in \mathcal{Y}^n} u^{\mathsf{T}} \left(\prod_{t=1}^{n} M_{\theta_t}(y_t)\right) \mathbf{1} = 1,$$

$$\frac{\partial}{\partial \theta_j} \sum_{y_1^n \in \mathcal{Y}^n} u^{\mathsf{T}} \left(\prod_{t=1}^{n} M_{\theta_t}(y_t)\right) \mathbf{1} = 0. \quad (7)$$

Based on Lemma 3.1, we note that the entropy rate exists for all $\theta \in D$ and focus on the derivative

$$g'_n(\theta^*) \overset{(a)}{=} -\frac{1}{n} \sum_{j=1}^{n} \frac{\partial}{\partial \theta_j} \sum_{y_1^n \in \mathcal{Y}^n} \pi_1^{\mathsf{T}} \left(\prod_{t=1}^{n} M_{\theta_t}(y_t)\right) \mathbf{1}$$

$$\cdot \left(\ln\left[C_j \pi_1^{\mathsf{T}}\left(\prod_{t=1}^{n} M_{\theta_t}(y_t)\right)\mathbf{1}\right] - \ln C_j\right)\Bigg|_{\theta_j = \theta^*}$$

$$\overset{(b)}{=} -\frac{1}{n} \sum_{j=1}^{n} \frac{\partial}{\partial \theta_j} \sum_{y_1^n \in \mathcal{Y}^n} \pi_1^{\mathsf{T}} \left(\prod_{t=1}^{n} M_{\theta_t}(y_t)\right)\mathbf{1} \cdot \ln\left[C_j \pi_1^{\mathsf{T}}\left(\prod_{t=1}^{n} M_{\theta_t}(y_t)\right)\mathbf{1}\right]\Bigg|_{\theta_j = \theta^*}$$

$$
\overset{(c)}{=} -\frac{1}{n} \sum_{j=1}^{n} \frac{\partial}{\partial \theta_j} \sum_{y_1^n \in \mathcal{Y}^n} \pi_1^{\mathrm{T}} M (y_1^{j-1}) M_{\theta_j} (y_j) M (y_{j+1}^n) \mathbf{1}
$$

$$
\cdot \ln \left[\frac{\pi_1^{\mathrm{T}} M (y_1^{j-1}) M_{\theta_j} (y_j) M (y_{j+1}^n) \mathbf{1}}{\left(\pi_1^{\mathrm{T}} M (y_1^{j-1}) \mathbf{1} \right) \left(\pi_{j+1}^{\mathrm{T}} M (y_{j+1}^n) \mathbf{1} \right)} \right] \Bigg|_{\theta_j = \theta^*},
$$

where (*a*) holds for arbitrary positive values $C_1, \ldots, C_n$, (*b*) follows because (7) implies that $\ln C_j$ gives no contribution if $\frac{\partial}{\partial \theta_j} C_j = 0$, and (*c*) follows from choosing

$$
C_j = \Pr \left(Y_1^{j-1} = y_1^{j-1} \right) \Pr \left(Y_{j+1}^n = y_{j+1}^n \right) = \left(\pi_1^{\mathrm{T}} M (y_1^{j-1}) \mathbf{1} \right) \left(\pi_{j+1}^{\mathrm{T}} M (y_{j+1}^n) \mathbf{1} \right).
$$

One subtlety is that $\pi_{j+1}^{\mathrm{T}} = \pi_j^{\mathrm{T}} \sum_{y \in \mathcal{Y}} M_{\theta_j} (y)$ is affected by θ_j. So, small changes in θ_j cause small changes in π_{j+1} and we must add the condition $\frac{\mathrm{d}}{\mathrm{d}\theta} \pi = \mathbf{0}$ to guarantee that $\frac{\partial}{\partial \theta_j} C_j = 0$. After adding this below condition, we may safely assume that $\pi_j = \pi$ for $j = 1, \ldots, n$. See Remark 3.3 for more details.

For Borel measurable sets $A \subseteq \mathcal{A}_0$ and $B \subseteq \left\{ u \in \mathcal{A}_0 \mid \pi^{\mathrm{T}} u = 1 \right\}$, the sets

$$
U_j(A) \triangleq \left\{ y_1^{j-1} \in \mathcal{Y}^{j-1} \;\middle|\; \alpha_j^{\mathrm{T}} = \frac{\pi^{\mathrm{T}} M (y_1^{j-1})}{\pi^{\mathrm{T}} M (y_1^{j-1}) \mathbf{1}} \in A \right\},
$$

$$
V_j(B) \triangleq \left\{ y_j^n \in \mathcal{Y}^{n-j-1} \;\middle|\; \beta_j = \frac{M (y_j^n) \mathbf{1}}{\pi^{\mathrm{T}} M (y_j^n) \mathbf{1}} \in B \right\}
$$

will be used to define the measures $\mu^{(j)}(A) \triangleq \Pr(Y_1^{j-1} \in U_j(A))$ and $\nu^{(j)}(B) \triangleq \Pr(Y_j^n \in V_j(B))$ for the forward/backward state probabilities. In this case, $\mu^{(j)}(\cdot), \nu^{(j)}(\cdot)$ are probability measures on $\mathcal{A}_0$ for the random variables α_j, β_j. Using these measures, we find that $g_n'(\theta^*)$ is given by

$$
-\frac{1}{n} \sum_{j=1}^{n} \frac{\partial}{\partial \theta_j} \sum_{y_1^n \in \mathcal{Y}^n} \overbrace{\pi^{\mathrm{T}} M (y_1^{j-1})}^{\alpha_j^{\mathrm{T}} \cdot \pi^{\mathrm{T}} M (y_1^{j-1}) \mathbf{1}} M_{\theta_j} (y_j) \overbrace{M (y_{j+1}^n) \mathbf{1}}^{\beta_{j+1} \cdot \pi^{\mathrm{T}} M (y_{j+1}^n) \mathbf{1}}
$$

$$
\times \ln \left[\frac{\overbrace{\pi^{\mathrm{T}} M (y_1^{j-1})}^{\alpha_j^{\mathrm{T}}}}{\pi^{\mathrm{T}} M (y_1^{j-1}) \mathbf{1}} M_{\theta_j} (y_j) \frac{\overbrace{M (y_{j+1}^n) \mathbf{1}}^{\beta_{j+1}}}{\pi^{\mathrm{T}} M (y_{j+1}^n) \mathbf{1}} \right] \Bigg|_{\theta_j = \theta^*}
$$

$$= -\frac{1}{n} \sum_{j=1}^{n} \int_{\mathcal{A}_0} \mu^{(j)}(\mathrm{d}\alpha) \int_{\mathcal{A}_0} \nu^{(j+1)}(\mathrm{d}\beta) \frac{\partial}{\partial \theta_j}$$

$$\times \sum_{y_j \in \mathcal{Y}} \alpha^{\mathrm{T}} M_{\theta_j}(y_j) \beta \ln \left(\alpha^{\mathrm{T}} M_{\theta_j}(y_j) \beta \right) \Bigg|_{\theta_j = \theta^*}$$

$$= -\frac{1}{n} \sum_{j=1}^{n} \int_{\mathcal{A}_0} \mu^{(j)}(\mathrm{d}\alpha) \int_{\mathcal{A}_0} \nu^{(j+1)}(\mathrm{d}\beta)$$

$$\times \sum_{y_j \in \mathcal{Y}} \left[\alpha^{\mathrm{T}} M'(y_j) \beta \ln \left(\alpha^{\mathrm{T}} M(y_j) \beta \right) + \alpha^{\mathrm{T}} M'(y_j) \beta \right].$$

All that is left is to compute the sum. If the HMP is almost-surely mixing, then the results of Section 2.4 show that measures converge weakly (i.e., $\mu^{(j)} \to \mu$ and $\nu^{(j)} \to \nu$). Moreover, Lemma A.2 in Appendix A.1 shows that the convergence rate is exponential. Therefore, most of the terms in the sum have essentially the same value. Like the LSR, we neglect terms within $(\ln n)^2$ of the block edge because their contribution is negligible as $n \to \infty$. The exponential convergence of the stationary measures also shows that the interior terms become equal at the superpolynomial rate $\gamma^{(\ln n)^2} = n^{\ln n \cdot \ln \gamma}$. Therefore, $f_n(\theta)$ and $f_n'(\theta)$ converge uniformly for all $\theta \in D$ and

$$\lim_{n \to \infty} -\frac{1}{n} \sum_{j=1}^{n} \int_{\mathcal{A}_0} \mu^{(j)}(\mathrm{d}\alpha) \int_{\mathcal{A}_0} \nu^{(j+1)}(\mathrm{d}\beta)$$

$$\times \sum_{y_j \in \mathcal{Y}} \left[\alpha^{\mathrm{T}} M'(y_j) \beta \ln \left(\alpha^{\mathrm{T}} M(y_j) \beta \right) + \alpha^{\mathrm{T}} M'(y_j) \beta \right]$$

converges to

$$\frac{\mathrm{d}}{\mathrm{d}\theta} H(\mathcal{Y}; \theta) \Big|_{\theta = \theta^*}$$

$$= -\int_{\mathcal{A}_0} \mu(\mathrm{d}\alpha) \int_{\mathcal{A}_0} \nu(\mathrm{d}\beta) \sum_{y \in \mathcal{Y}} \left[\alpha^{\mathrm{T}} M'(y) \beta \ln \left(\alpha^{\mathrm{T}} M(y) \beta \right) + \alpha^{\mathrm{T}} M'(y) \beta \right]. \quad (8)$$

Finally, the last term in (8) is shown to be zero in Lemma 3.4. □

Remark 3.3. The necessity of the condition $\frac{\mathrm{d}}{\mathrm{d}\theta} \pi = 0$ in Theorem 3.2 can be a bit subtle. This is because the π-term in many equations (e.g., $\pi^{\mathrm{T}} M (y_{j+1}^n) \mathbf{1}$) actually represents the state distribution at a particular time (e.g., time $j + 1$). The indices are dropped after the first few steps because the underlying Markov chain is stationary and the state distribution is independent of time. For example, the proof liberally uses the assumption that

$$\Pr(Y_{j+1}^n = y_{j+1}^n) = \sum_{q,q' \in Q} \Pr(Q_{j+1} = q) \Pr(Q_{n+1} = q', Y_{j+1}^n = y_{j+1}^n | Q_{j+1} = q)$$

$$= \pi M (y_{j+1}^n) \mathbf{1},$$

where the last step clearly requires that $\Pr(Q_{j+1} = q) = \pi(q)$. Moreover, this is not simply a problem with the proof. The author has applied the formula from Theorem 3.2 to a Markov chain (where the true entropy-rate derivative is well known) and shown that the two expressions become equal only if $\frac{d}{d\theta} \pi = \mathbf{0}$.

Lemma 3.4. *The following properties of the forward/backward Blackwell measures will be useful:*

$$\int_{\mathcal{A}_0} \mu(d\alpha)\alpha = \pi,$$

$$\int_{\mathcal{A}_0} \nu(d\beta)\beta = \mathbf{1},$$

$$\int_{\mathcal{A}_0} \mu(d\alpha) \sum_{y \in \mathcal{Y}} \alpha^T M (y)\beta = 1,$$

$$\int_{\mathcal{A}_0} \nu(d\beta) \sum_{y \in \mathcal{Y}} \alpha^T M (y)\beta = 1,$$

$$\int_{\mathcal{A}_0} \mu(d\alpha) \int_{\mathcal{A}_0} \nu(d\beta) \sum_{y \in \mathcal{Y}} \alpha^T M'(y)\beta = 0.$$

Proof. The proof is deferred to the appendix. □

3.4 Behavior of the entropy rate in the high-noise regime

Suppose that the domain of θ includes a "high-noise" point θ^* where the channel output provides no information about the channel state. In this case, the forward/backward Blackwell measures become singletons on $\pi, \mathbf{1}$ and the entropy rate $H(\mathcal{Y};\theta)$ converges to the single-letter entropy $H(Y;\theta)$ as $\theta \to \theta^*$. In the high-noise regime, one can also evaluate the derivative from Theorem 3.2 in closed form and extend the formula to the second derivative. In this section, we compare the expansions of $H(\mathcal{Y};\theta)$ and $H(Y;\theta)$.

First, we consider the single-letter entropy

$$H(Y;\theta) = -\sum_{y\in\mathcal{Y}} \Pr(Y_t = y) \log\left(\Pr(Y_t = y)\right)$$

$$= -\sum_{y\in\mathcal{Y}} \pi^{\mathrm{T}} M_\theta(y) \mathbf{1} \log(\pi^{\mathrm{T}} M_\theta(y)\mathbf{1}),$$

where π is the stationary distribution of the underlying Markov chain as a function of θ.

Lemma 3.5. *Under the assumption that $\frac{\mathrm{d}}{\mathrm{d}\theta}\pi = \mathbf{0}$ for all $\theta \in D$, the first derivative with respect to θ of the single-letter entropy is given by*

$$\frac{\mathrm{d}}{\mathrm{d}\theta} H(Y;\theta) = -\sum_{y\in\mathcal{Y}} \pi^{\mathrm{T}} M'_\theta(y)\mathbf{1}\log\left(\pi^{\mathrm{T}} M_\theta(y)\mathbf{1}\right).$$

Under the same assumption, the second derivative with respect to θ is given by

$$\frac{\mathrm{d}^2}{\mathrm{d}\theta^2} H(Y;\theta) = -\sum_{y\in\mathcal{Y}} \frac{\left(\pi^{\mathrm{T}} M'_\theta(y)\mathbf{1}\right)^2}{\pi^{\mathrm{T}} M_\theta(y)\mathbf{1}} - \sum_{y\in\mathcal{Y}} \pi^{\mathrm{T}} M''_\theta(y)\mathbf{1}\log\left(\pi^{\mathrm{T}} M_\theta(y)\mathbf{1}\right). \quad (9)$$

Proof. In particular, the first derivative is given by

$$\frac{\mathrm{d}}{\mathrm{d}\theta} H(Y;\theta) = -\frac{\mathrm{d}}{\mathrm{d}\theta}\sum_{y\in\mathcal{Y}} \pi^{\mathrm{T}} M_\theta(y)\mathbf{1}\log\left(\pi^{\mathrm{T}} M_\theta(y)\mathbf{1}\right)$$

$$= -\sum_{y\in\mathcal{Y}} \left(\left(\frac{\mathrm{d}}{\mathrm{d}\theta}\pi^{\mathrm{T}}\right) M_\theta(y)\mathbf{1} + \pi^{\mathrm{T}} M'_\theta(y)\mathbf{1}\right) \log\left(\pi^{\mathrm{T}} M_\theta(y)\mathbf{1}\right)$$

$$-\frac{\mathrm{d}}{\mathrm{d}\theta}\sum_{y\in\mathcal{Y}} \pi^{\mathrm{T}} M_\theta(y)\mathbf{1}$$

$$= -\sum_{y\in\mathcal{Y}} \pi^{\mathrm{T}} M'_\theta(y)\mathbf{1}\log\left(\pi^{\mathrm{T}} M_\theta(y)\mathbf{1}\right)$$

because $\frac{\mathrm{d}}{\mathrm{d}\theta}\pi = \mathbf{0}$ and $\sum_{y\in\mathcal{Y}} \pi^{\mathrm{T}} M_\theta(y)\mathbf{1} = 1$ for all θ. Since $\frac{\mathrm{d}}{\mathrm{d}\theta}\pi = \mathbf{0}$ for all $\theta \in D$, the second derivative is given by

$$\frac{\mathrm{d}^2}{\mathrm{d}\theta^2} H(Y;\theta) = -\frac{\mathrm{d}}{\mathrm{d}\theta}\sum_{y\in\mathcal{Y}} \pi^{\mathrm{T}} M'_\theta(y)\mathbf{1}\log\left(\pi^{\mathrm{T}} M_\theta(y)\mathbf{1}\right)$$

$$= -\sum_{y\in\mathcal{Y}} \pi^{\mathrm{T}} M''_\theta(y)\mathbf{1}\log\left(\pi^{\mathrm{T}} M_\theta(y)\mathbf{1}\right) - \sum_{y\in\mathcal{Y}} \frac{\left(\pi^{\mathrm{T}} M'_\theta(y)\mathbf{1}\right)^2}{\pi^{\mathrm{T}} M_\theta(y)\mathbf{1}}. \qquad \square$$

Now, we consider closed-form evaluation of Theorem 3.2. Since the first derivative is often zero at $\theta = \theta^*$, we are fortunate that a new formula for the second derivative can also be evaluated in closed form.

Theorem 3.6. *If there is a function $s(y)$, a $\theta^* \in D$, and a matrix P such that* $\lim_{\theta \to \theta^*} M(y) = s(y)P$ *for all $y \in \mathcal{Y}$, then*

$$\frac{\mathrm{d}}{\mathrm{d}\theta} H(\mathcal{Y};\theta)\Big|_{\theta=\theta^*} = -\sum_{y \in \mathcal{Y}} \pi^{\mathrm{T}} M'(y) \mathbf{1} \ln(s(y))$$

and

$$\frac{\mathrm{d}^2}{\mathrm{d}\theta^2} H(\mathcal{Y};\theta)\Bigg|_{\theta=\theta^*} = -\sum_{y \in \mathcal{Y}} \pi^{\mathrm{T}} M''(y) \mathbf{1} \ln(s(y)) - \sum_{y \in \mathcal{Y}} \frac{\left(\pi^{\mathrm{T}} M'(y) \mathbf{1}\right)^2}{\pi^{\mathrm{T}} M(y) \mathbf{1}}. \quad (10)$$

Proof. The proof is deferred to the appendix. $\qquad\square$

3.5 HMP example: a binary Markov-1 source with BSC noise

Consider the HMP defined by a binary Markov-1 source observed through a BSC(ε). The two-state Markov process is defined by $\Pr(Q_{t+1} = j \mid Q_t = i) = p_{ij}$ with stationary distribution $\Pr(Q_t = i) = \pi(i)$, and $\pi(0) = 1 - \pi(1) = \frac{1-p_{11}}{2-p_{00}-p_{11}}$. The output of the HMP is simply the observation of the state through a BSC or more specifically

$$h_{i,j}(y) = \begin{cases} 1 - \varepsilon & \text{if } y = i, \\ \varepsilon & \text{otherwise.} \end{cases}$$

The entropy rate of this process was considered earlier using a range of techniques [30, 31, 17, 44]. Now, we will consider the entropy rate of this process as $\varepsilon \to \frac{1}{2}$ (i.e., in the high-noise regime). This special case was also treated earlier and very similar results were obtained using different methods in [19, 18, 32].

Since we are interested in the high-noise regime, we start by analyzing the system using the upper bound $H(\mathcal{Y}) \leq H(Y)$. This gives

$$H(Y) = -\sum_{y \in \mathcal{Y}} \Pr(Y = y) \ln(\Pr(Y = y)),$$

where

$$\Pr(Y = 0) = \pi(0)p_{00}(1-\varepsilon) + \pi(0)p_{01}\varepsilon + \pi(1)p_{10}(1-\varepsilon) + \pi(1)p_{11}\varepsilon,$$

$$\Pr(Y = 1) = \pi(0)p_{00}\varepsilon + \pi(0)p_{01}(1-\varepsilon) + \pi(1)p_{10}\varepsilon + \pi(1)p_{11}(1-\varepsilon).$$

Using the Taylor expansion of $H(Y;\theta)$ around $\theta = \frac{1}{2} - \varepsilon$, we find that

$$H(\mathcal{Y}) \le H(Y;\theta) = \ln 2 - \frac{4(p_{00}^2 - p_{11}^2)}{(2 - p_{00} - p_{11})^2} \frac{\theta^2}{2} + O\left(\theta^4\right). \tag{11}$$

To calculate this expansion exactly for $H(\mathcal{Y})$, we apply Theorem 3.6. The conditions of the theorem are satisfied because

$$M_\theta(y) = \begin{cases} \begin{bmatrix} p_{00}(1-\varepsilon) & p_{01}\varepsilon \\ p_{10}(1-\varepsilon) & p_{11}\varepsilon \end{bmatrix} & \text{if } y=0, \\[12pt] \begin{bmatrix} p_{00}\varepsilon & p_{01}(1-\varepsilon) \\ p_{10}\varepsilon & p_{11}(1-\varepsilon) \end{bmatrix} & \text{if } y=1, \end{cases}$$

$$= \begin{cases} \frac{1}{2}\begin{bmatrix} p_{00} & p_{01} \\ p_{10} & p_{11} \end{bmatrix} + \theta \begin{bmatrix} p_{00} & -p_{01} \\ p_{10} & -p_{11} \end{bmatrix} & \text{if } y=0, \\[12pt] \frac{1}{2}\begin{bmatrix} p_{00} & p_{01} \\ p_{10} & p_{11} \end{bmatrix} - \theta \begin{bmatrix} p_{00} & -p_{01} \\ p_{10} & -p_{11} \end{bmatrix} & \text{if } y=1, \end{cases}$$

implies that $M_\theta(0) = M_\theta(1)$ at $\theta = 0$ (i.e., $\varepsilon = \frac{1}{2}$). Computing (10), which is simplified by the symmetry of $M_\theta(y)$ and the fact that $M_\theta''(y)$ is the zero matrix, gives

$$\left. \frac{\mathrm{d}^2}{\mathrm{d}\theta^2} H(\mathcal{Y};\theta) \right|_{\theta=0} = -2 \frac{\left(\begin{bmatrix} \frac{1-p_{11}}{2-p_{00}-p_{11}} & \frac{1-p_{00}}{2-p_{00}-p_{11}} \end{bmatrix} \begin{bmatrix} p_{00} & -p_{01} \\ p_{10} & -p_{11} \end{bmatrix} \begin{bmatrix} 1 \\ 1 \end{bmatrix} \right)^2}{\frac{1}{2}\begin{bmatrix} \frac{1-p_{11}}{2-p_{00}-p_{11}} & \frac{1-p_{00}}{2-p_{00}-p_{11}} \end{bmatrix} \begin{bmatrix} p_{00} & p_{01} \\ p_{10} & p_{11} \end{bmatrix} \begin{bmatrix} 1 \\ 1 \end{bmatrix}}$$

$$= -\frac{4(p_{00}^2 - p_{11}^2)}{(2 - p_{00} - p_{11})^2}. \tag{12}$$

Since $H(\mathcal{Y};0) = \ln 2$, this implies that the upper bound is tight with respect to the first nonzero term in the high-noise expansion.

3.6 Further example: a conditionally Gaussian HMP

Consider an HMP where the output distribution, conditioned on the state of the underlying Markov chain, is Gaussian. For example, this could be the output of an ISI channel with additive white Gaussian noise (AWGN). Suppose that the Gaussian associated with the transition from state i to state j has mean $\theta \cdot m_{ij}$

and variance 1; then this implies that $h_{ij}(y) = (1/\sqrt{2\pi})e^{-(y-\theta m_{ij})^2/2}$. Since the HMP loses state dependence as $\theta \to 0$, we first consider the derivatives with respect to θ of the single-letter entropy

$$H(Y;\theta) = -\int_{-\infty}^{\infty} \pi^{\mathrm{T}} M_{\theta}(y) \mathbf{1} \log\left(\pi^{\mathrm{T}} M_{\theta}(y) \mathbf{1}\right) \mathrm{d}y.$$

In this case, the stationary distribution does not depend on θ, so translating Lemma 3.5 to the continuous-alphabet case gives

$$\frac{\mathrm{d}}{\mathrm{d}\theta} H(Y;\theta)\Big|_{\theta=0} = -\lim_{\theta \to 0} \int_{-\infty}^{\infty} \pi^{\mathrm{T}} M_{\theta}'(y) \mathbf{1} \log\left(\pi^{\mathrm{T}} M_{\theta}(y) \mathbf{1}\right) \mathrm{d}y$$

$$= -\lim_{\theta \to 0} \int_{-\infty}^{\infty} \sum_{i,j \in \mathcal{Q}} \pi(i) p_{ij} \frac{1}{\sqrt{2\pi}} e^{-(y-\theta m_{ij})^2/2} m_{ij}(y - \theta m_{ij})$$

$$\times \log\left(\sum_{k,l \in \mathcal{Q}} \pi(k) p_{kl} \frac{1}{\sqrt{2\pi}} e^{-(y-\theta m_{kl})^2/2}\right) \mathrm{d}y$$

$$= -\int_{-\infty}^{\infty} \sum_{i,j \in \mathcal{Q}} \pi(i) p_{ij} \frac{1}{\sqrt{2\pi}} e^{-y^2/2} m_{ij} y \log\left(\frac{1}{\sqrt{2\pi}} e^{-y^2/2}\right) \mathrm{d}y$$

$$= -\int_{-\infty}^{\infty} \sum_{i,j \in \mathcal{Q}} \pi(i) p_{ij} \frac{1}{\sqrt{2\pi}} e^{-y^2/2} m_{ij} \left[y \log\left(\frac{1}{\sqrt{2\pi}}\right) - \frac{y^3}{2}\right] \mathrm{d}y$$

$$= 0,$$

because the odd moments of a zero-mean Gaussian are zero. Likewise, the formula for the second derivative (9) can be translated into

$$\frac{\mathrm{d}^2}{\mathrm{d}\theta^2} H(Y;\theta) = -\int_{-\infty}^{\infty} \pi^{\mathrm{T}} M_{\theta}''(y) \mathbf{1} \log\left(\pi^{\mathrm{T}} M_{\theta}(y) \mathbf{1}\right) \mathrm{d}y - \int_{-\infty}^{\infty} \frac{\left(\pi^{\mathrm{T}} M_{\theta}'(y) \mathbf{1}\right)^2}{\pi^{\mathrm{T}} M_{\theta}(y) \mathbf{1}} \mathrm{d}y.$$

The second term T_2 of the expression for $\frac{\mathrm{d}^2}{\mathrm{d}\theta^2} H(Y;\theta)\big|_{\theta=0}$ is given by

$$T_2 = -\lim_{\theta \to 0} \int_{-\infty}^{\infty} \frac{\left(\sum_{i,j \in \mathcal{Q}} \pi(i) p_{ij} \frac{1}{\sqrt{2\pi}} e^{-(y-\theta m_{ij})^2/2} m_{ij}(y - \theta m_{ij})\right)^2}{\sum_{i,j \in \mathcal{Q}} \pi(i) p_{ij} \frac{1}{\sqrt{2\pi}} e^{-(y-\theta m_{ij})^2/2}} \mathrm{d}y$$

$$= -\int_{-\infty}^{\infty} \frac{\left(\sum_{i,j \in \mathcal{Q}} \pi(i) p_{ij} \frac{1}{\sqrt{2\pi}} e^{-y^2/2} m_{ij} y\right)^2}{\frac{1}{\sqrt{2\pi}} e^{-y^2/2}} \mathrm{d}y$$

$$= -\left(\sum_{i,j \in \mathcal{Q}} \pi(i) p_{ij} m_{ij} \right)^2 \int_{-\infty}^{\infty} \frac{1}{\sqrt{2\pi}} e^{-y^2/2} y^2 \, dy$$

$$= -\left(\sum_{i,j \in \mathcal{Q}} \pi(i) p_{ij} m_{ij} \right)^2.$$

Using the fact that

$$\pi^{\mathrm{T}} M_{\theta}''(y) \mathbf{1} = \sum_{i,j \in \mathcal{Q}} \pi(i) p_{ij} \frac{1}{\sqrt{2\pi}} e^{-(y-\theta m_{ij})^2/2} m_{ij}^2 \left[(y - \theta m_{ij})^2 - 1 \right],$$

we can write the first term T_1 of the expression for $\frac{d^2}{d\theta^2} H(Y;\theta)\big|_{\theta=0}$ as

$$T_1 = -\lim_{\theta \to 0} \int_{-\infty}^{\infty} \pi^{\mathrm{T}} M_{\theta}''(y) \mathbf{1} \log \left(\pi^{\mathrm{T}} M_{\theta}(y) \mathbf{1} \right) dy$$

$$= -\int_{-\infty}^{\infty} \sum_{i,j \in \mathcal{Q}} \pi(i) p_{ij} \frac{1}{\sqrt{2\pi}} e^{-y^2/2} m_{ij}^2 \left(y^2 - 1 \right) \log \left(\frac{1}{\sqrt{2\pi}} e^{-y^2/2} \right) dy$$

$$\overset{(a)}{=} \frac{1}{2} \sum_{i,j \in \mathcal{Q}} \pi(i) p_{ij} m_{ij}^2 \int_{-\infty}^{\infty} \frac{1}{\sqrt{2\pi}} e^{-y^2/2} \left(y^4 - y^2 \right) dy$$

$$= \sum_{i,j \in \mathcal{Q}} \pi(i) p_{ij} m_{ij}^2,$$

where (a) follows from the fact that the fourth moment of a standard Gaussian is 3.

Comparing Lemma 3.5 with Theorem 3.6 shows that the first two terms in the expansion of $H(Y;\theta)$ match the first two terms in the expansion of $H(\mathcal{Y};\theta)$ at $\theta = 0$. Therefore, we have

$$\frac{d^2}{d\theta^2} H(\mathcal{Y};\theta)\bigg|_{\theta=0} = \frac{d^2}{d\theta^2} H(Y;\theta)\bigg|_{\theta=0} = \sum_{i,j \in \mathcal{Q}} \pi(i) p_{ij} m_{ij}^2 - \left(\sum_{i,j \in \mathcal{Q}} \pi(i) p_{ij} m_{ij} \right)^2.$$

$$\tag{13}$$

4 Application: high-noise capacity expansions for FSCs

4.1 The derivative of capacity for an FSC

Now, we will use the previous result to compute the derivative of the capacity. The mutual information $I(X;Y)$ between the random variables X and Y is defined by $I(X;Y) \triangleq H(Y) - H(Y|X)$, where the conditional entropy is defined by $H(Y|X) \triangleq H(X,Y) - H(X)$. Since the mutual information depends on the input distribution, the capacity is defined to be the supremum of the mutual information over all input distributions [11]. Therefore, some care must be taken when expressing the derivative of the capacity in terms of the derivative of the mutual information.

Consider a family of FSCs whose entropy rate is differentiable with respect to some parameter θ. Let the input distribution be Markov with memory m (e.g., defined by the vector $\vec{P}$ containing $|\mathcal{X}|^{m+1}$ values) and the optimal input distribution be $\vec{P}(\theta)$. In this case, we let the mutual information rate be $\mathcal{I}(\theta, \vec{P})$ and the Markov-m capacity be $\mathcal{C}(\theta) = \mathcal{I}(\theta, \vec{P}(\theta))$.

Lemma 4.1. *The derivative of the Markov-m capacity is given by*

$$\frac{\mathrm{d}}{\mathrm{d}\theta}\mathcal{C}(\theta) = \frac{\mathrm{d}}{\mathrm{d}\theta}\mathcal{I}(\theta, \vec{P}(\theta)) = \mathcal{I}'_\theta(\theta, \vec{P}(\theta)), \tag{14}$$

where $\mathcal{I}'_\theta(\theta, \vec{P}(\theta))$ is the derivative (with respect to θ) of the mutual information rate evaluated at the capacity-achieving input distribution for θ.

Proof. Expanding the derivative of $\mathcal{C}(\theta)$ in terms of $\mathcal{I}'_\theta(\theta, \vec{P})$ and the gradient vector $\mathcal{I}'_P(\theta, \vec{P})$ (with respect to input distribution) gives

$$\mathrm{d}\mathcal{I}(\theta, \vec{P}) = \mathcal{I}'_\theta(\theta, \vec{P})\mathrm{d}\theta + \mathcal{I}'_P(\theta, \vec{P}) \cdot \mathrm{d}\vec{P}.$$

The optimality of $\vec{P}(\theta)$ implies that $\mathcal{I}'_P(\theta, \vec{P}(\theta)) \cdot \mathrm{d}\vec{P} = 0$ for any $\mathrm{d}\vec{P}$ satisfying $\mathrm{d}\vec{P} \cdot \mathbf{1} = 0$ (i.e., the sum of $\vec{P}(\theta)$ is a constant). So, the derivative of the capacity is the derivative of the mutual information rate and we have (14). $\square$

Corollary 4.2. *If there is a "high-noise" point $\theta^* \in D$ where the Markov-m capacity satisfies $\mathcal{C}(\theta^*) = 0$ and $\mathcal{C}'(\theta^*) = 0$, then*

$$\frac{\mathrm{d}^2}{\mathrm{d}\theta^2}\mathcal{C}(\theta)\bigg|_{\theta=\theta^*} = \mathcal{I}''_\theta(\theta, \vec{P}(\theta)),$$

where $\mathcal{I}''_\theta(\theta, \vec{P}(\theta))$ is the second derivative (with respect to θ) of the mutual information rate evaluated at the capacity-achieving input distribution for θ.

Proof. First, we write the second derivative as

$$\frac{\mathrm{d}^2}{\mathrm{d}\theta^2}\mathcal{C}(\theta)\Big|_{\theta=\theta^*} = \lim_{\theta\to\theta^*}\frac{\mathrm{d}}{\mathrm{d}\theta}\mathcal{I}'_\theta(\theta,\vec{P}(\theta))$$

$$= \mathcal{I}''_\theta(\theta^*,\vec{P}(\theta^*)) + \lim_{\theta\to\theta^*}\left[\frac{\mathrm{d}}{\mathrm{d}\vec{P}}\mathcal{I}'_\theta(\theta,\vec{P})\right]_{\vec{P}=\vec{P}(\theta^*)}\cdot\vec{P}'(\theta^*).$$

Now, recall that $\mathcal{I}'_\theta(\theta^*,\vec{P}(\theta^*)) = 0$ and suppose that the second term is positive. In this case, a small change in $\vec{P}$ in the direction $\vec{P}'(\theta^*)$ must give an $\mathcal{I}'_\theta(\theta^*,\vec{P}) > 0$. But, this contradicts the fact that

$$0 = \mathcal{C}'(\theta^*) \geq \max_{\vec{P}}\mathcal{I}'_\theta(\theta^*,\vec{P}).$$

Therefore, the second term must be zero. $\qquad\square$

If the domain of θ includes a "high-noise" point θ^* where the channel output provides no information about the channel state, then Theorem 3.6 shows that the first two θ-derivatives of the entropy rate $H(\mathcal{Y};\theta)$ can be calculated at $\theta = \theta^*$. In fact, one also sees that they match the first two θ-derivatives of the single-letter entropy $H(Y;\theta)$ at $\theta = \theta^*$. Using Lemma 4.1 and Corollary 4.2, we see that these derivatives also equal the derivative of the Markov-m capacity in this case. But this equality holds for all m, so we can take a limit to see that it must hold also for the true capacity [10]. Even without this, however, we can use the fact that $H(\mathcal{Y};\theta) \leq H(Y;\theta)$ to upper bound the maximum entropy rate over all input distributions.

4.2 FSC example: a BSC with an RLL constraint

Consider the FSC defined by the BSC(ε) with a $(0, 1)$ run-length-limited (RLL) constraint [23]. This is a standard binary symmetric channel with a constraint that the input cannot have two ones in a row (e.g., this requires a two-state input process). The two-state input process is defined by $\Pr(X_{t+1} = j \mid X_t = i) = p_{ij}$ with $p_{11} = 0$, $\Pr(X_t = i) = \pi(i)$, and $\pi(0) = 1 - \pi(1) = 1/(2 - p_{00})$.

The mutual information rate between the input and output satisfies

$$I(\mathcal{X};\mathcal{Y}) = H(\mathcal{Y}) - H(\mathcal{Y}|\mathcal{X})$$

$$\leq H(Y_i) - h(\varepsilon),$$

where $h(\varepsilon) = -\varepsilon\ln\varepsilon - (1-\varepsilon)\ln(1-\varepsilon)$ is the binary entropy function in nats. Now, we can let $\theta = \frac{1}{2} - \varepsilon$ and combine the entropy-rate expansion from

(11) with the fact that $h(\frac{1}{2} - \theta) = \ln 2 - 2\theta^2 + O(\theta^4)$. The resulting high-noise expansion for the upper bound is

$$I(\mathcal{X}; \mathcal{Y}) \leq \frac{8(1 - p_{00})}{(2 - p_{00})^2} \theta^2 + O\left(\theta^4\right).$$

Notice that the leading coefficient achieves a unique maximum value of $2/\ln 2$ at $p_{00} = 0$. Since this upper bound only depends on the single-letter probabilities, it cannot be increased by extending the memory of the input process.

To see that this rate is achievable, we apply Theorem 3.6 to our system. Taking the result from (12), we find that

$$
\begin{aligned}
I(\mathcal{X}; \mathcal{Y}) &= H(\mathcal{Y}) - H(\mathcal{Y}|\mathcal{X}) \\
&= \left(1 - \frac{2p_{00}^2}{(2 - p_{00})^2} \theta^2 + o\left(\theta^2\right)\right) - \left(1 - 2\theta^2 + O\left(\theta^4\right)\right) \\
&= \frac{8(1 - p_{00})}{(2 - p_{00})^2} \theta^2 + o\left(\theta^2\right).
\end{aligned}
$$

So, the leading term of the actual expansion matches the upper bound.

From a coding perspective, this result implies that we should choose our Shannon random codebook to be sequences with mostly alternating 01 patterns and an occasional 00 pattern (i.e., it occurs with probability $p_{00} \to 0$). It is also worth mentioning that this constraint costs nothing when the noise is large because the slope of the expansion matches the slope of the unconstrained BSC as $p_{00} \to 0$.

4.3 FSC example: intersymbol-interference channels in AWGN

Consider a family of finite-memory ISI channels parameterized by θ. Let the time-t output Y_t be a Gaussian whose mean is given by θ times a deterministic function of the current input and the previous k inputs. Under these conditions, the output process is a conditionally Gaussian HMP, with state $Q_t = (X_{t-1}, \ldots, X_{t-k})$, as defined in Section 3.6. Moreover, the conditional entropy rate $H(\mathcal{Y}|\mathcal{X})$ only depends on the noise variance, which can be taken to be 1 without loss of generality. Therefore, θ-derivatives of the mutual information rate, $I(\mathcal{X}; \mathcal{Y}) = H(\mathcal{Y}) - H(\mathcal{Y}|\mathcal{X})$, depend only on θ-derivatives of the entropy rate $H(\mathcal{Y})$.

Let the mean of the output process induced by a state transition $Q_t = i$ to $Q_{t+1} = j$ be m_{ij}. One can explore the high-noise regime by keeping the noise variance fixed to 1 and letting $\theta \to 0$. In this case, one can combine (13) and

Corollary 4.2 to see that

$$
\mathcal{C}(\theta) = \frac{\theta^2}{2}\left[\sum_{i,j\in\mathcal{Q}}\pi(i)p_{ij}m_{ij}^2 - \left(\sum_{i,j\in\mathcal{Q}}\pi(i)p_{ij}m_{ij}\right)^2\right] + n\left(\theta^2\right)
$$

The first term in this expansion can be optimized over the input distribution p_{ij}, but there are a few caveats. Let $e_{ij} = \pi(i)p_{ij}$ be the edge occupancy probabilities that satisfy $\sum_{i,j\in\mathcal{Q}} e_{ij} = 1$; then stationarity of the underlying Markov chain implies that $\sum_j(e_{ij} - e_{ji}) = 0$. One also finds that not all state transitions are valid, but setting $e_{ij} = 0$ if $(i,j) \notin \mathcal{V}$ gives the following convex[4] optimization problem with linear constraints:

$$
\text{maximize} \quad \sum_{i,j\in\mathcal{Q}} e_{ij}m_{ij}^2 - \left(\sum_{i,j\in\mathcal{Q}} e_{ij}m_{ij}\right)^2
$$

$$
\text{subject to} \quad \sum_{i,j\in\mathcal{Q}} e_{ij} = 1,
$$

$$
\sum_{j\in\mathcal{Q}}(e_{ij} - e_{ji}) = 0 \quad \forall i.
$$

A similar result is given in [40] for linear ISI channels with balanced inputs (i.e., a zero-mean input). In this case, the $\sum e_{ij}m_{ij}$ term is zero and the optimization problem is reduced to finding the maximum mean-weight cycle in a directed graph with edge weights m_{ij}^2. The formula above generalizes the previous result to nonlinear ISI channels and eliminates the zero-mean input requirement.

5 Connection to the formula of Vontobel *et al.*

The results of this article are closely related to an observation by Vontobel *et al.* [41] that the first part of the generalized Blahut–Arimoto algorithm [11] for FSCs actually computes the derivative of the mutual information [41]. Their result is somewhat different because it considers derivatives with respect to the edge occupancy probabilities $\pi(i)p_{ij}$ rather than the observation probabilities. Their approach is also dissimilar because the answer is given exactly for finite blocks rather than focusing on the asymptotically long blocks and the

---

[4] The objective function is actually concave, but one can negate the objective and minimize instead.

forward/backward stationary measures. Moreover, the result in this article does not apply to changes in the HMP which change the stationary distribution π, while the derivative result in [41] focuses exclusively on changes in the edge occupancy probabilities.

Ideally, one would have a unified treatment of the derivative, with respect to changes in both the edge occupancy probabilities $\pi(i)p_{ij}$ and the observation probabilities, of the entropy rate of an FSC. Indeed, a simple formula, in terms of forward/backward stationary measures, can be cobbled together by translating the derivative formula in [41] to stationary measures and combining this with Theorem 3.2. To clarify the connection, their result is shown first in terms of conditional density functions for α and β. Paraphrasing their result, in terms of the derivative of the edge occupancy probabilities

$$\Delta_{ij} = \frac{\mathrm{d}}{\mathrm{d}\theta}\pi(i)p_{ij}\big|_{\theta=0},$$

gives

$$\frac{\mathrm{d}}{\mathrm{d}\theta}H(\mathcal{X}|\mathcal{Y};\theta)\big|_{\theta=0} = -\sum_{i,j\in\mathcal{Q}}\Delta_{ij}\int_{\mathcal{A}_0\times\mathcal{A}_0}f_{\alpha|Q_t}(\alpha|i)f_{\beta|Q_{t+1}}(\beta|j)$$

$$\times\sum_{y\in\mathcal{Y}}h_{ij}(y)\ln\frac{\alpha(i)M_{ij}(y)\beta(j)}{\sum_{k\in\mathcal{Q}}\alpha(i)M_{ik}(y)\beta(k)}\,\mathrm{d}\alpha\,\mathrm{d}\beta.$$

One can decompose this formula to see that the term Δ_{ij} gives the change in the edge occupancy probability, the term $f_{\alpha|Q_t}(\alpha|i)f_{\beta|Q_{t+1}}(\beta|j)f_{Y|Q_tQ_{t+1}}(y|i,j)$ is the probability of α,β,y given the transition, and the logarithmic term gives the contribution to $H\left(Q_{t+1}=j|Q_t=i,Y_{-\infty}^{\infty}\right)$ for this α,β,y.

Next, we modify this expression to use unconditional α,β distributions with

$$\frac{\mathrm{d}}{\mathrm{d}\theta}H(\mathcal{X}|\mathcal{Y};\theta)\big|_{\theta=0} \overset{(a)}{=} -\sum_{i,j\in\mathcal{Q}}\Delta_{ij}\int_{\mathcal{A}_0\times\mathcal{A}_0}\frac{\mu_i(\mathrm{d}\alpha)}{\pi(i)}\cdot\frac{\nu_j(\mathrm{d}\beta)}{\pi(j)}$$

$$\times\sum_{y\in\mathcal{Y}}h_{ij}(y)\ln\frac{\alpha(i)M_{ij}(y)\beta(j)}{\sum_{k\in\mathcal{Q}}\alpha(i)M_{ik}(y)\beta(k)}$$

$$\overset{(b)}{=} -\sum_{i,j\in\mathcal{Q}}\Delta_{ij}\int_{\mathcal{A}_0\times\mathcal{A}_0}\frac{\mu(\mathrm{d}\alpha)\alpha(i)}{\pi(i)}\cdot\frac{\nu(\mathrm{d}\beta)\beta(j)\pi(j)}{\pi(j)}$$

$$\times\sum_{y\in\mathcal{Y}}h_{ij}(y)\ln\frac{\alpha(i)M_{ij}(y)\beta(j)}{\sum_{k\in\mathcal{Q}}\alpha(i)M_{ik}(y)\beta(k)}$$

$$\overset{(c)}{=} - \sum_{i,j \in \mathcal{Q}} \Delta_{ij} \int_{\mathcal{A}_0 \times \mathcal{A}_0} \mu(\mathrm{d}\alpha)\nu(\mathrm{d}\beta)$$

$$\times \sum_{y \in \mathcal{Y}} \frac{\alpha(i)M_{ij}(y)\beta(j)}{\pi(i)p_{ij}} \ln \frac{\alpha(i)M_{ij}(y)\beta(j)}{\sum_{k \in \mathcal{Q}} \alpha(i)M_{ik}(y)\beta(k)},$$

where (a) holds because $\mu_i(\mathrm{d}\alpha)/\pi(i)$ is the conditional density of α given the true state is i and $\nu_j(\mathrm{d}\beta)/\pi(j)$ is the conditional density of β given the true state is j, (b) follows from Lemmas 2.21 and 2.25, and (c) follows from $M_{ij}(y) = p_{ij}h_{ij}(y)$. Finally, using $H(\mathcal{Y};\theta) = H(\mathcal{X};\theta) - H(\mathcal{X}|\mathcal{Y};\theta) + H(\mathcal{Y}|\mathcal{X};\theta)$ and

$$\frac{\mathrm{d}}{\mathrm{d}\theta} H(\mathcal{X};\theta)\Big|_{\theta=0} = - \sum_{i,j \in \mathcal{Q}} \Delta_{ij} \ln p_{ij},$$

$$\frac{\mathrm{d}}{\mathrm{d}\theta} H(\mathcal{Y}|\mathcal{X};\theta)\Big|_{\theta=0} = - \sum_{i,j \in \mathcal{Q}} \Delta_{ij} \sum_{y \in \mathcal{Y}} h_{ij}(y) \ln h_{ij}(y),$$

we find that

$$\frac{\mathrm{d}}{\mathrm{d}\theta} H(\mathcal{Y};\theta)\Big|_{\theta=0}$$

is given by

$$- \sum_{i,j \in \mathcal{Q}} \Delta_{ij} \int_{\mathcal{A}_0 \times \mathcal{A}_0} \mu(\mathrm{d}\alpha)\nu(\mathrm{d}\beta)$$

$$\times \sum_{y \in \mathcal{Y}} \left[h_{ij}(y) \ln M_{ij}(y) + \frac{\alpha(i)M_{ij}(y)\beta(j)}{\pi(i)p_{ij}} \ln \frac{\alpha(i)M_{ij}(y)\beta(j)}{\sum_{k \in \mathcal{Q}} \alpha(i)M_{ik}(y)\beta(k)} \right].$$

It is straightforward to combine this with Theorem 3.2, though the final expression is even more unwieldy.

6 Conclusions

This article considers the derivative of the entropy rate for general hidden Markov processes and derives a closed-form expression for this derivative in the high-noise limit. An application is presented relating to the achievable information rates of finite-state channels. Again, a closed-form expression is derived for the high-noise limit. Two examples of interest are considered.

First, transmission over a BSC under a (0, 1) RLL constraint is treated and the capacity-achieving input distribution is derived in the high-noise limit. Second, an intersymbol interference channel in AWGN is considered and the capacity is derived in the high-noise limit.

Acknowledgments. The author would like to thank an anonymous reviewer for pointing out a number of errors and inconsistencies in the article. He is also grateful to Pascal Vontobel for his excellent comments on an earlier draft. This work also benefited from interesting discussions with Brian Marcus and is a natural extension of past work with Paul H. Siegel and Joseph B. Soriaga.

This research was supported in part by the National Science Foundation under Grant No. 0747470.

A Technical details

A.1 Lemmas for Theorem 3.2

Lemma A.1. *Consider the function $F(\alpha, \beta) = -\alpha^{\mathrm{T}} M' \beta \log (\alpha^{\mathrm{T}} M \beta)$ where M is a nonnegative matrix and M' is a real matrix. This function is Lipschitz continuous with respect to $\|\cdot\|_1$ on $(\alpha, \beta) \in \mathcal{P}_\delta \times \mathcal{B}_\delta$, where $\mathcal{B}_\delta = \{u \in \mathcal{A}_\delta \mid \pi^{\mathrm{T}} \beta = 1\}$, $\eta = \min_i \pi(i) > 0$, and $\delta > 0$. This implies that*

$$\left| F(\alpha, \beta) - F(\alpha', \beta) \right| \le L_\alpha \left\| \alpha - \alpha' \right\|_1,$$

$$\left| F(\alpha, \beta) - F(\alpha, \beta') \right| \le L_\beta \left\| \beta - \beta' \right\|_1,$$

$$\left| F(\alpha, \beta) - F(\alpha', \beta') \right| \le L_\alpha \left\| \alpha - \alpha' \right\|_1 + L_\beta \left\| \beta - \beta' \right\|_1,$$

where $c = \delta^2 \sum_{i,j} M_{ij}$ and

$$L_\alpha = \|M\|_1 \frac{1}{\eta} \log \frac{1}{c} + \|M'\|_1 \|M\|_1 \frac{1}{\eta^2 c},$$

$$L_\beta = \|M\|_\infty \log \frac{1}{c} + \|M'\|_\infty \|M\|_\infty \frac{1}{c}.$$

Proof. Let $G : \mathbb{R}^m \to \mathbb{R}$ be any function that is differentiable on a convex set $D \subseteq \mathbb{R}^m$. Then, the mean value theorem of vector calculus implies that

$$G(y) - G(x) = G'(x + t(y - x))^{\mathrm{T}} (y - x)$$

for some $t \in [0, 1]$. Applying Hölder's inequality allows one to upper bound the Lipschitz constant with respect to $\|\cdot\|_1$ and gives the upper bound

$$G(y) - \bar{G}(x) \le \sup_{t \in [0,1]} \|G''(x + t(y-x))\|_\infty \|x - y\|_1$$

$$\le \sup_{z \in D} \|G'(z)\|_\infty \|x - y\|_1.$$

Since $F(\alpha, \beta)$ is differentiable with respect to α, we can bound the Lipschitz constant L_α with

$$
\begin{aligned}
L_\alpha &= \sup_{\alpha \in \mathcal{P}_\delta} \sup_{\beta \in \mathcal{B}} \sup_{\|u\|_\infty \le 1} \left| u^\mathsf{T} M' \beta \log \frac{1}{\alpha^\mathsf{T} M \beta} - \alpha^\mathsf{T} M' \beta \frac{u^\mathsf{T} M \beta}{\alpha^\mathsf{T} M \beta} \right| \\
&\overset{(a)}{\le} \sup_{\alpha \in \mathcal{P}_\delta} \sup_{\beta \in \mathcal{B}} \sup_{\|u\|_\infty \le 1} \left[|u^\mathsf{T} M' \beta| \log \frac{1}{c} + |\alpha^\mathsf{T} M' \beta| \, |u^\mathsf{T} M \beta| \frac{1}{c} \right] \\
&\overset{(b)}{\le} \|M\|_1 \|\beta\|_1 \log \frac{1}{c} + \|M'\|_1 \|\beta\|_1 \|M\|_1 \|\beta\|_1 \frac{1}{c} \\
&\overset{(c)}{\le} \|M\|_1 \frac{1}{\eta} \log \frac{1}{c} + \|M'\|_1 \|M\|_1 \frac{1}{\eta^2 c},
\end{aligned}
\tag{A.1}
$$

where (a) follows from $\alpha^\mathsf{T} M \beta \ge c$ with $c = \delta^2 \sum_{i,j} M_{ij}$, (b) follows from $|x^\mathsf{T} M y| \le \|x\|_\infty \|M\|_1 \|y\|_1$, and (c) follows from $\|\beta\|_1 \le \eta^{-1}$, which holds because $\pi^\mathsf{T} \beta = 1$.

Likewise, $F(\alpha, \beta)$ is differentiable with respect to β and we can bound the Lipschitz constant L_β with

$$
\begin{aligned}
L_\beta &= \sup_{\alpha \in \mathcal{P}_\delta} \sup_{\beta \in \mathcal{B}} \sup_{\|u\|_\infty \le 1} \left| \alpha^\mathsf{T} M' u \log \frac{1}{\alpha^\mathsf{T} M \beta} - \alpha^\mathsf{T} M' \beta \frac{\alpha^\mathsf{T} M u}{\alpha^\mathsf{T} M \beta} \right| \\
&\overset{(a)}{\le} \sup_{\alpha \in \mathcal{P}_\delta} \sup_{\beta \in \mathcal{B}} \sup_{\|u\|_\infty \le 1} \left[|\alpha^\mathsf{T} M' u| \log \frac{1}{c} + |\alpha^\mathsf{T} M' \beta| \, |\alpha^\mathsf{T} M u| \frac{1}{c} \right] \\
&\overset{(b)}{\le} \|M\|_\infty \log \frac{1}{c} + \|M'\|_\infty \|M\|_\infty \frac{1}{c},
\end{aligned}
\tag{A.2}
$$

where (a) is the same as above and (b) follows from $|x^\mathsf{T} M y| \le \|x\|_1 \|M\|_\infty \|y\|_\infty$. $\qquad \square$

Lemma A.2. *If the HMP is ϵ-primitive for $\epsilon > 0$, then for some $\gamma < 1$ and $C < \infty$ we have*

$$\sum_{y \in \mathcal{Y}} E \left[\alpha^{\mathrm{T}} M'(y)\beta \log(\alpha^{\mathrm{T}} M(y)\beta) - \alpha_j^{\mathrm{T}} M'(y)\beta_{j+1} \log(\alpha_j^{\mathrm{T}} M(y)\beta_{j+1}) \right]$$

$$\leq 2\overline{L}_\alpha C \gamma^{j-1} + 2\overline{L}_\beta C \gamma^{n-j+1},$$

where $c(y) = \delta^2 \sum_{i,j} [M(y)]_{ij}$ and

$$\overline{L}_\alpha = \sum_{y \in \mathcal{Y}} \left[\|M(y)\|_1 \frac{1}{\eta} \log \frac{1}{c(y)} + \|M'(y)\|_1 \|M(y)\|_1 \frac{1}{\eta^2 c(y)} \right],$$

$$\overline{L}_\beta = \sum_{y \in \mathcal{Y}} \left[\|M(y)\|_\infty \log \frac{1}{c(y)} + \|M'(y)\|_\infty \|M(y)\|_\infty \frac{1}{c(y)} \right].$$

The expectation assumes that α, β are drawn from their respective stationary distributions while α_j, β_{j+1} are drawn from the distributions implied by an arbitrary initialization of α_1, β_{n+1}.

Proof. Since the HMP is ϵ-primitive for $\epsilon > 0$, there is a δ such that $\min_i \alpha_i > \delta$ and $\min_i \beta_i > \delta$ on the entire support of α, β. It also follows that $\eta = \min_i \pi(i) > 0$. Now, consider the function $F_y(\alpha, \beta) = -\alpha^{\mathrm{T}} M'(y)\beta \log(\alpha^{\mathrm{T}} M(y)\beta)$. Under these conditions, Lemma A.1 shows that this function is Lipschitz continuous with respect to $\|\cdot\|_1$ on the support of α, β with Lipschitz constants $L_\alpha(y)$ and $L_\beta(y)$ defined by generalizing (A.1) and (A.2). Therefore, we can write

$$\sum_{y \in \mathcal{Y}} E_{\alpha,\beta} \left[F_y(\alpha, \beta) - F_y(\alpha_j, \beta_{j+1}) \right]$$

$$\leq \sum_{y \in \mathcal{Y}} E_{\alpha,\beta} \left[L_\alpha(y) \|\alpha - \alpha_j\|_1 + L_\beta(y) \|\beta - \beta_{j+1}\|_1 \right]$$

$$\overset{(a)}{\leq} \sum_{y \in \mathcal{Y}} E_{\alpha,\beta} \left[L_\alpha(y) 2d(\alpha, \alpha_j) + L_\beta(y) 2d(\beta, \beta_{j+1}) \right]$$

$$\overset{(b)}{\leq} \sum_{y \in \mathcal{Y}} L_\alpha(y) 2C \gamma^{j-1} + \sum_{y \in \mathcal{Y}} L_\beta(y) 2C \gamma^{n-j+1},$$

where (a) follows from Lemma 2.4 and (b) follows from Lemma 2.12 because the HMP is ϵ-primitive. $\square$

A.2 Proof of Lemma 3.4

Proof. The first two results follow from Lemmas 2.21 and 2.25. Substituting and integrating gives

$$\int_{\mathcal{A}_0} \mu(\mathrm{d}\alpha)\alpha(q) = \int_{\mathcal{A}_0} \underbrace{\mu_q(\mathrm{d}\alpha)}_{\mathrm{Pr}(Q=q,\alpha\in\mathrm{d}\alpha)} = \mathrm{Pr}(Q=q)$$

and

$$\int_{\mathcal{A}_0} \nu(\mathrm{d}\beta)\beta(q) = \int_{\mathcal{A}_0} \frac{1}{\pi(q)} \underbrace{\nu_q(\mathrm{d}\beta)}_{\mathrm{Pr}(Q=q,\beta\in\mathrm{d}\beta)} = 1.$$

Using the fact that

$$\sum_{y\in\mathcal{Y}} M(y) = P,$$

we can evaluate the third and fourth results with

$$\int_{\mathcal{A}_0} \mu(\mathrm{d}\alpha)\alpha^{\mathrm{T}} \sum_{y\in\mathcal{Y}} M(y)\beta = \pi^{\mathrm{T}} P\beta = \pi^{\mathrm{T}}\beta = 1$$

and

$$\alpha^{\mathrm{T}} \sum_{y\in\mathcal{Y}} M(y) \int_{\mathcal{A}_0} \nu(\mathrm{d}\beta)\beta = \alpha^{\mathrm{T}} P\mathbf{1} = \alpha^{\mathrm{T}}\mathbf{1} = 1.$$

Finally, the fifth result follows from

$$\frac{\mathrm{d}}{\mathrm{d}\theta} \int_{\mathcal{A}_0} \mu(\mathrm{d}\alpha)\alpha^{\mathrm{T}} \sum_{y\in\mathcal{Y}} M_\theta(y) \int_{\mathcal{A}_0} \nu(\mathrm{d}\beta)\beta = \frac{\mathrm{d}}{\mathrm{d}\theta}\pi^{\mathrm{T}} P_\theta\mathbf{1} = \frac{\mathrm{d}}{\mathrm{d}\theta}\mathbf{1} = 0.$$

$\square$

A.3 Proof of Theorem 3.6

Proof. First, we point out that $\lim_{\theta\to\theta^*} M_\theta(y) = s(y)P$ implies that output symbols provide no state information at $\theta = \theta^*$ so that $H(\mathcal{Y};\theta^*) = H(Y_1;\theta^*)$. This also implies that, at $\theta = \theta^*$, the forward and backward Blackwell measures are Dirac measures, $\mu(A) = 1_A(\pi)$ and $\nu(B) = 1_B(\mathbf{1})$, concentrated on $\pi, \mathbf{1}$. By

Theorem 3.2, the derivative of the entropy rate is uniformly continuous on D and we have

$$\lim_{\theta \to \theta^*} \frac{\mathrm{d}}{\mathrm{d}\theta} H(\mathcal{Y};\theta) = - \lim_{\theta \to \theta^*} E_{\alpha,\beta}\left[\sum_{y \in \mathcal{Y}} \alpha^{\mathrm{T}} M_\theta'(y)\beta \ln\left(\alpha^{\mathrm{T}} M_\theta(y)\beta\right)\right]$$

$$= -\sum_{y \in \mathcal{Y}} \pi^{\mathrm{T}} M'(y)\mathbf{1} \ln(s(y)) - \pi^{\mathrm{T}}\left(\sum_{y \in \mathcal{Y}} M'(y)\right)\mathbf{1} \ln\left(\pi^{\mathrm{T}} P\mathbf{1}\right)$$

$$\overset{(a)}{=} -\sum_{y \in \mathcal{Y}} \pi^{\mathrm{T}} M'(y)\mathbf{1} \ln(s(y)),$$

where (a) holds because $\pi^{\mathrm{T}} P\mathbf{1} = 1$.

For the second derivative, we apply the derivative shortcut a second time by noting that

$$g_n''(\theta) = \sum_{i=1}^{n}\sum_{j=1}^{n} \frac{\partial}{\partial \theta_i}\frac{\partial}{\partial \theta_j} g_n(\theta).$$

Applying this to $g_n(\theta_1,\ldots,\theta_n)$ for the entropy rate gives

$$g_n''(\theta^*)$$

$$= -\frac{1}{n}\sum_{i=1}^{n}\sum_{j=1}^{n} \frac{\partial}{\partial \theta_i}\frac{\partial}{\partial \theta_j} \sum_{y_1^n \in \mathcal{Y}^n} \pi^{\mathrm{T}}\left(\prod_{t=1}^{n} M_{\theta_t}(y_t)\right)\mathbf{1}$$

$$\cdot \log\left[\pi^{\mathrm{T}}\left(\prod_{t=1}^{n} M_{\theta_t}(y_t)\right)\mathbf{1}\right]\Bigg|_{(\theta_1,\ldots,\theta_n)=(\theta^*,\ldots,\theta^*)}$$

$$\overset{(a)}{=} -\frac{1}{n}\sum_{i=1}^{n} \frac{\partial}{\partial \theta_i}\sum_{j=1}^{n}\sum_{y_1^n \in \mathcal{Y}^n} \pi^{\mathrm{T}} M(y_1^{j-1}) M_{\theta_j}'(y_j) M(y_{j+1}^n)\mathbf{1}$$

$$\cdot \log\left[\pi^{\mathrm{T}}\left(\prod_{t=1}^{n} M_{\theta_t}(y_t)\right)\mathbf{1}\right]\Bigg|_{(\theta_1,\ldots,\theta_n)=(\theta^*,\ldots,\theta^*)}$$

$$- \frac{1}{n}\sum_{i=1}^{n} \frac{\partial}{\partial \theta_i}\sum_{j=1}^{n}\frac{\partial}{\partial \theta_j}\sum_{y_1^n \in \mathcal{Y}^n} \pi^{\mathrm{T}}\left(\prod_{t=1}^{n} M_{\theta_t}(y_t)\right)\mathbf{1}\Bigg|_{(\theta_1,\ldots,\theta_n)=(\theta^*,\ldots,\theta^*)} \tag{A.3}$$

$$= -\frac{1}{n}\sum_{j=1}^{n}\sum_{y_1^n\in\mathcal{Y}^n}\pi^{\mathrm{T}}M(y_1^{j-1})M''(y_j)M(y_{j+1}^n)\mathbf{1}\cdot\log\left[\pi^{\mathrm{T}}\left(\prod_{t=1}^{n}M(y_t)\right)\mathbf{1}\right] \qquad (T_1)$$

$$-\frac{1}{n}\sum_{j=1}^{n}\sum_{y_1^n\in\mathcal{Y}^n}\frac{\left(\pi^{\mathrm{T}}M(y_1^{j-1})M'(y_j)M(y_{j+1}^n)\mathbf{1}\right)^2}{\pi^{\mathrm{T}}\left(\prod_{t=1}^{n}M(y_t)\right)\mathbf{1}} \qquad (T_2)$$

$$-\frac{2}{n}\sum_{j=1}^{n}\sum_{i=1}^{j-1}\sum_{y_1^n\in\mathcal{Y}^n}\pi^{\mathrm{T}}M(y_1^{i-1})M'(y_i)M(y_{i+1}^{j-1})M'(y_j)M(y_{j+1}^n)\mathbf{1}$$

$$\cdot\log\left[\pi^{\mathrm{T}}\left(\prod_{t=1}^{n}M_{\theta_t}(y_t)\right)\mathbf{1}\right] \qquad (T_3)$$

$$-\frac{2}{n}\sum_{j=1}^{n}\sum_{i=1}^{j-1}\sum_{y_1^n\in\mathcal{Y}^n}\frac{\left(\pi^{\mathrm{T}}M(y_1^{i-1})M'(y_i)M(y_{i+1}^n)\mathbf{1}\right)\left(\pi^{\mathrm{T}}M(y_1^{j-1})M'(y_j)M(y_{j+1}^n)\mathbf{1}\right)}{\pi^{\mathrm{T}}\left(\prod_{t=1}^{n}M(y_t)\right)\mathbf{1}},$$
$$\qquad (T_4)$$

where the term labeled (A.3) is zero because it equals $-(1/n)\mathrm{d}^2\mathbf{1}/\mathrm{d}^2\theta$. Using the term labels in the equation (i.e., T_1, T_2, ...), we see that $g_n''(\theta^*) = T_1 + T_2 + T_3 + T_4$, where the terms T_1, T_2 are associated with $i = j$, and the terms T_3, T_4 are associated with $i \neq j$. Using this decomposition, we can reduce each term separately.

For the first term, $M(y) = s(y)P$ implies that

$$T_1 = -\frac{1}{n}\sum_{j=1}^{n}\sum_{y_1^n\in\mathcal{Y}^n}\pi^{\mathrm{T}}M(y_1^{j-1})M''(y_j)M(y_{j+1}^n)\mathbf{1}\cdot\log\left[\pi^{\mathrm{T}}\left(\prod_{t=1}^{n}M(y_t)\right)\mathbf{1}\right]$$

$$= -\frac{1}{n}\sum_{j=1}^{n}\sum_{y_1^n\in\mathcal{Y}^n}\frac{s(y_1^n)}{s(y_j)}\pi^{\mathrm{T}}M''(y_j)\mathbf{1}\cdot\log\left(s(y_1^n)\right)$$

$$= -\frac{1}{n}\sum_{j=1}^{n}\sum_{y_1^n\in\mathcal{Y}^n}\left(\frac{s(y_1^n)}{s(y_j)}\pi^{\mathrm{T}}M''(y_j)\mathbf{1}\cdot\log\left(s(y_j)\right)\right.$$

$$\left.+\frac{s(y_1^n)}{s(y_j)}\pi^{\mathrm{T}}M''(y_j)\mathbf{1}\sum_{k=1,k\neq j}^{n}\log\left(s(y_k)\right)\right)$$

$$\overset{(a)}{=}-\frac{1}{n}\sum_{j=1}^{n}\left(\sum_{y_j\in\mathcal{Y}}\pi^{\mathrm{T}}M''(y_j)\mathbf{1}\cdot\log\left(s(y_j)\right)+0\right)$$

$$= -\sum_{y\in\mathcal{Y}}\pi^{\mathrm{T}}M''(y)\mathbf{1}\cdot\log\left(s(y)\right),$$

where (a) follows from the fact that

$$\sum_{y_j \in \mathcal{Y}} \frac{s(y_1^n)}{s(y_j)} \pi^{\mathsf{T}} M''(y_j) \mathbf{1} \sum_{k=1, k \neq j}^{n} \log(s(y_k))$$

$$= \left(\prod_{i=1, i \neq j}^{n} s(y_i) \right) \left(\sum_{k=1, k \neq j}^{n} \log(s(y_k)) \right) \sum_{y_j \in \mathcal{Y}} \pi^{\mathsf{T}} M''(y_j) \mathbf{1} = 0.$$

For the second term, $M(y) = s(y)P$ implies that

$$T_2 = -\frac{1}{n} \sum_{j=1}^{n} \sum_{y_1^n \in \mathcal{Y}^n} \frac{\left(\pi^{\mathsf{T}} M(y_1^{j-1}) M'(y_j) M(y_{j+1}^n) \mathbf{1} \right)^2}{\pi^{\mathsf{T}} \left(\prod_{t=1}^{n} M(y_t) \right) \mathbf{1}}$$

$$= -\frac{1}{n} \sum_{j=1}^{n} \sum_{y_1^n \in \mathcal{Y}^n} \frac{s(y_1^n)^2}{s(y_1^n) s(y_j)^2} \left(\pi^{\mathsf{T}} M'(y_j) \mathbf{1} \right)^2$$

$$= -\frac{1}{n} \sum_{j=1}^{n} \sum_{y_j \in \mathcal{Y}} \frac{\left(\pi^{\mathsf{T}} M'(y_j) \mathbf{1} \right)^2}{s(y_j)}$$

$$= -\sum_{y \in \mathcal{Y}} \frac{\left(\pi^{\mathsf{T}} M'(y) \mathbf{1} \right)^2}{s(y)}$$

$$= -\sum_{y \in \mathcal{Y}} \frac{\left(\pi^{\mathsf{T}} M'(y) \mathbf{1} \right)^2}{\pi^{\mathsf{T}} M(y) \mathbf{1}}.$$

For the third term, we notice first that $\sum_{y \in \mathcal{Y}} M'(y) = 0$ implies that

$$\sum_{y_i, y_j, y_k \in \mathcal{Y}^n} \pi^{\mathsf{T}} M'(y_i) P^{j-i-1} M'(y_j) \mathbf{1} \cdot \log(s(y_k)) = 0$$

if either $i \neq k$ or $j \neq k$. This gives

$$T_3 = -\frac{2}{n} \sum_{j=1}^{n} \sum_{i=1}^{j-1} \sum_{y_1^n \in \mathcal{Y}^n} \pi^{\mathsf{T}} M(y_1^{i-1}) M'(y_i) M(y_{i+1}^{j-1}) M'(y_j) M(y_{j+1}^n) \mathbf{1}$$

$$\cdot \log \left[\pi^{\mathsf{T}} \left(\prod_{t=1}^{n} M_{\theta_t}(y_t) \right) \mathbf{1} \right]$$

$$= -\frac{2}{n} \sum_{j=1}^{n} \sum_{i=1}^{j-1} \sum_{y_1^n \in \mathcal{Y}^n} \frac{s(y_1^n)}{s(y_i)s(y_j)} \pi^T M'(y_i) P^{j-i-1} M'(y_j) \mathbf{1} \cdot \log\left(s(y_1^n)\right)$$

$$= -\frac{2}{n} \sum_{k=1}^{n} \sum_{j=1}^{n} \sum_{i=1}^{j-1} \sum_{y_i,y_j,y_k \in \mathcal{Y}^n} \pi^T M'(y_i) P^{j-i-1} M'(y_j) \mathbf{1} \cdot \log\left(s(y_k)\right)$$

$$= 0$$

because $i < j$.

For the fourth term, we have

$$T_4 = -\frac{2}{n} \sum_{j=1}^{n} \sum_{i=1}^{j-1} \sum_{y_1^n \in \mathcal{Y}^n} \frac{\left(\pi^T M(y_1^{i-1}) M'(y_i) M(y_{i+1}^n) \mathbf{1}\right) \left(\pi^T M(y_1^{j-1}) M'(y_j) M(y_{j+1}^n) \mathbf{1}\right)}{\pi^T \left(\prod_{t=1}^{n} M(y_t)\right) \mathbf{1}}$$

$$= -\frac{2}{n} \sum_{j=1}^{n} \sum_{i=1}^{j-1} \sum_{y_1^n \in \mathcal{Y}^n} \frac{s(y_1^n)^2}{s(y_1^n) s(y_i) s(y_j)} \left(\pi^T M'(y_i) \mathbf{1}\right) \left(\pi^T M'(y_j) \mathbf{1}\right)$$

$$= -\frac{2}{n} \sum_{j=1}^{n} \sum_{i=1}^{j-1} \sum_{y_i,y_j \in \mathcal{Y}^n} \left(\pi^T M'(y_i) \mathbf{1}\right) \left(\pi^T M'(y_j) \mathbf{1}\right)$$

$$= 0$$

because $\sum_{y \in \mathcal{Y}} M'(y) = 0$.   □

References

[1] D. Arnold and H. Loeliger. *On the information rate of binary-input channels with memory*. In Proc. IEEE Int. Conf. Communications, Helsinki, Finland, June 2001, pp. 2692–2695

[2] D. Arnold, H. A. Loeliger, P. O. Vontobel, A. Kavčić, and W. Zeng. *Simulation-based computation of information rates for channels with memory*. IEEE Trans. Inf. Theory **52**(8) (2006) 3498–3508

[3] D. M. Arnold. *Computing Information Rates of Finite-state Models with Application to Magnetic Recording*. PhD thesis, Swiss Federal Institute of Technology, Zurich, 2003

[4] R. G. Bartle and D. R. Sherbert. *Introduction to Real Analysis*, 3rd edn. Wiley, New York, 1999

[5] L. E. Baum and T. Petrie. *Statistical inference for probabilistic functions of finite state Markov chains*. Ann. Math. Statist. **37** (1966) 1554–1563

[6] P. Billingsley. *Convergence of Probability Measures*, 2nd edn. Wiley Interscience, New York, 1999

[7] J. Birch. *Approximations for the entropy for functions of Markov chains*. Ann. Math. Statist. **33**(3) (1962) 930–938

[8] D. Blackwell. *Entropy of functions of finite-state Markov chains*. In Trans. First Prague Conf. Information Theory, Statistical Decision Functions, Random Processes, 1957, pp. 13–20

[9] D. Blackwell, L. Breiman, and A. J. Thomasian. *Proof of Shannon's transmission theorem for finite-state indecomposable channels*. Ann. Math. Statist. **29** (1958) 1209–1220

[10] J. Chen and P. H. Siegel. *Markov processes asymptotically achieve the capacity of finite-state intersymbol interference channels*. IEEE Trans. Inf. Theory **54**(3) (2008) 1295–1303

[11] T. M. Cover and J. A. Thomas. *Elements of Information Theory*. Wiley Interscience, New York, 1991

[12] Y. Ephraim and N. Merhav. *Hidden Markov processes*. IEEE Trans. Inf. Theory **48** (2002) 1518–1569

[13] H. Furstenberg and H. Kesten. *Products of random matrices*. Ann. Math. Statist. **31** (1960) 457–469

[14] H. Furstenberg and Y. Kifer. *Random matrix products and measures on projective spaces*. Israel J. Math. **46**(1) (1983) 12–32

[15] R. G. Gallager. *Information Theory and Reliable Communication*. Wiley, New York, 1968

[16] A. J. Goldsmith and P. P. Varaiya. *Capacity, mutual information, and coding for finite-state Markov channels*. IEEE Trans. Inf. Theory **42**(3) (1996) 868–886

[17] G. Han and B. Marcus. *Analyticity of entropy rate of hidden Markov chains*. IEEE Trans. Inf. Theory **52**(12) (2006) 5251–5266

[18] G. Han and B. Marcus. *Asymptotics of noisy constrained channel capacity*. In Proc. IEEE Int. Symp. Information Theory, Nice, France, June 2007, pp. 991–995

[19] G. Han and B. Marcus. *Derivatives of entropy rate in special families of hidden Markov chains*. IEEE Trans. Inf. Theory **53**(7) (2007) 2642–2652

[20] T. E. Harris. *On chains of infinite order*. Pacific J. Math. **5**(1) (1955) 707–724

[21] T. Holliday, A. Goldsmith, and P. Glynn. *Capacity of finite state channels based on Lyapunov exponents of random matrices*. IEEE Trans. Inf. Theory **52**(8) (2006) 3509–3532

[22] F. Jelinek. *Continuous speech recognition by statistical methods*. Proc. IEEE **64**(4) (1976) 532–556

[23] A. Kavčić. *On the capacity of Markov sources over noisy channels*. In Proc. IEEE Global Telecommunications Conf., San Antonio, TX, November 2001, pp. 2997–3001

[24] A. Krogh, M. Brown, I. Mian, K. Sjolander, and D. Haussler. *Hidden Markov models in computational biology: Applications to protein modeling*. J. Mol. Biol. **235**(5) (1994) 1501–1531

[25] F. Le Gland and L. Mevel. *Basic properties of the projective product with application to products of column-allowable nonnegative matrices*. Math. Control Signals Syst. **13**(1) (2000) 41–62

[26] F. Le Gland and L. Mevel. *Exponential forgetting and geometric ergodicity in hidden Markov models*. Math. Control Signals Syst. **13**(1) (2000) 63–93

[27] C. Méasson, A. Montanari, T. J. Richardson, and R. L. Urbanke. *The generalized area theorem and some of its consequences*. IEEE Trans. Inf. Theory **55**(11) (2009) 4793–4821

[28] C. Méasson, A. Montanari, and R. L. Urbanke. *Maxwell construction: the hidden bridge between iterative and maximum a posteriori decoding.* IEEE Trans. Inf. Theory **54**(12) (2008) 5277–5307

[29] M. Mushkin and I. Bar-David. *Capacity and coding for Gilbert–Elliott channels.* IEEE Trans. Inf. Theory **35**(6) (1989) 1277–1290

[30] E. Ordentlich and T. Weissman. *New bounds on the entropy rate of hidden Markov processes.* In Proc. IEEE Information Theory Workshop, San Antonio, TX, October 2004, pp. 117–122

[31] E. Ordentlich and T. Weissman. *Approximations for the entropy rate of a hidden Markov process.* In Proc. IEEE Int. Symp. Information Theory, Adelaide, Australia, September 2005, pp. 2198–2202

[32] E. Ordentlich and T. Weissman. *On the optimality of symbol-by-symbol filtering and denoising.* IEEE Trans. Inf. Theory **52**(1) (2006) 19–40

[33] V. I. Oseledec. *A multiplicative ergodic theorem. Lyapunov characteristic numbers for dynamical systems.* Trans. Moscow Math. Soc. (1968) 197–231

[34] Y. Peres. *Domains of analytic continuation for the top Lyapunov exponent.* Ann. Inst. H. Poincaré Prob. Statist. **28**(1) (1992) 131–148

[35] T. Petrie. *Probabilistic functions of finite state Markov chains.* Ann. Math. Statist. **40**(1) (1969) 97–115

[36] H. D. Pfister. *On the Capacity of Finite State Channels and the Analysis of Convolutional Accumulate-m Codes.* PhD thesis, University of California, San Diego, CA, 2003

[37] H. D. Pfister, J. B. Soriaga, and P. H. Siegel. *On the achievable information rates of finite state ISI channels.* In Proc. IEEE Global Telecommunications Conf., San Antonio, TX, November 2001, pp. 2992–2996

[38] D. Ruelle. *Analyticity properties of the characteristic exponents of random matrix products.* Adv. Math. **32** (1979) 68–80

[39] E. Seneta. *Non-Negative Matrices: An Introduction to Theory and Applications,* 2nd edn. Wiley, New York, 1981

[40] J. B. Soriaga, H. D. Pfister, and P. H. Siegel. *On the low-rate Shannon limit for binary intersymbol interference channels.* IEEE Trans. Commun. **51**(12) (2003) 1962–1964

[41] P. O. Vontobel, A. Kavčić, D. M. Arnold, and H. A. Loeliger. *A generalization of the Blahut–Arimoto algorithm to finite-state channels.* IEEE Trans. Inf. Theory **54**(5) (2008) 1887–1918

[42] A. Ziv. *Relative distance – an error measure in round-off error analysis.* Math. Comput. **39**(160) (1982) 563–569

[43] O. Zuk, E. Domany, I. Kanter, and M. Aizenman. *From finite-system entropy to entropy rate for a hidden Markov process.* IEEE Signal Process. Lett. **13**(9) (2006) 517–520

[44] O. Zuk, I. Kanter, and E. Domany. *The entropy of a binary hidden Markov process.* J. Statist. Phys. **121**(3) (2005) 343–360

7
Computing entropy rates for hidden Markov processes

MARK POLLICOTT

Mathematics Institute, University of Warwick, Coventry CV4 7AL, UK
E-mail address: mpollic@maths.warwick.ac.uk

Abstract. In this article we want to show how certain analytic techniques from dynamical systems, and more particularly thermodynamics, can be used to give new explicit formulae for entropy rates for certain hidden Markov processes. As a byproduct, the method gives potentially very accurate numerical approximations to the entropy rate.

1 Introduction

We want to describe an approach to studying entropy rates for certain hidden Markov processes. Our motivation for studying this problem comes from a previous approach to Lyapunov exponents for random matrix products, which we shall also briefly describe. We were first introduced to the connection between Lyapunov exponents and entropy rates by the article of Jacquet *et al.* [8]. However, there is a close analogy which probably dates back as far as the work of Furstenberg [3] and Blackwell [1]. They studied Lyapunov exponents and entropy rates, respectively, by considering associated stationary measures. For simplicity, we shall restrict ourselves to the specific case of binary symmetric channels with noise. However, there is scope for generalizing this method to more general settings.

Our aim is to present new explicit formulae for the entropy rates and, thus, by suitable approximations, give an algorithm for the explicit computation. The usual techniques for studying entropy rates tend to give algorithms which give exponential convergence (reflecting the use of positive matrices and associated

Entropy of Hidden Markov Processes and Connections to Dynamical Systems: Papers from the Banff International Research Station Workshop, ed. B. Marcus, K. Petersen, and T. Weissman. Published by Cambridge University Press. © Cambridge University Press 2011.

transfer operators). The techniques we describe typically lead to a faster super-exponential convergence. This is characteristic of the analytic methods (rooted in a beautiful circle of ideas initiated by Grothendieck, Ruelle, and Cvitanovic; see [4] and [15]) that we employ.

To motivate some of the key ideas, we will first consider the somewhat simpler setting of integrals of analytic functions on the unit interval. There are, of course, many classical approaches to estimating integrals in terms of a finite evenly spaced set of points. We describe yet another approximation (from [9]) which illustrates our theme. We then progress to the more subtle case of describing Lyapunov exponents for random products of positive matrices and, finally, entropy rates for hidden Markov processes.

Whereas the treatment for Lyapunov exponents is a summary of our earlier work, and we refer the reader to [14] for complete details and proofs, the results on entropy rates for hidden Markov processes do not appear elsewhere. However, since the proofs are a variant on those to be found in the other cases, we present a summary of the approach and the main ideas in the final section. The underlying framework is based on ideas from thermodynamic formalism, which we could loosely describe as the study of dynamical systems, with a dash of ergodic theory and statistical physics, in the great tradition of Sinai, Bowen, and Ruelle.

2 A little motivation: integration on the unit interval

We want to motivate the main work in terms of the more classical problem of approximating integrals on the unit interval. We refer the reader to [9] for more details.

Given an (analytic) function $f : [0, 1] \to \mathbb{R}$, we can always approximate its integral by an average over a finite set of points. For example, given $n \geq 1$ we can naively write

$$\int_0^1 f(x)\, dx \approx \frac{1}{n} \sum_{k=1}^n f\left(\frac{k}{n}\right).$$

The implied error term here is typically of order $O(1/n)$, which for the purposes of numerical approximation is not particularly efficient. However, using exactly the same information on the values of the function, but arranged in a different way, it is possible to improve on this. Indeed, there are various classical methods which use different combinations of the same numerical values $f(k/n)$, $k = 0, \ldots, n$, to give better approximations to the integral. These include the classical Newton–Cotes approximations [2]. The best known of these is probably Simpson's rule:

$$\int_0^1 f \, dx \approx \frac{1}{3n} \left(f(0) + 2 \sum_{k=1}^{n/2-1} f\left(\frac{2k}{n}\right) + 4 \sum_{k=1}^{n/2} f\left(\frac{2k-1}{n}\right) + f(1) \right),$$

where the error is only of order $O(1/n^4)$.

In general, these particular classical methods typically give polynomial errors, i.e., estimates for which the error is $O(n^{-k})$ as $n \to \infty$ for some $k > 0$. Let us consider a closely related problem where we replace the points

$$\{k/n : 1 \le j \le n\}$$

by the following sequences of finite sets.

Notation. For each $N \ge 1$, we define a family of points

$$P_N = \left\{ \frac{j}{(2^m - 1)} : 0 \le j \le 2^m - 1, \quad 1 \le m \le N \right\}.$$

The set P_N has cardinality $|P_N| \approx 2^{N+1}$.

Let us now consider the following natural question.

Question 2.1. Is there a method for combining the values $\{f(x) : x \in P_N\}$ so as to get a good estimate for the integral?

Of course, by "good" we essentially mean the best estimate we can manage. To proceed, it is convenient to introduce the following notation.

Definition 2.2. We define three quantities:

(1) for $n \ge 0$, we define

$$e_n = (-1)^n \frac{2^{-n(n+1)/2}}{\prod_{i=1}^n (1 - 2^{-i})}$$

(for example, $e_0 = -1$, $e_1 = -1$, $e_2 = \frac{1}{3}$, $e_3 = -\frac{1}{21}$, $e_4 = \frac{1}{315}$, etc.);

(2) for each $1 \le m \le N$, we define

$$\beta_m(f) = \frac{1}{2^m} \sum_{j=0}^{2^m-1} f\left(\frac{j}{2^m - 1}\right);$$

(3) finally, for each $N \geq 1$, we define

$$\mu_N(f) = \frac{\sum_{m=1}^{N} \frac{e_{N-m}}{1-2^{-m}} \beta_m(f)}{\sum_{m=1}^{N} \frac{e_{N-m}}{1-2^{-m}}}.$$

In particular, we see from the above definitions that $\mu_N(f)$ depends only on the values f takes on the points in P_N. Of course, the values $\beta_m(f)$ are already approximations themselves to the integral of f. However, the next result shows that the reorganization of these values in the definition of $\mu_N(f)$ leads to a better approximation to this integral.

Theorem 2.3. *For any analytic function* $f : [0,1] \to \mathbb{R}$,

$$\left| \int_0^1 f(x)\,dx - \mu_N(f) \right| = O\left(\exp(-cN^2)\right) \quad \text{as } N \to \infty$$

for some $c > 0$.

Full details of the proof can be found in [9]. However, for the present we want to stress that it is important to assume that f is real analytic. In fact, in the present context the value $c > 0$ can be chosen arbitrarily close to $\log 2$.

Since $P_N = O(2^{N+1})$, the above theorem tells us that the error is super-polynomial in the number of points $|P_N|$, i.e., for some $C > 0$,

$$\left| \int_0^1 f(x)\,dx - \mu_N(f) \right| = O\left(\exp(-C \log^2 |P_N|)\right).$$

It is illuminating to recall a particular example from [9].

Example 2.4. We can apply the theorem to $\int_0^1 f(x)\,dx$, where

$$f(x) = \frac{5\pi}{2}(e^\pi - 2)^{-1} e^{\pi x} \cos(\pi x/2).$$

The integral is actually equal to 1. The following table gives the approximations $\mu_N(f)$ to the integral.

N	$\mu_N(f)$
1	0.18575506891852380346423780644
2	−0.84112428438320560388180161679.2
3	0.40608775333324283787989060678
4	1.09233276774560006235284301088
5	0.99681510352795871656533973381
6	1.00004673478255155995818417282
7	0.99999977398633338017700990934
8	0.99999999982780643498244012823
9	1.00000000000326837809455213367
10	0.99999999999999360196233983786
11	1.00000000000000000436159826786
12	0.99999999999999999989026376489
13	1.00000000000000000000000742107229
14	1.00000000000000000000000000513617960
15	0.99999999999999999999999999999993061845
16	0.99999999999999999999999999999993061845283932
17	0.99999999999999999999999999999999999999997039178592326

Remark 2.5. Well-known, and often more accurate, alternatives to Newton–Cotes approximations are those of quadrature, using unequally spaced points. Of these, the best known is probably Gauss–Legendre quadrature, where the points are chosen to be zeros of the Legendre polynomials. Even in the present case of analytic functions $f : [0,1] \to \mathbb{R}$, the best estimates of the error term for either Newton–Cotes or Gauss–Legendre quadrature are exponential [10].

We now turn to the next problem: the computation of Lyapunov exponents.

3 A little more motivation: Lyapunov exponents

In this section we want to consider the analogous case of the largest Lyapunov exponent for a fixed pair of $d \times d$ positive matrices $\{A_1, A_2\}$ chosen with respect to an independent and identically distributed (i.i.d.), or Bernoulli, distribution.

Let $||\cdot||$ denote any norm on $d \times d$ matrices. In the particular case that $d = 2$ and we have the $(\frac{1}{2}, \frac{1}{2})$-Bernoulli measure, the definition takes the following form.

Definition 3.1. The Lyapunov exponent λ is given by the limit

$$\lambda := \lim_{n \to +\infty} \frac{1}{n} \sum_{i_1,\dots,i_n} \frac{1}{2^n} \log ||A_{i_1} \cdots A_{i_n}||,$$

where the summation is over the 2^n possible choices $i_1,\dots,i_n \in \{1,2\}$.

By a famous result of Kesten and Furstenberg one has the following pointwise estimate with respect to the Bernoulli measure $\mu = (\frac{1}{2}, \frac{1}{2})^{\mathbb{Z}^+}$.

Proposition 3.2. Furstenberg–Kesten [3]. *For almost every $\underline{i} \in \Sigma$ (with respect to the measure μ), one has*

$$\lambda = \lim_{n \to +\infty} \frac{1}{n} \log \|A_{i_1} \cdots A_{i_n}\|.$$

We have chosen to formulate the results in this simple case in order to simplify the notation. However, analogous results hold for any finite set of $d \times d$ matrices chosen with respect to a Markov measure.

Standing assumption. Henceforth, we shall always assume that the matrices are all positive (i.e., each of the entries of each of the matrices A_1, A_2 is strictly positive).

We now need to introduce some notation.

Definition 3.3. Given a finite string $\underline{i} = (i_1, i_2, \ldots, i_n) \in \{1, 2\}^n$, let $|\underline{i}| = n$ denote its length. Let $\sigma \underline{i} = (i_2, \ldots, i_n, i_1)$ denote the cyclic permutation given by shifting the terms one place to the left (modulo one).

Let $A_{\underline{i}} = A_{i_1} \cdots A_{i_n}$ denote the 2×2 matrix coming from matrix multiplication of the matrices indexed by the terms in $\underline{i}$. Let $\lambda_{\underline{i}}$ and $x_{\underline{i}}$ be the maximal positive eigenvalue and eigenvector (by the Perron–Frobenius theorem), i.e., $A_{\underline{i}} x_{\underline{i}} = \lambda_{\underline{i}} x_{\underline{i}}$.

We are now in a position to introduce the key ingredient in the analysis of the Lyapunov exponents.

Definition 3.4. For $j = 1, 2$ and $t \in \mathbb{R}$, we define a complex function

$$d^{(j)}(z, t) = \exp\left(-\sum_{n=1}^{\infty} \frac{z^n}{2^n n} \sum_{|\underline{i}|=n} \frac{1}{(1 - \lambda_{\underline{i}}^{-2})} \left(\prod_{l=0}^{n-1} \frac{\|A_j x_{\sigma^l \underline{i}}\|}{\|x_{\sigma^l \underline{i}}\|} \right)^t \right),$$

which converges to an analytic function for $|z|$ small.

We can now state the following key technical result.

Proposition 3.5. *For each $j = 1, 2$, the function $d^{(j)}(z, t)$ is analytic for $z \in \mathbb{C}$ and $t \in \mathbb{R}$, and we can write*

$$\lambda = \frac{\frac{\partial d^{(1)}}{\partial t}(1, 0)}{\frac{\partial d^{(1)}}{\partial z}(1, 0)} + \frac{\frac{\partial d^{(2)}}{\partial t}(1, 0)}{\frac{\partial d^{(2)}}{\partial z}(1, 0)}.$$

Moreover, expanding $d^{(j)}(z,t) = 1 + \sum_{n=1}^{\infty} a_n^{(j)}(t) z^n$, *we have*

(1) *the coefficients* $a_n^{(j)}(t)$ *depend only on the maximal eigenvalues and eigenvectors of the matrices* $A_{\underline{i}}$, *with* $|\underline{i}| \leq n$; *and*

(2) *we can bound* $|a_n^{(j)}(t)| = O\left(\exp\left(-Cn^2\right)\right)$.

An explicit formula for $a_n^{(j)}(t)$ comes from expanding the expression for $d^{(j)}(z,t)$ as a power series in z. Despite the slightly complicated formulation of the above result, it leads to the following very simple conclusion.

Theorem 3.6. *There are approximations* λ_n *to* λ *defined in terms of the maximal eigenvalues and eigenvectors of the* 2^{n+1} *matrices* $A_{\underline{i}}$, *with* $|\underline{i}| \leq n$. *Moreover, for some* $C > 0$.

$$|\lambda - \lambda_n| = O\left(\exp\left(-Cn^2\right)\right).$$

In particular, the above theorem means that $\lambda_n \to \lambda$ faster than any exponential. We can deduce this theorem from the preceding proposition by making the explicit choice

$$\lambda_n = \frac{\sum_{k=1}^{n} \frac{\partial a_k^{(1)}}{\partial t}(0)}{\sum_{k=1}^{n}(k+1)a_k^{(1)}(0)} + \frac{\sum_{k=1}^{n} \frac{\partial a_k^{(2)}}{\partial t}(0)}{\sum_{k=1}^{n}(k+1)a_n^{(2)}(0)}$$

for each $n \geq 1$.

Let us illustrate this method with an example.

Example 3.7. If we consider the matrices

$$A_1 = \begin{pmatrix} 2 & 1 \\ 1 & 1 \end{pmatrix} \quad \text{and} \quad A_2 = \begin{pmatrix} 3 & 1 \\ 2 & 1 \end{pmatrix},$$

then using the complex functions to estimate λ we get the following table.

n	λ_n
1	1.1435949601546489930611282560219921476826
2	1.1432978985074534413937646485571388968329
3	1.1433110787994660471763348416564186089168
4	1.1433110350856466164727559958382071786676
5	1.1433110351029501308232209336496360457362
6	1.1433110351029492458371384694231808633421
7	1.1433110351029492458432518595030145277475
8	1.1433110351029492458432518536555875134112
9	1.1433110351029492458432518536555882994025

For $n = 9$, this is empirically accurate to the 40 decimal places presented and $\lambda_{n=9}$ gives an approximation

$$\lambda = 1.1433110351029492458432518536555882991025\ldots$$

to the Lyapunov exponent λ. This is presented to the number of decimal places to which it empirically appears to be accurate (i.e., by comparison with the other terms $\{\lambda_n\}$ in the approximation).

Remark 3.8. The standard approximations to the Lyapunov exponent are known to have exponential errors (cf. [13, pages 146–147]). In essence, the convergence is based on the spectral gap for the transfer operator associated to the actions of the matrices on projective space (which is a contraction with respect to the standard Birkhoff metric). In fact, one would probably not expect faster convergence, due to the effect of the rest of the spectrum of this operator.

It is also interesting to consider a simple family of pairs of matrices.

Example 3.9. We consider the following families of matrices.

$$A_1 = \begin{pmatrix} 2 & 1 \\ 1 & 1 \end{pmatrix} \text{ and } A_2 = \begin{pmatrix} \cos(\theta) & \sin(\theta) \\ -\sin(\theta) & \cos(\theta) \end{pmatrix} \begin{pmatrix} 2 & 1 \\ 1 & 1 \end{pmatrix} \begin{pmatrix} \cos(\theta) & -\sin(\theta) \\ \sin(\theta) & \cos(\theta) \end{pmatrix}$$

for any $\theta \in (-0.5, 0.5)$, say. The Lyapunov exponent satisfies

$$\lambda(\theta) \leq \lambda(0) = \log\left(\frac{3 + \sqrt{5}}{2}\right) = 0.962424\ldots.$$

In a neighborhood of $\theta = 0$, we have a second-order approximation

$$\lambda(\theta) = \lambda(0) - \left(\frac{\sqrt{5}}{3 + \sqrt{5}}\right)\theta^2 + O(\theta^3),$$

where $\sqrt{5}/(3 + \sqrt{5}) = 0.427491\ldots$. We can plot $\lambda(\theta)$ (and the approximation). The solid line in Figure 1 is the graph of $\lambda(\theta)$ and the dashed line is that of the second-order approximation.

Finally, we move on to the main topic of this article: entropy rates for hidden Markov processes.

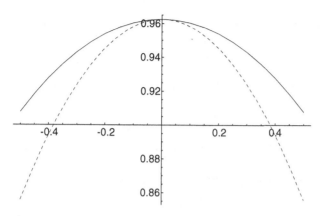

Figure 1

4 Hidden Markov processes and entropy rates

Let $Y = \{Y_n\}_n$ denote a process on the finite set of states $\{1,2,\ldots,k\}$. Let μ_Y be a Markov probability measure on Y, associated to a stochastic matrix P. Let $Z = \{Z_n\}_n$ denote a second process on the states $\{1,2,\ldots,m\}$, with $m < k$. Then given a map $\Psi : \{1,2,\ldots,k\} \to \{1,2,\ldots,m\}$ between the state spaces we can consider the corresponding map $\Phi : Y \to Z$, where $Z_n = \Psi(Y_n)$. We can consider the factor measure $\mu_Z = \Phi(\mu_X)$, and then the associated measure corresponds to a hidden Markov process.

In the language of digital communication, one might describe μ_Z as the output process obtained when passing a finite-state Markov chain through a noisy channel.

We shall concentrate on the following basic example.

Example 4.1. Let $\{X_n\}$ be a Markov process with associated stochastic matrix

$$\begin{pmatrix} p & 1-p \\ 1-p & p \end{pmatrix},$$

where $0 < p < 1$. At time n a binary symmetric channel with cross-over probability p can be characterized by the equation $Z_n = X_n \oplus E_n$, where

(1) $X_n \in \{0,1\}$ denotes the binary input; and
(2) $E_n \in \{0,1\}$ denotes the i.i.d. binary noise with probability vector $(1-\epsilon, \epsilon)$ and $\oplus$ denotes addition modulo 2.

In particular, $Z = \{Z_n\}$ represents the corrupted output (due to the noise).

The product $Y = X \times E$ defined by $Y_n = (X_n, E_n)$ is again a Markov process. In this case, it is easy to see that the associated stochastic matrix is

$$\begin{pmatrix} p(1 \; \epsilon) & p\epsilon & (1-p)(1-\epsilon) & (1-p)\epsilon \\ p(1-\epsilon) & p\epsilon & (1-p)(1-\epsilon) & (1-p)\epsilon \\ (1-p)(1-\epsilon) & (1-p)\epsilon & p(1-\epsilon) & p\epsilon \\ (1-p)(1-\epsilon) & (1-p)\epsilon & p(1-\epsilon) & p\epsilon \end{pmatrix}.$$

We can let Z be a process on the four states $\{(0,0),(0,1),(1,0),(1,1)\}$ and define $\Phi : Y \to Z$ using the maps $\Psi(0,0) = \Psi(1,1) = 0$ and $\Psi(0,1) = \Psi(1,0) = 1$. We want to consider the factor measure $\mu_Z = \Psi\mu_Y$ on Z.

Unfortunately, the factor measures μ_Z will not typically be Markov measures. Therefore, it is useful to consider natural characteristics, of which one of the most important is the following.

Definition 4.2. The *entropy rate* $H(Y)$ of the process $Z = \{Z_n\}$ (with the associated measure μ_Z) is defined by $H(Z) = \lim_{n \to +\infty}(1/n)H_n(Z)$, where

$$H_n(Z) = - \sum_{z_0,\dots,z_{n-1}} \mu_Z[z_0,\dots,z_{n-1}] \log \mu_Z[z_0,\dots,z_{n-1}].$$

In the special case that μ_Y is a Markov measure corresponding to a stochastic matrix $P = P_Y$, with right eigenvector p, say, then the entropy (rate) is well known to be

$$H(Y) = - \sum_{i,j} p_i P_{ij} \log P_{ij}.$$

Remark 4.3. Although we have concentrated on a specific example, this is part of a more general class of examples which we briefly recall. More generally, assume that Z is a process on m states and for each state $j \in \{1,2,\dots,m\}$ we can associate the submatrix P_a of Y by deleting the rows and columns indexed by states from $\{1,2,\dots,k\}$ which do not map to j.

Let $\Delta = \{(x_1,\dots,x_m) : x_1 + \dots + x_m = 1, x_1,\dots,x_m \geq 0\}$ denote the $(m-1)$-dimensional simplex and let Δ_j be the subset that comes by setting to zero all of the coordinates which do not project to j. We then define $f_j : \Delta \to \mathbb{R}^m$ by $f_j(\underline{x}) = \underline{x}^T P_j / \underline{x}^T P_j \underline{1}$, where $\underline{x} = (x_1,\dots,x_n)$ and $\underline{1} = (1,\dots,1)$.

Given $z = \{z_n\}$, we let $x_n = p(y_n | z_n)$. In particular, we have $x_{n+1} = f_{z_{n+1}}(x_n)$, i.e., we have an iterated function scheme on Δ described in terms of the maps f_j.

5 Formulae for entropy rates

We might like to try to adapt the methodology for computing integrals and Lyapunov exponents to give an expression for the entropy rate of some hidden Markov processes. Indeed, part of the motivation comes from the existing connection between entropy rates and Lyapunov exponents, as described in [8]. However, rather than approaching the entropy rate by first expressing it in terms of a Lyapunov exponent, which we then, in turn, analyze as in Section 3, we shall take an approach based on estimating integrals (with respect to the Blackwell measure).

Fix a choice of $\epsilon > 0$ and $0 < p < 1$. Consider the linear fractional transformations $f_1 : \mathbb{R}^+ \to \mathbb{R}^+$ and $f_2 : \mathbb{R}^+ \to \mathbb{R}^+$ defined by

$$f_1(x) = \frac{(1-\epsilon)}{\epsilon} \frac{px+(1-p)}{(1-p)x+p}$$

and

$$f_2(x) = \frac{\epsilon}{(1-\epsilon)} \frac{(1-p)x+p}{px+(1-p)}.$$

We can also associate weight functions $r_1 : \mathbb{R}^+ \to [0,1]$ and $r_2 : \mathbb{R}^+ \to [0,1]$ defined by

$$r_1(x) = \frac{((1-\epsilon)p+\epsilon(1-p))x+(1-\epsilon)(1-p)+\epsilon p}{x+1},$$

$$r_2(x) = \frac{(\epsilon p+(1-\epsilon)(1-p))x+\epsilon(1-p)+(1-\epsilon)p}{x+1}.$$

We can use the maps f_1, f_2 and weights r_1, r_2 to define an important measure on $\mathbb{R}^+$.

Definition 5.1. We define the (Blackwell) probability measure μ by the identity

$$\mu(A) = \int_{f_1^{-1}A} r_1(x)\,d\mu(x) + \int_{f_2^{-1}A} r_2(x)\,d\mu(x).$$

This measure was introduced in Blackwell's seminal paper from 1957 [1]. The principal complication in studying it is the implicit nature of its definition.

Proposition 5.2. *The entropy rate of a binary symmetric channel with cross-over probability $\epsilon > 0$ and input Markov chain with transition matrix*

$$\begin{pmatrix} p & 1-p \\ 1-p & p \end{pmatrix}$$

is given by

$$H = - \int (r_1(x) \log r_1(x) + r_2(x) \log r_2(x)) \, d\mu(x).$$

By analogy with the approach for Lyapunov exponents, we might try to estimate the integral for the entropy rate using complex functions defined in terms of

(1) the fixed points $x = x_{\underline{i}}$ of the maps $f_{i_n} \cdots f_{i_1} : \mathbb{R}^+ \to \mathbb{R}^+$, where $\underline{i} = (i_1, \ldots, i_n) \in \{1,2\}^n$ and $n \geq 1$; and
(2) the weights

$$r_{i_1}(x_{\underline{i}}) r_{i_2}(f_{i_1} x_{\underline{i}}) r_{i_3}(f_{i_2} f_{i_1} x_{\underline{i}}) \cdots r_{i_n}(f_{i_{n-1}} \cdots f_{i_1} x_{\underline{i}}).$$

As in the computation of the Lyapunov exponents, we can define functions which will be used to give expressions for the entropy rate. For a fixed $j \in \{1,2\}$, let us denote

$$r_i^{(j,t)}(z) = r_i(z) \exp\left(-tr_j(f_i z) \log r_j(f_i z)\right)$$

for $t \in (-\epsilon, \epsilon)$ and $i \in \{1,2\}$.

Definition 5.3. For $j \in \{1,2\}$ and $t \in \mathbb{R}$, we define a complex function

$$d^{(j)}(z,t) = \exp\left(-\sum_{n=1}^{\infty} \frac{z^n}{n} Z_n^{(j)}(t)\right),$$

where

$$Z_n^{(j)}(t) := \sum_{|\underline{i}|=n} \frac{r_{i_1}^{(j,t)}(x_{\underline{i}}) r_{i_2}^{(j,t)}(f_{i_1} x_{\underline{i}}) r_{i_3}^{(j,t)}(f_{i_2} f_{i_1} x_{\underline{i}}) \cdots r_{i_n}^{(j,t)}(f_{i_{n-1}} \cdots f_{i_1} x_{\underline{i}})}{1 - \lambda_{\underline{i}}},$$

which converges to an analytic function for $|z|$ sufficiently small (e.g., $|z| < \frac{1}{2}$).

We can now state the following key technical result.

Proposition 5.4. *For each $j = 1, 2$, the function $d^{(j)}(z,t)$ is analytic for $z \in \mathbb{C}$ and $t \in \mathbb{R}$. Moreover, expanding*

$$d^{(j)}(z,t) = 1 + \sum_{n=1}^{\infty} a_n^{(j)}(t) z^n,$$

we have

(1) *the coefficients* $a_n^{(j)}(t)$ *depend only on the maximal eigenvalues and eigenvectors of the matrices* $A_{\underline{i}}$, *with* $|\underline{i}| \leq n$;

(2) *we can bound* $|a_n^{(j)}(t)| = O\left(\exp\left(-Cn^2\right)\right)$; *and*

(3) *we can write*

$$H = \frac{\frac{\partial d^{(1)}(1,t)}{\partial t}|_{t=0}}{\frac{\partial d^{(1)}(z,0)}{\partial z}|_{z=1}} + \frac{\frac{\partial d^{(2)}(1,t)}{\partial t}|_{t=0}}{\frac{\partial d^{(2)}(z,0)}{\partial z}|_{z=1}}.$$

An explicit formula for $a_n^{(j)}(t)$ comes from expanding the expression for $d^{(j)}(z,t)$ as a power series in z, i.e.,

$$d^{(j)}(z,t) = 1 + \left(-\sum_{n=1}^{\infty} \frac{z^n}{n} Z_n^{(j)}(t)\right) + \frac{1}{2}\left(-\sum_{n=1}^{\infty} \frac{z^n}{n} Z_n^{(j)}(t)\right)^2 + \cdots.$$

For example,

$$a_1^{(j)}(t) = -Z_1^{(j)}(t),$$

$$a_2^{(j)}(t) = \frac{1}{2}\left(Z_1^{(j)}(t)\right)^2 - \frac{1}{2}Z_2^{(j)}(t),$$

$$a_3^{(j)}(t) = -\frac{1}{6}\left(Z_1^{(j)}(t)\right)^3 + \frac{1}{2}\left(Z_1^{(j)}(t)Z_2^{(j)}(t)\right) - \frac{1}{6}Z_3^{(j)}(t),$$

$$\cdots.$$

Despite the slightly complicated nature of the above result, it leads to the following very simple conclusion.

Theorem 5.5. *There are approximations* H_n *to* H *defined in terms of the maximal eigenvalues of the* 2^{n+1} *matrices* $A_{\underline{i}}$ *and weightings of fixed points, with* $|\underline{i}| \leq n$. *Moreover, for some* $C > 0$,

$$|H - H_n| = O\left(\exp\left(-Cn^2\right)\right).$$

In particular, the above theorem means that H_n converges to H faster than any exponential. We can deduce this theorem from the preceding proposition, by making the explicit choice

$$H_n = \frac{\sum_{k=1}^{n} \frac{\partial a_k^{(1)}}{\partial t}|_{t=0}}{\sum_{k=1}^{n}(k+1)a_k^{(1)}(0)} + \frac{\sum_{k=1}^{n} \frac{\partial a_k^{(2)}}{\partial t}|_{t=0}}{\sum_{k=1}^{n}(k+1)a_n^{(2)}(0)}$$

for each $n \geq 1$. In practice, the values of H_n can be easily evaluated by any computer program that does basic manipulations applied to the Nth-order differential expansion for $\exp\left(-\sum_{n=1}^{N}(z^n/n)Z_n^{(j)}(t)\right)$, for N suitably large.

Remark 5.6. The approach described in [8] for computing H involves directly estimating the Lyapunov exponents and therefore we would expect it to have, at best, exponential convergence.

6 Examples

For definiteness, let us consider the specific case where we choose $p = 0.3$. By applying Theorem 5.5, we get the following numerical approximations to the entropy rate H for three different values of ϵ.

Example 6.1. Consider the case where $\epsilon = 0.1$. The first 12 approximations to the entropy rate H are given in the following table.

n	nth approximation H_n to H
1	0.6695762188924017222059823716665738904
2	0.6584363094376296050449308000119378473
3	0.6592182569002402462990204371102245291
4	0.6592124235327986432500897457195334
5	0.659212415370513825990945266340317516
6	0.6592124153801376031854319151545553251
7	0.6592124153800641822555891372061903
8	0.6592124153800641884649812690053861
9	0.659212415380064188468453625846998960
10	0.659212415380064188468453654488609212
11	0.659212415380064188468453654486913548
12	0.659212415380064188468453654486913549

For comparison, the more standard approach with the same data for $n = 12$ gives an estimate for H of

$$0.6592124153800641884641649864973743024,$$

which appears to be accurate to somewhat fewer decimal places (i.e., 20 places as opposed to 35).

Example 6.2. Consider the case where $\epsilon = 0.05$. The first 12 approximations to the entropy rate H are given in the following table.

n	nth approximation H_n to H
1	0.649716380502570388049059721652168093125382270609931577
2	0.638872831505531868118185758445930726424597836500913661
3	0.639321291425218666038301968402827811797492170748918523
4	0.639319149549044371892644590643994790558375746801300467
5	0.639319145726601645487482153395710461092700918828406479
6	0.639319145725759026714798759361581543207516822294029079
7	0.639319145725759203984276521261981115706640032193321810
8	0.639319145725759203981818636575956471133699225028879025
9	0.639319145725759203981818724863800016293366611661269640
10	0.639319145725759203981818724864560454891091372750356015
11	0.639319145725759203981818724864560448411778119333427618
12	0.639319145725759 20398181872486456044841178095043229234611

For comparison, the more standard approach with the same data for $n = 12$ gives an estimate

$$0.639319145725759203981818717621507182709597356320832719422,$$

which again appears to be accurate to somewhat fewer decimal places (i.e., 25 places as opposed to 55).

Example 6.3. Consider the case where $\epsilon = 0.01$. The approximation for $n = 12$ to the entropy rate H is

$$0.6174008375150805894156894641082285725424274376957975286126518405501560992511.$$

For comparison, the more standard approach with the same data for $n = 12$ gives an estimate

$$0.6174008375150805894156894641082285725424251522027617335829260119008480796 0,$$

which again appears to be accurate to somewhat fewer decimal places (i.e., 40 places as opposed to 75).

We can use either method to plot the entropy rate $H = H(\epsilon)$, for the fixed value $p = 0.3$, say, as a function of ϵ, as in Figure 2.

Remark 6.4. A close inspection of the proof (described in Section 7 below) gives rigorous error estimates of $|H - H_n|$. Similar estimates for Lyapunov exponents can be found in [14].

Remark 6.5. The entropy rate depends on the values of the derivatives of the function $d^{(j)}(z, t)$, which in turn have an analytic dependence on p and ϵ.

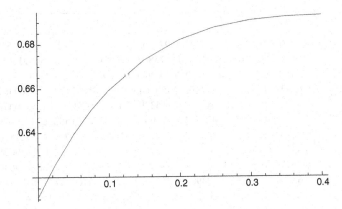

Figure 2

This allows us to recover the analytic dependence of $H = H(p, \epsilon)$ on both p and ϵ [7]. This is not very surprising, since in both cases the observation is rooted in perturbation theory for the transfer operator. A little more work (using the implicit function theorem) would give explicit formulae for the derivatives $(\partial^l H / \partial \epsilon)|_{\epsilon=0}$ in terms of the fixed points of the contractions.

Remark 6.6. In practice, rather than finding the fixed points $x_{\underline{i}}$ for the projective maps $f_{\underline{i}} = f_{i_{n-1}} \cdots f_{i_1} : \mathbb{R}^+ \to \mathbb{R}^+$, it is computationally easier to consider the unique positive eigenvector $v_{\underline{i}}$ for the matrix product $B(\underline{i}) := B(i_n)B(i_{n-1}) \cdots B(i_1)$, where $i_1, \ldots, i_n \in \{1, 2\}$,

$$B(1) = \begin{pmatrix} (1 - \epsilon)p & (1 - \epsilon)(1 - p) \\ \epsilon p & (1 - \epsilon)p \end{pmatrix},$$

and

$$B(2) = \begin{pmatrix} \epsilon p & \epsilon(1 - p) \\ (1 - \epsilon)(1 - p) & (1 - \epsilon)p \end{pmatrix}.$$

Given $x, y > 0$, we readily see that if we write

$$\begin{pmatrix} x' \\ y' \end{pmatrix} = B(j) \begin{pmatrix} x \\ y \end{pmatrix}, \quad \text{where } j = 1, 2,$$

then the corresponding projective action is

$$\frac{x}{y} \mapsto \frac{x'}{y'} = f_j \left(\frac{x}{y} \right).$$

In particular, if $v = \begin{pmatrix} v_1 \\ v_2 \end{pmatrix}$ is the positive eigenvector of $B(\underline{i})$, then we can write the fixed point as $x_{\underline{i}} = v_1/v_2$. Moreover, writing $B(\underline{i}) = \begin{pmatrix} \alpha & \beta \\ \gamma & \delta \end{pmatrix}$, we see that $f_{\underline{i}}(x) = (\alpha x + \beta)/(\gamma x + \delta)$ and $f_{\underline{i}}'(x) = \det B(\underline{i})/(\gamma x + \delta)$ and, therefore, $\lambda_{\underline{i}} = \det B(\underline{i})/(\gamma x_{\underline{i}} + \delta)^2$.

7 Overview of the proof of Theorem 5.5

The proof of Theorem 5.5 (or, more particularly, Proposition 5.4) follows the same general principles as that of Theorems 2.3 and 3.6, which appear elsewhere [9, 14]. In the present context, it seems more appropriate merely to sketch the steps needed in the proof.

As before, the key ingredient is to consider a suitable bounded linear transfer operator $\mathcal{L}$ defined on analytic functions on a suitable open set $U \subset \mathbb{C}$. The proof again draws on ideas from the classical work of Ruelle in 1976 [15] (based on even earlier work of Grothendieck in 1955 [5]). Ultimately, we need to connect the entropy rate to the complex functions $d^{(j)}(z, t)$, which play a role roughly akin to characteristic polynomials for computing eigenvalues of matrices. The role of such matrices is replaced by (Ruelle) transfer operators defined on spaces of (analytic) functions.

The strategy of the proof of Theorem 2.3 is as follows:

Step 1: introduce transfer operators and their determinants;

Step 2: relate the entropy rate to (weighted) transfer operators; and

Step 3: relate the transfer operators to the determinants $d^{(j)}(z, t)$ and show that the terms in the expansion of $z \mapsto d^{(j)}(z, t)$ converge quickly.

We now elaborate on each of these three steps.

7.1 Step 1: transfer operators and determinants

By definition, we see that the images of the contractions satisfy

$$\overline{f_1(\mathbb{R}^+)} \cup \overline{f_2(\mathbb{R}^+)} = \left[\frac{(1-\epsilon)(1-p)}{\epsilon p}, \frac{(1-\epsilon)p}{\epsilon(1-p)} \right] \cup \left[\frac{\epsilon p}{(1-\epsilon)(1-p)}, \frac{\epsilon(1-p)}{(1-\epsilon)p} \right],$$

which is a closed bounded subset of $\mathbb{R}^+$. In particular, we can choose an open neighborhood $U \subset \mathbb{C}$ of $\overline{f_1(\mathbb{R}^+)} \cup \overline{f_2(\mathbb{R}^+)}$ such that the extensions of the linear fractional maps to the complex plane satisfy $\overline{f_1(U)}, \overline{f_2(U)} \subset U$.

Let $C^\omega(U)$ denote the Hilbert space of analytic functions on U which are square integrable, with the inner product $\langle f_1, f_2 \rangle = \int_U f_1(z) \overline{f_2(z)} \, d\mathrm{vol}(z)$ [15].

Definition 7.1. We define a bounded linear operator $\mathcal{L}: C^\omega(U) \to C^\omega(U)$ by

$$\mathcal{L}w(z) = r_1(z)w(f_1 z) + r_2(z)w(f_2 z), \quad z \in U,$$

called the *(Ruelle) transfer operator.*

Crucially, this operator is of trace class (on the space of analytic functions $C^\omega(U)$) [15, 9]. In particular, $\mathrm{tr}(\mathcal{L})$ (and $\mathrm{tr}(\mathcal{L}^n)$, for each $n \geq 1$) is finite. Moreover, each $\mathrm{tr}(\mathcal{L}^n)$ can be explicitly expressed in terms of

(1) the unique fixed points $x = x_{\underline{i}}$ of the maps $f_{\underline{i}} := f_{i_n} \cdots f_{i_1}: \mathbb{R}^+ \to \mathbb{R}^+$, where $\underline{i} = (i_1, \ldots, i_n) \in \{1, 2\}^n$ and $n \geq 1$;
(2) the weights of these fixed points of the form

$$r_{\underline{i}} := r_{i_1}(x_{\underline{i}}) r_{i_2}(f_{i_1} x_{\underline{i}}) r_{i_3}(f_{i_2} f_{i_1} x_{\underline{i}}) \cdots r_{i_n}(f_{i_{n-1}} \cdots f_{i_1} x_{\underline{i}}); \text{ and}$$

(3) the weights of the fixed points of the form

$$\lambda_{\underline{i}} = f'_{\underline{i}}(x_{\underline{i}}).$$

More precisely, we can write the following lemma.

Lemma 7.2. *For each $n \geq 1$,*

$$\mathrm{tr}(\mathcal{L}^n) = \sum_{|\underline{i}|=n} \frac{r_{\underline{i}}}{(1 - \lambda_{\underline{i}})}.$$

Proof. We first claim that for each $n \geq 1$, we can write

$$\mathcal{L}^n w(z) = \sum_{|\underline{i}|=n} r_{i_1}(z) r_{i_2}(f_{i_1} z) r_{i_3}(f_{i_2} f_{i_1} z) \cdots r_{i_n}(f_{i_{n-1}} \cdots f_{i_1} z) w(f_{i_n} \cdots f_{i_1} z),$$

where the summation is over the 2^n strings $\underline{i} = (i_1, \ldots, i_n) \in \{1, 2\}^n$. To see this, we observe the inductive step

$$\mathcal{L}^{n+1} w(z) = \mathcal{L}\left(\mathcal{L}^n w\right)(z)$$

$$= \mathcal{L}\left(\sum_{|\underline{i}|=n} r_{i_1}(\cdot) r_{i_2}(f_{i_1} \cdot) \cdots r_{i_n}(f_{i_{n-1}} \cdots f_{i_1} \cdot) w(f_{i_n} \cdots f_{i_1} \cdot)\right)(z)$$

$$= \sum_{i_0} \sum_{|\underline{i}|=n} r_{i_1}(f_{i_0} z) r_{i_2}(f_{i_1} f_{i_0} z) \cdots r_{i_n}(f_{i_{n-1}} \cdots f_{i_1} f_{i_0} z) w(f_{i_n} \cdots f_{i_1} f_{i_0} z),$$

which after relabeling gives the required expression for $\mathcal{L}^{n+1}w$. If we view this as the sum of 2^n individual linear maps

$$\mathcal{L}_{\underline{i}}w(z) = r_{i_1}(z)r_{i_2}(f_{i_1}z)r_{i_3}(f_{i_2}f_{i_1}z)\cdots r_{i_n}(f_{i_{n-1}}\cdots f_{i_1}z)w(f_{i_n}\cdots f_{i_1}z),$$

then these are each of trace class, and $\operatorname{tr}(\mathcal{L}^n) = \sum_{|\underline{i}|=n}\operatorname{tr}(\mathcal{L}_{\underline{i}})$. Since each trace $\operatorname{tr}(\mathcal{L}_{\underline{i}})$ is a sum of eigenvalues, we can consider the eigenvalue equations $\mathcal{L}_{\underline{i}}\psi_{\underline{i}} = \alpha\psi_{\underline{i}}$ evaluated at the fixed point $f_{\underline{i}}x_{\underline{i}} = x_{\underline{i}}$. If the eigenfunction $\psi_{\underline{i}}$ is nonzero at $x_{\underline{i}}$, then we deduce that $\alpha = r_{\underline{i}}$; but, on the other hand, if the eigenfunction $\psi_{\underline{i}}$ has a zero at $x_{\underline{i}}$ of order $k \geq 1$, then we can differentiate the eigenvalue equation k times and evaluate at $x_{\underline{i}}$ to deduce that $\alpha = \lambda_{\underline{i}}^k r_{\underline{i}}$. Moreover, it is easy to show that all eigenvalues occur (with unit multiplicity), and then

$$\operatorname{tr}(\mathcal{L}_{\underline{i}}) = r_{\underline{i}}\left(\sum_{m=0}^{\infty}\lambda_{\underline{i}}^m\right) = \frac{r_{\underline{i}}}{1-\lambda_{\underline{i}}},$$

cf. [12]. Summing over the 2^n strings $\underline{i}$ such that $|\underline{i}| = n$ gives the claimed result. $\qquad\square$

We can now write the determinant in the following way:

$$d(z) := \exp\left(-\sum_{n=1}^{\infty}\frac{z^n}{n}\operatorname{tr}(\mathcal{L}^n)\right) = \exp\left(-\sum_{n=1}^{\infty}\frac{z^n}{n}\sum_{|\underline{i}|=n}\frac{r_{\underline{i}}}{(1-\lambda_{\underline{i}})}\right), \quad (1)$$

and by work of Ruelle [15] we have the following proposition.

Proposition 7.3. *The function $d(z)$ is entire as a function of z. Moreover, we can expand*

$$d(z) = 1 + \sum_{n=1}^{\infty}a_n z^n, \quad \text{where } |a_n| = O(e^{-cn^2})$$

for some $c > 0$.

7.2 Step 2: weighted transfer operators and entropy rates

The maximal eigenvalue 1 for $\mathcal{L}$ corresponds to an eigenfunction which is identically 1 and has an eigenprojection μ, i.e., $\mathcal{L}^*\mu = \mu$. From the definition of the Blackwell measure μ, we see that for any interval $[a,b]$ we have

$$\int \chi_{[a,b]}\,\mathrm{d}\mu(x) = \int \mathcal{L}\chi_{[a,b]}\,\mathrm{d}\mu(x).$$

Thus, by approximation, we see that the projection $\mathcal{L}^*\mu = \mu$ is precisely the Blackwell measure (in Definition 4.2), cf. [6, 7].

We recall that

$$H = -\int (r_1(x)\log r_1(x) + r_2(x)\log r_2(x))\,d\mu(x). \tag{2}$$

We want to define a family of modified versions of the transfer operators.

Definition 7.4. The bounded linear operator $\mathcal{L}_t^{(j)} : C^\omega(U) \to C^\omega(U)$ defined by

$$\mathcal{L}_t^{(j)}w(z) = r_1(z)e^{-tr_j(f_1z)\log r_j(f_1z)}w(f_1z) + r_2(z)e^{-tr_j(f_2z)\log r_j(f_2z)}w(f_2z),$$

where $z \in U$, $t \in (-\epsilon,\epsilon)$, say, and $j \in \{1,2\}$ (with further restrictions on the domain U, as necessary, to ensure that f_1, f_2 are analytic on U) is called a *weighted (Ruelle) transfer operator*.

By standard perturbation theory [11], the maximal eigenvalue $\lambda_t^{(j)}$ for $\mathcal{L}_t^{(j)}$ has an analytic dependence on $t \in (-\epsilon,\epsilon)$, providing $\epsilon > 0$ is sufficiently small.

The significance of this is explained by the following characterization of the entropy rates.

Lemma 7.5. *The entropy rate is given by*

$$H = \frac{\partial\lambda_t^{(1)}}{\partial t}\bigg|_{t=0} + \frac{\partial\lambda_t^{(2)}}{\partial t}\bigg|_{t=0}.$$

Proof. This comes from differentiating the identity $\mathcal{L}_t^{(j)}h_t^{(j)} = \lambda_t^{(j)}h_t^{(j)}$, for $j = 1,2$, to get

$$\mathcal{L}_t^{(j)}\left(-(r_j\log r_j)h_t^{(j)} + \frac{\partial h_t^{(j)}}{\partial t}\right) = \frac{\partial\lambda_t^{(j)}}{\partial t}h_t^{(j)} + \lambda_t^{(j)}\frac{\partial h_t^{(j)}}{\partial t}.$$

Letting $t = 0$ and applying μ to both sides gives, for $j = 1,2$,

$$-\int r_j\log r_j\,d\mu = \frac{\partial\lambda_t^{(j)}}{\partial t}\bigg|_{t=0}.$$

The result then follows by adding the expressions for $j = 1,2$ and applying (2). $\square$

Remark 7.6. The identity (2) offers an alternative simpler, albeit less rapidly convergent, approach to estimating H. Since $\mathcal{L}^*\mu = \mu$ and 1 is an eigenvalue of maximal modulus, we have that for any fixed $z_0 \in U$

$$H = - \lim_{n \to +\infty} \left(\mathcal{L}^n \left(r_1(\cdot) \log r_1(\cdot) \right)(z_0) + \mathcal{L}^n \left(r_2(\cdot) \log r_2(\cdot) \right)(z_0) \right)$$

$$= - \lim_{n \to +\infty} \sum_{|\underline{i}|=n} r_{i_1}(z_0) r_{i_2}(f_{i_1} z_0) \cdots r_{i_n}(f_{i_{n-1}} \cdots f_{i_1} z_0)$$

$$\times \left(\sum_{j=1}^{2} r_j(f_{\underline{i}} z_0) \log r_j(f_{\underline{i}} z_0) \right).$$

The convergence is just exponential, the rate being given by the spectral gap of $\mathcal{L}$ (i.e., the modulus of the second eigenvalue). However, it is useful in giving an alternative way to check the estimates of H in Theorem 5.5.

7.3 Step 3: transfer operators and determinants

We can define a family of *determinants* (generalizing slightly those in (1)) by

$$d^{(j)}(z,t) := \exp \left(- \sum_{n=1}^{\infty} \frac{z^n}{n} \operatorname{tr} \left((\mathcal{L}_t^{(j)})^n \right) \right).$$

By modifying Lemma 7.2, we can find explicit values for the traces. The operator $(\mathcal{L}_t^{(j)})^n$ takes the form

$$(\mathcal{L}^{(j)})^n w(z) = \sum_{|\underline{i}|=n} \frac{r_{\underline{i}} \times \left(e^{-t(r_j \log r_j)} \right)_{\underline{i}}}{1 - \lambda_{\underline{i}}},$$

where

$$\left(e^{-t(r_j \log r_j)} \right)_{\underline{i}}$$
$$= e^{-tr_j(f_{i_1} x_{\underline{i}}) \log r_j(f_{i_1} x_{\underline{i}})} e^{-tr_j(f_{i_1} x_{\underline{i}}) \log r_j(f_{i_1} x_{\underline{i}})} \cdots e^{-tr_j(f_{i_n} \cdots f_{i_1} x_{\underline{i}}) \log r_j(f_{i_n} \cdots f_{i_1} x_{\underline{i}})}.$$

The analogue of Proposition 7.3 is the following (cf. [15]).

Proposition 7.7. *The function $d^{(j)}(z,t)$ is entire in both variables. Moreover, we can expand*

$$d^{(j)}(z,t) = 1 + \sum_{n=1}^{\infty} a_n^{(j)}(t) z^n, \quad \text{where } |a_n(t)| = O(e^{-cn^2})$$

for some $c > 0$.

Furthermore, the zeros of $z \mapsto d^{(j)}(z,t)$ are described in terms of the eigenvalues of the weighted transfer operators.

Lemma 7.8. *There is a zero for $d^{(j)}(z,t)$ of the form $z := z^{(j)}(t) = 1/\lambda_t^{(j)}$.*

Thus, by the implicit function theorem,

$$
-\frac{\partial d^{(j)}(1,t)}{\partial t}\bigg|_{t=0}\left(\frac{\partial d^{(j)}(z,0)}{\partial z}\bigg|_{z=1}\right)^{-1} = \frac{\partial z^{(j)}(t)}{\partial t}\bigg|_{t=0} = -\frac{\partial \lambda_t^{(j)}}{\partial t}\bigg|_{t=0}.
$$

Finally, using Lemma 7.5, we can write

$$
H = \frac{\partial \lambda_t^{(1)}}{\partial t}\bigg|_{t=0} + \frac{\partial \lambda_t^{(2)}}{\partial t}\bigg|_{t=0} = \frac{\frac{\partial d^{(1)}}{\partial t}(1,0)}{\frac{\partial d^{(1)}}{\partial z}(1,0)} + \frac{\frac{\partial d^{(2)}}{\partial t}(1,0)}{\frac{\partial d^{(2)}}{\partial z}(1,0)},
$$

as required.

References

[1] D. Blackwell. *The entropy of functions of finite-state Markov chains.* In Trans. First Prague Conf. Information Theory, 1957, pp. 13–20

[2] P. Davis and P. Rabinowitz. *Numerical Integration.* Blaisdell, Waltham, MA, 1967

[3] H. Furstenberg. *Noncommuting random products.* Trans. Amer. Math. Soc. **108** (1963) 377–428

[4] R. Gharavi and V. Ananatharam. *An upper bound for the largest Lyapunov exponent of a Markovian product of nonnegative matrices.* Theor. Comput. Sci. **332** (2005) 543–557

[5] A. Grothendieck. *Produits tensoriels topologiques et espaces nucléaires.* Mem. Amer. Math. Soc. **16** (1955) 1–140

[6] G. Han and B. Marcus. *Derivatives of entropy rate in special families of hidden Markov chains.* IEEE Trans. Inf. Theory **53** (2007) 2642–2652

[7] G. Han and B. Marcus. *Analyticity of entropy rate of hidden Markov chains.* IEEE Trans. Inf. Theory **52**(12) (2006) 5251–5266

[8] P. Jacquet, G. Seroussi, and W. Szpankowski. *On the entropy of a hidden Markov process.* In Proc. Data Compression Conf., Snowbird, UT, 2004 (http://www.cs.purdue.edu/homes/spa/papers/dcc04-hmm.ps)

[9] O. Jenkinson and M. Pollicott. *A dynamical approach to accelerating numerical integration with equidistributed points.* Tr. Mat. Inst. Steklov. Din. Sist. Optim. **256** (2007) 290–304

[10] N. Kambo. *Error of the Newton–Cotes and Gauss–Legendre quadrature formulas.* Math. Comput. **24** (1970) 261–269

[11] T. Kato. *Perturbation Theory for Linear Operators.* Grundlehren der Mathematischen Wissenschaften **132**. Springer, Berlin, 1976

[12] D. Mayer. *On a ζ function related to the continued fraction transformation.* Bull. Soc. Math. Fr. **104** (1976) 195–203

[13] Y. Peres. *Domains of analytic continuation for the top Lyapunov exponent.* Ann. Inst. H. Poincaré **28** (1992) 131–148

[14] M. Pollicott. *Maximal Lyapunov exponents for random matrix products.* Invent. Math. **181** (2010) 209–224

[15] D. Ruelle. *Zeta-functions for expanding maps and Anosov flows.* Invent. Math. **34** (1976) 231–242

8

Factors of Gibbs measures for full shifts

MARK POLLICOTT

Mathematics Institute, University of Warwick,
Coventry CV4 7AL, UK
E-mail address: masdbl@warwick.ac.uk

THOMAS KEMPTON

Mathematics Institute, University of Warwick,
Coventry CV4 7AL, UK
E-mail address: t.kempton@warwick.ac.uk

Abstract. We study the images of Gibbs measures under one-block factor maps on full shifts and the changes in the variations of the corresponding potential functions.

1 Introduction

In this article we consider the images of invariant measures under simple factor maps. Let $\sigma_1 : \Sigma_1 \to \Sigma_1$ and $\sigma_2 : \Sigma_2 \to \Sigma_2$ denote two full shifts, on finite alphabets A and B, respectively. Assume that $\Pi : \Sigma_1 \to \Sigma_2$ is a one-block factor map (i.e., a semiconjugacy satisfying $\Pi\sigma_1 = \sigma_2\Pi$, where $(\Pi(x))_0 = \pi(x_0)$ for a surjective map $\pi : A \to B$). Furthermore, let us consider a σ_1-invariant probability measure μ_{ψ_1} which is a Gibbs measure for a potential function $\psi_1 : \Sigma_1 \to \mathbb{R}$. In particular, the image $\nu = \pi_*\mu_{\psi_1}$ will be a σ_2-invariant probability measure.

A very natural question would be to ask under which hypotheses on ψ_1 would ν be a Gibbs measure for a potential $\psi_2 : \Sigma_2 \to \mathbb{R}$ (i.e., $\nu = \mu_{\psi_2}$), and how would the regularity of ψ_1 be reflected in that of ψ_2.

In the particular case that ψ_1 is locally constant, and thus μ_{ψ_1} is a generalized Markov measure, it was shown by Chazottes and Ugalde [2] that ν is a Gibbs measure for a Hölder continuous function $\psi_2 : \Sigma_2 \to \mathbb{R}$ (i.e., $\nu = \mu_{\psi_2}$), although not necessarily still a generalized Markov measure. One of our main results is the following.

Entropy of Hidden Markov Processes and Connections to Dynamical Systems: Papers from the Banff International Research Station Workshop, ed. B. Marcus, K. Petersen, and T. Weissman. Published by Cambridge University Press. © Cambridge University Press 2011.

Theorem 1.1. *Assume that* $\sum_{n=0}^{\infty} n \operatorname{var}_n(\psi_1) < +\infty$; *then* $\sum_{n=0}^{\infty} \operatorname{var}_n(\psi_2) < +\infty$.

The method we describe can be extended to the case of suitable subshifts of finite type and factor maps. These results appear in Chapter 7 in this book.

In Section 2, we recall some basic properties of Gibbs measures. In Section 3, we discuss a construction of the potential function ψ_2. In Section 4, we present the proof of a key step in this construction (Proposition 3.2). Finally, in Section 5, we present and prove the main results.

After completing this work, Ugalde informed the authors of his contemporaneous work with Chazottes. In [3] they have proved related results using a somewhat different method. In particular, in [3] it is shown that if $\sum_{n=0}^{\infty} n^{2+\epsilon} \operatorname{var}_n(\psi_1) < +\infty$ then $\sum_{n=0}^{\infty} \operatorname{var}_n(\psi_2) < +\infty$. However, the methods there appear to give sharper bounds on the actual rates of decay of the terms $\operatorname{var}_n(\psi_2)$ as $n \to +\infty$ than we can obtain.

2 Gibbs measures and equilibrium states

We begin with some general definitions and results. Given $k \geq 2$, let $\Sigma = \{1,\ldots,k\}^{\mathbb{Z}^+}$ denote a full shift space, with the shift map $\sigma : \Sigma \to \Sigma$ defined by $(\sigma \underline{w})_n = w_{n+1}$. For each $n \geq 0$, we define the nth-level variation

$$\operatorname{var}_n(\psi) := \sup\{|\psi(x) - \psi(y)| : x_i = y_i, 0 \leq i \leq n-1\}$$

of $\psi : \Sigma \to \mathbb{R}$. Continuity of the function ψ corresponds to $\operatorname{var}_n(\psi) \to 0$ as $n \to +\infty$.

We say that ψ has *summable variation* if $\sum_{n=0}^{\infty} \operatorname{var}_n(\psi) < +\infty$. Stronger still, we say that ψ is *Hölder continuous* if there exists $0 < \theta < 1$ such that

$$||\psi||_{\theta} := \sup_{n \geq 1} \left\{ \frac{\operatorname{var}_n(\psi)}{\theta^n} \right\} < +\infty.$$

Given any continuous function $\psi : \Sigma \to \mathbb{R}$ on a subshift of finite type Σ, we can define the *pressure* by

$$P(\psi) = \sup_{\mu \in \mathcal{M}} \left\{ h(\mu) + \int \psi d\mu \right\},$$

where the supremum is over the space $\mathcal{M}$ of all σ-invariant probability measures. This is equivalent to other standard definitions, using the variational principle [5].

A measure which realizes this supremum is called an *equilibrium state*. Any continuous function has at least one equilibrium state, and every invariant probability measure is the equilibrium state for some continuous potential (cf. [4]). If ψ has summable variation, then there is a unique equilibrium state μ_ψ. Given $x_0, \ldots, x_{n-1} \in \{1, \ldots, k\}$, we denote

$$[x_0, \ldots, x_{n-1}] = \{\underline{w} \in \Sigma : w_i = x_i \text{ for } 0 \le i \le n-1\}.$$

We have the following alternative characterization (cf. [1]).

Lemma 2.1. *If ψ has summable variation, then the following are equivalent:*

(1) *A σ-invariant probability measure μ is the unique equilibrium state for ψ.*
(2) *There exist $C_1, C_2 > 0$ such that*

$$C_1 \le \frac{\mu[w_0, \ldots, w_{n-1}]}{\exp(\psi^n(\underline{w}) - nP(\psi))} \le C_2 \tag{1}$$

for all $\underline{w} \in \Sigma$.

We shall refer to the unique invariant measure satisfying the Bowen–Gibbs inequality (1) as a *Gibbs measure* (in the sense of Bowen [1]).

Since we are primarily interested in measures, rather than functions, we can replace ψ by $\psi - P(\psi)$ (and still have $\mu_{\psi_1} = \mu_{\psi_1 - P(\psi_1)}$), and so in the sequel we shall assume, without loss of generality, that $P(\psi) = 0$.

Remark 2.2. For the two-sided shift $\sigma : \Sigma \to \Sigma$ on $\Sigma = \{1, \ldots, k\}^{\mathbb{Z}}$ and a function $\psi : \Sigma \to \mathbb{R}$, we define $\mathrm{var}_n(\psi) := \sup\{|\psi(x) - \psi(y)| : x_i = y_i, |i| \le n-1\}$. If $\sum_{n=0}^{\infty} n^2 \, \mathrm{var}_n(\psi) < +\infty$, we can add a coboundary to obtain a function $\psi'(x)$ which depends only on $(x_n)_{n=0}^{\infty}$ and for which $\sum_{n=0}^{\infty} n \, \mathrm{var}_n(\psi') < +\infty$. Then Theorem 1.1, for example, applies.

3 Constructing the potential ψ_2

If μ_{ψ_1} is a Gibbs measure for a continuous function ψ_1 (satisfying $P(\psi_1) = 0$), then we write[1]

$$\nu[z_0, \ldots, z_n] = \sum_{x_0, \ldots, x_n} \mu_{\psi_1}[x_0, \ldots, x_n] \asymp \sum_{x_0, \ldots, x_n} e^{\psi_1^{n+1}(x)},$$

---

[1] Here $\asymp$ means that the ratio of the two sides is bounded above and below (away from zero) independently of n.

with each summation being over finite strings $x_0, \ldots, x_n$ from Σ_1 for which $\pi(x_i) = z_i$, for $i = 0, \ldots, n$, and any $x \in [x_0, \ldots, x_{n-1}]$. This motivates the construction of the potential function via the limit of a sequence of functions in $C(\Sigma_2, \mathbb{R})$.

We begin by fixing, for the moment, a sequence $\underline{w} \in \Sigma_2$.

Definition 3.1. We consider a sequence of functions $(u_n(\underline{z}))_{n=1}^{\infty}$ in $C(\Sigma_2, \mathbb{R})$ defined by

$$u_n(\underline{z}) = \frac{\sum_{\underline{x} = x_0 \cdots x_n} \exp(\psi_1^{n+1}(\underline{xw}))}{\sum_{\underline{x'} = x_1 \cdots x_n} \exp(\psi_1^n(\underline{x'w}))},$$

where $\underline{z} \in \Sigma_2$ for $n \geq 2$, $u_1(\underline{z}) = \sum_{x_0} \exp(\psi_1(x_0 \underline{w}))$, and

(1) $\psi_1^n = \sum_{i=0}^{n-1} \psi_1 \circ \sigma^i$ and $\psi_1^{n+1} = \sum_{i=0}^{n} \psi_1 \circ \sigma^i$;
(2) each summation is over finite strings from Σ_1 for which x_i projects to z_i for $i = 0, \ldots, n$; and
(3) $\underline{xw} \in \Sigma_1$ denotes the concatenation of words to give the sequence $(x_0, \ldots, x_n, w_0, w_1, \ldots)$.

It is clear that $u_n(\underline{z})$ depends only on $z_0, \ldots, z_n$, i.e., $u_n : \Sigma_2 \to \mathbb{R}$ is a locally constant function.

The sequence $(u_n(\underline{z}))_{n=1}^{\infty}$ has an explicit dependence on the sequence $\underline{w}$. When appropriate, we will change the notation to reflect the $\underline{w}$ dependence by writing $u_{\underline{w},n}(\underline{z})$. We need the following result.

Proposition 3.2. *The sequence* $\{u_{\underline{w},n}(\underline{z})\}$ *converges uniformly to a continuous function* $u(\underline{z}) > 0$ *which is independent of* $\underline{w}$.

We will return to the proof of this proposition in the next section. Before stating a corollary, we present a lemma we need in its proof.

Lemma 3.3. *If* μ_{ψ_1} *is a Gibbs measure with continuous potential* ψ_1*, then there exists* $C > 0$ *such that for any* $\underline{x} = x_0, \ldots, x_n$*,* $\underline{w}, \underline{w}' \in \Sigma_1$*,* $n \in \mathbb{N}$*, we have*

$$\frac{\exp(\psi_1^{n+1}(\underline{xw}))}{\exp(\psi_1^{n+1}(\underline{xw'}))} \leq C.$$

Proof. By definition, there exist $C_1, C_2 > 0$ such that

$$C_1 \exp(\psi_1^{n+1}(\underline{xw})) \leq \mu_{\psi_1}[x_0, \ldots, x_n] \leq C_2 \exp(\psi_1^{n+1}(\underline{xw'})) \tag{2}$$

for all $\underline{w} \in \Sigma$, and thus

$$\frac{\exp(\psi_1^{n+1}(x\underline{w}))}{\exp(\psi_1^{n+1}(x\underline{w}'))} \leq \frac{C_2}{C_1}.$$

The result follows with $C = C_2/C_1$. $\square$

Corollary 3.4. *The measure* $\nu = \Pi_* \mu_{\psi_1}$ *is a Gibbs measure for* $\psi_2(\underline{z}) := \log u(\underline{z})$ *(i.e.,* $\nu = \mu_{\psi_2}$*).*

Proof. Fix $n \geq 1$. We can write

$$\psi_2^n(\underline{z}) = \lim_{m \to +\infty} \log u_{\underline{w},m}(\underline{z}) + \cdots + \lim_{m \to +\infty} \log u_{\underline{w},m}(\sigma^n \underline{z})$$

$$= \lim_{m \to +\infty} \log \left(\frac{\sum_{\underline{x}=x_0 \cdots x_m} \exp(\psi_1^{m+1}(x\underline{w}))}{\sum_{\overline{x}=x_{n+1} \cdots x_m} \exp(\psi_1^{m-n}(\overline{x}\underline{w}))} \right).$$

Moreover, for $m > n$ we can factor the summation

$$\sum_{\underline{x}=x_0 \cdots x_m} \exp(\psi_1^{m+1}(x\underline{w}))$$

$$= \sum_{\underline{x}'=x_0 \cdots x_n} \sum_{\overline{x}=x_{n+1} \cdots x_m} \exp(\psi_1^{n+1}(\underline{x}'\overline{x}\underline{w})) \exp(\psi_1^{m-n}(\overline{x}\underline{w}))$$

and then bound

$$\frac{1}{C} \sum_{\underline{x}'=x_0 \cdots x_n} \exp(\psi_1^{n+1}(\underline{x}'\underline{w})) \leq \underbrace{\frac{\sum_{\underline{x}=x_0 \cdots x_m} \exp(\psi_1^{m+1}(x\underline{w}))}{\sum_{\overline{x}=x_{n+1} \cdots x_m} \exp(\psi_1^{m-n}(\overline{x}\underline{w}))}}_{=\exp(\sum_{i=0}^n \log u_{\underline{w},m}(\sigma^i \underline{z}))}$$

$$\leq C \sum_{\underline{x}'=x_0 \cdots x_n} \exp(\psi_1^{n+1}(\underline{x}'\underline{w})),$$

where C is as in the previous lemma.

Since μ_{ψ_1} is a Gibbs measure for ψ_1, we can apply (2), and then summing over strings with $\pi(x_i) = z_i$, for $i = 0, \ldots, n-1$, and letting $m \to +\infty$ gives

$$\frac{C_1}{C} \leq \frac{\nu[z_0 \cdots z_{n-1}]}{\exp(\psi_2^n(\underline{z}))} \leq C_2 C.$$

It then follows from the definitions that ν is a Gibbs measure for ψ_2. $\square$

4 Proof of Proposition 3.2

We return to the proof postponed from the previous section.

Definition 4.1. Let $s > n$, $\bar{x} = x_{n+1} \cdots x_s$ be a given finite string and $\underline{w} \in \Sigma$. We define

$$P^{(s,n)}(\bar{x}, \underline{w}) = \frac{\sum_{\underline{x} = x_1 \cdots x_n} \exp(\psi_1^n(\underline{x}\bar{x}\underline{w}))}{\sum_{\underline{x}' = x_1 \cdots x_s} \exp(\psi_1^n(\underline{x}'\underline{w}))}$$

(where $\pi(x_i) = z_i$ for $i = 1, \ldots, s$).

We require the following lemma.

Lemma 4.2. *For $s > n$, we have the identity*

$$u_{\underline{w},s}(\underline{z}) = \sum_{\bar{x} = x_{n+1} \cdots x_s} u_{\bar{x}\underline{w},n}(\underline{z}) P^{(s,n)}(\bar{x}, \underline{w}),$$

where $\underline{z} \in \Sigma_2$ (and $\pi(x_i) = z_i$ for $i = n+1, \ldots, s$).

Proof. By definition, the *numerator* of $u_{\underline{w},s}(\underline{z})$ is

$$\sum_{\underline{x} = x_0 \cdots x_s} \exp(\psi_1^{s+1}(\underline{x}\underline{w})) = \sum_{\underline{x} = x_0 \cdots x_n} \sum_{\bar{x} = x_{n+1} \cdots x_s} \exp(\psi_1^{n+1}(\underline{x}\bar{x}\underline{w})) \exp(\psi_1^{s-n}(\bar{x}\underline{w}))$$

(where $\pi(x_i) = z_i$ for $i = 0, \ldots, s$) and we have used $\psi_1^{s+1}(\bar{x}\underline{x}\underline{w}) = \psi_1^{n+1}(\underline{x}\bar{x}\underline{w}) + \psi_1^{s-n}(\bar{x}\underline{w})$. We can further rewrite this as

$$\sum_{\bar{x} = x_{n+1} \cdots x_s} \underbrace{\left(\frac{\sum_{\underline{x} = x_0 \cdots x_n} \exp(\psi_1^{n+1}(\underline{x}\bar{x}\underline{w}))}{\sum_{\underline{x}' = x_1 \cdots x_n} \exp(\psi_1^n(\underline{x}'\bar{x}\underline{w}))} \right)}_{u_{\bar{x}\underline{w},n}(\underline{z})}$$

$$\times \underbrace{\left(\sum_{\underline{x}' = x_1 \cdots x_n} \exp(\psi_1^n(\underline{x}'\bar{x}\underline{w})) \right) \exp(\psi_1^{s-n}(\bar{x}\underline{w}))}_{\sum_{\underline{x}' = x_1 \cdots x_n} \exp(\psi_1^{s+1}(\underline{x}'\bar{x}\underline{w}))}.$$

We can now divide by the *denominator* of $u_s(\underline{z})$ to get the result. □

Corollary 4.3. *We have*

$$\frac{u_{\underline{w},s}(\underline{z})}{u_{\underline{w}',s}(\underline{z})} = \frac{\sum_{\bar{x} = x_{n+1} \cdots x_s} u_{\bar{x}\underline{w},n}(\underline{z}) P^{(s,n)}(\bar{x}, \underline{w})}{\sum_{\bar{x} = x_{n+1} \cdots x_s} u_{\bar{x}\underline{w}',n}(\underline{z}) P^{(s,n)}(\bar{x}, \underline{w}')}.$$

The following special case is illustrative.

Example 4.4. (Markov case) Assume that ψ_1 depends on only the first two coordinates. Let $s = n+2$ and then observe from the definitions that $u_{x_n x_{n+1} \underline{w}, n}(\underline{z})$ is independent of $\underline{w}$. In particular,

$$\frac{u_{w_1 w_2, n+2}(\underline{z})}{u_{w_1' w_2', n+2}(\underline{z})} = \frac{\sum_{x_n, x_{n+1}} u_{x_n x_{n+1}, n}(\underline{z}) P^{(n+2, n)}(x_n x_{n+1}, w_1 w_2)}{\sum_{x_n, x_{n+1}} u_{x_n x_{n+1}, n}(\underline{z}) P^{(n+2, n)}(x_n x_{n+1}, w_1' w_2')}.$$

In this case, Lemma 4.2 corresponds to a linear action by a strictly positive matrix, which contracts the simplex, and so the images converge at an exponential rate to a fixed line.

It remains to use the corollary to complete the proof of Proposition 3.2. We require the following lemma.

Lemma 4.5. *There exists* $c > 0$ *such that* $P^{(s,n)}(\overline{x}, \underline{w}) / P^{(s,n)}(\overline{x}, \underline{w}') \geq c$ *for all* $\overline{x} = x_n \cdots x_s$ *and* $\underline{w}, \underline{w}' \in \Sigma$.

Proof. Since $\underline{x}\overline{x}\underline{w}$ and $\underline{x}\overline{x}\underline{w}'$ agree in s places, we see that

$$|\psi_1^n(\underline{x}\overline{x}\underline{w}) - \psi_1^n(\underline{x}\overline{x}\underline{w}')| \leq \log\left(\frac{C_2}{C_1}\right),$$

and the result follows with $c = (C_1/C_2)^2$. $\square$

For each $n \geq 0$ and $\underline{z} \in \Sigma_2$, we denote

$$\lambda_n = \lambda_n(\underline{z}) := \sup\left\{ \frac{u_{\underline{w}, n}(\underline{z})}{u_{\underline{w}', n}(\underline{z})} : \underline{w}, \underline{w}' \in \Sigma_2 \right\} \geq 1.$$

Lemma 4.6. *The sequence* $\lambda_0 \geq \lambda_1 \geq \lambda_2 \geq \cdots \geq \lambda_n \geq \cdots \geq 1$ *is a monotone decreasing sequence. Furthermore, the intervals* $[\inf_{\underline{w}} u_{\underline{w}, n}(\underline{z}), \sup_{\underline{w}} u_{\underline{w}, n}(\underline{z})]$ *form a nested sequence in* n.

Proof. Fix $n \geq 1$. By definition,

$$u_{\underline{w}, n+1}(\underline{z}) = \frac{\sum_{\underline{x} = x_0 \cdots x_n} \sum_{x_{n+1}} \exp(\psi_1^{n+1}(\underline{x} x_{n+1} \underline{w})) \exp(\psi_1(x_{n+1} \underline{w}))}{\sum_{\underline{x}' = x_1 \cdots x_n} \sum_{x_{n+1}} \exp(\psi_1^n(\underline{x}' x_{n+1} \underline{w})) \exp(\psi_1(x_{n+1} \underline{w}))}$$

$$\leq \max_{x_{n+1}} \underbrace{\frac{\sum_{\underline{x} = x_0 \cdots x_n} \exp(\psi_1^{n+1}(\underline{x} x_{n+1} \underline{w}))}{\sum_{\underline{x}' = x_1 \cdots x_n} \exp(\psi_1^n(\underline{x}' x_{n+1} \underline{w}))}}_{u_{x_{n+1} \underline{w}, n}(\underline{z})}.$$

The inequality here comes from the fact that

$$\frac{\sum_{x_{n+1}} a(x_{n+1})}{\sum_{x_{n+1}} b(x_{n+1})} \le \max_{x_{n+1}} \frac{a(x_{n+1})}{b(x_{n+1})}.$$

Similarly, $\min_{x'_{n+1}} u_{x'_{n+1}\underline{w}',n}(\underline{z}) \le u_{\underline{w},n+1}(\underline{z})$. Thus, if $\underline{w}, \underline{w}' \in \Sigma_2$, then

$$\frac{u_{\underline{w},n+1}(\underline{z})}{u_{\underline{w}',n+1}(\underline{z})} \le \frac{u_{x_n\underline{w},n}(\underline{z})}{u_{x'_n\underline{w}',n}(\underline{z})} \le \lambda_n.$$

Taking the supremum over $\underline{w}, \underline{w}' \in \Sigma_2$ gives $\lambda_{n+1} \le \lambda_n$. $\qquad\square$

The following inequality is central to our analysis.

Lemma 4.7. *Let* $s > n$. *Then*

$$\lambda_s \le c \exp\left(\sum_{k=s-n}^{s} \mathrm{var}_k(\psi_1)\right) + (1-c)\lambda_n. \tag{3}$$

Proof. If $\lambda_n \le \exp\left(\sum_{k=s-n}^{s} \mathrm{var}_k(\psi_1)\right)$, then

$$\lambda_s = c\lambda_s + (1-c)\lambda_s \le c \exp\left(\sum_{k=s-n}^{s} \mathrm{var}_k(\psi_1)\right) + (1-c)\lambda_n,$$

as required. Therefore, we can assume that $\lambda_n \ge \exp\left(\sum_{k=s-n}^{s} \mathrm{var}_k(\psi_1)\right)$.

For ease of notation, we denote $b_{s,n} := \exp\left(\sum_{k=s-n}^{s} \mathrm{var}_k(\psi_1)\right)$. We want to combine Corollary 4.3 and Lemma 4.6. For any $\underline{w}, \underline{w}' \in \Sigma$, we can bound

$$\frac{u_{\underline{w},s}(\underline{z})}{u_{\underline{w}',s}(\underline{z})}$$

$$= \frac{\sum_{\bar{x}=x_{n+1}\cdots x_s} c u_{\bar{x}\underline{w},n}(\underline{z}) P^{(s,n)}(\bar{x},\underline{w}) + (1-c) u_{\bar{x}\underline{w},n}(\underline{z}) P^{(s,n)}(\bar{x},\underline{w})}{\sum_{\bar{x}=x_{n+1}\cdots x_s} c u_{\bar{x}\underline{w}',n}(\underline{z}) P^{(s,n)}(\bar{x},\underline{w}) + (P^{(s,n)}(\bar{x},\underline{w}') - c P^{(s,n)}(\bar{x},\underline{w})) u_{\bar{x}\underline{w}',n}(\underline{z})}$$

$$\le \frac{\sum_{\bar{x}=x_{n+1}\cdots x_s} c\left(b_{s,n} \min_{\bar{x}'}\{u_{\bar{x}'\underline{w}',n}(\underline{z})\}\right) P^{(s,n)}(\bar{x},\underline{w}) + (1-c) \max_{\bar{x}}\{u_{\bar{x}\underline{w},n}(\underline{z})\} P^{(s,n)}(\bar{x},\underline{w})}{\sum_{\bar{x}=x_{n+1}\cdots x_s} c \min_{\bar{x}'}\{u_{\bar{x}'\underline{w}',n}(\underline{z})\} P^{(s,n)}(\bar{x},\underline{w}) + (P^{(s,n)}(\bar{x},\underline{w}') - c P^{(s,n)}(\bar{x},\underline{w})) \min_{\bar{x}'}\{u_{\bar{x}'\underline{w}',n}(\underline{z})\}}$$

$$= \frac{\sum_{\bar{x}=x_{n+1}\cdots x_s} c b_{s,n} P^{(s,n)}(\bar{x},\underline{w}) + (1-c) \frac{\max_{\bar{x}}\{u_{\bar{x}\underline{w},n}(\underline{z})\}}{\min_{\bar{x}'}\{u_{\bar{x}'\underline{w}',n}(\underline{z})\}} P^{(s,n)}(\bar{x},\underline{w})}{\sum_{\bar{x}=x_{n+1}\cdots x_s} c P^{(s,n)}(\bar{x},\underline{w}) + (P^{(s,n)}(\bar{x},\underline{w}') - c P^{(s,n)}(\bar{x},\underline{w}))}$$

$$\le c b_{s,n} + (1-c) \frac{\max_{\bar{x}}\{u_{\bar{x}\underline{w},n}(\underline{z})\}}{\min_{\bar{x}'}\{u_{\bar{x}'\underline{w}',n}(\underline{z})\}}$$

$$\le c b_{s,n} + (1-c)\lambda_n,$$

as required. For the second line here we decreased the denominator by replacing $u_{\bar{x}\underline{w}',n}(\underline{z})$ with $\min_{\bar{x}'}\{u_{\bar{x}'\underline{w}',n}(\underline{z})\}$ and increased the numerator by replacing $u_{\bar{x}\underline{w},n}(\underline{z})$ with $\max_{\bar{x}}\{u_{\bar{x}\underline{w},n}(\underline{z})\}$ and $b_{s,n}\min_{\bar{x}'}\{u_{\bar{x}'\underline{w}',n}(\underline{z})\}$. □

Choosing $s = 2n$ gives the following corollary.

Corollary 4.8. *For any $n \geq 1$,*

$$\lambda_{2n} \leq \underbrace{c\exp\left(\sum_{k=n}^{2n}\mathrm{var}_k(\psi_1)\right)}_{b_{2n,n}} + (1-c)\lambda_n.$$

In particular, since $b_{2n,n} \to 1$ as $n \to +\infty$, it is clear that $\lim_{n\to+\infty}\lambda_n = 1$. Moreover, the regularity of u, and thus of ψ_2, is determined by the speed of convergence.

To complete the proof of Proposition 3.2, we need the following simple lemma.

Lemma 4.9. *We have*

(1) *for any $\underline{w} \in \Sigma_2$, the limit $u(\underline{z}) = \lim_{n\to+\infty}u_{\underline{w},n}(\underline{z})$ exists; and*
(2) *$\psi_2 := \log(u(\underline{z}))$ is continuous in $\underline{z} \in \Sigma_2$.*

Proof. For the first part, it suffices to use the previous observation that $\lim_{n\to+\infty}\lambda_n = 1$.

For the second part, we observe that if $\underline{z}_1, \underline{z}_2 \in \Sigma_2$ agree in n terms, then for any $\underline{w}$ we have $u_n(\underline{z}_1) = u_n(\underline{z}_2)$. Thus,

$$\left|\frac{u(\underline{z}_1)}{u(\underline{z}_2)}\right| \leq \sup\left\{\frac{u_{\underline{w},n}(\underline{z}_1)}{u_{\underline{w}'n}(\underline{z}_2)} : \underline{w},\underline{w}' \in \Sigma_2\right\} = \lambda_n,$$

and again the result follows from $\lim_{n\to+\infty}\lambda_n = 1$. □

We now have that $\psi_2 = \log u$ is a well-defined continuous function which is a potential for ν, which completes the proof of Proposition 3.2.

Remark 4.10. There are more general hypotheses that one might consider, including, for example, the Walters condition, but it is not immediately clear how the methods in this article can be adapted to that case. However, observe that any potential ψ_1 satisfying the Bowen–Gibbs inequality must necessarily be uniformly continuous. Since the only condition on ψ_1 that we have used is that it is uniformly continuous, Proposition 3.2 shows that if μ_{ψ_1} is a Gibbs measure, then $\nu := \Pi(\mu_{\psi_1})$ is a Gibbs measure.

5 Regularity of the potential ψ_2

In this section, we will consider the regularity properties of $\psi_2 := \log u$. The following is our main result.

Theorem 5.1. *Let* $\kappa \geq 0$. *If* $\sum_{n=0}^{\infty} n^{\kappa+1} \operatorname{var}_n(\psi_1) < \infty$, *then*

$$\sum_{n=0}^{\infty} n^{\kappa} \operatorname{var}_n(\psi_2) < \infty.$$

Proof. Let $0 < c < 1$ be as in Lemma 4.5. Choose $1 < \beta < 1/(1-c)$ and an integer $M > 1$ sufficiently large that $\alpha := \beta(1-c)(1+1/M)(1-1/M)^{-\kappa} < 1$. Let us denote $a_n = \log \lambda_n$ and recall the trivial inequality $1 + x \leq \exp(x) \leq 1 + \beta x$ for $x > 0$ sufficiently small. Thus, providing N_0 is sufficiently large, we can deduce from (3) that for any $n > N_0$

$$1 + a_n \leq \exp(a_n) \leq c \exp\left(\sum_{m=[n/M]}^{n} \operatorname{var}_m(\psi_1)\right) + (1-c)\exp(a_{n-[n/M]})$$

$$\leq c + (1-c) + \beta c \sum_{m=[n/M]}^{n} \operatorname{var}_m(\psi_1) + \beta(1-c)a_{n-[n/M]}.$$

Hence, for any $N > N_0$,

$$\sum_{n=N_0}^{N} n^{\kappa} a_n \leq \beta c \sum_{n=N_0}^{N} n^{\kappa} \sum_{m=[n/M]}^{n} \operatorname{var}_m(\psi_1) + \beta(1-c) \sum_{n=N_0}^{N} n^{\kappa} a_{n-[n/M]}$$

(where $[\cdot]$ denotes the integer part).

We can bound

$$\sum_{n=N_0}^{N} n^{\kappa} \sum_{m=[n/M]}^{n} \operatorname{var}_m(\psi_1) \leq M^{\kappa} \sum_{n=N_0}^{N} \sum_{m=[n/M]}^{n} m^{\kappa} \operatorname{var}_m(\psi_1)$$

$$\leq M^{\kappa+1} \sum_{n=N_0}^{N} n^{\kappa+1} \operatorname{var}_n(\psi_1)$$

and

$$\sum_{n=N_0}^{N} n^\kappa a_{n-[n/M]}$$

$$\leq \frac{1}{(1-1/M)^\kappa} \sum_{n=N_0}^{N} (n-[n/M])^\kappa a_{n-[n/M]}$$

$$\leq \frac{1}{(1-1/M)^\kappa} \left(\sum_{m=[N_0-N_0/M]}^{[N-N/M]+1} m^\kappa a_m + \sum_{\substack{N_0 \leq n \leq N \\ M \mid n+1}} (n-[n/M])^\kappa a_{n-[n/M]} \right)$$

$$\leq \frac{(1+1/M)}{(1-1/M)^\kappa} \sum_{m=N_0}^{N} m^\kappa a_m + O(1),$$

where we have used

$$\sum_{\substack{N_0 \leq n \leq N \\ M \mid n+1}} (n-[n/M])^\kappa a_{n-[n/M]} \leq \frac{1}{M} \sum_{m=N_0-[N_0/M]}^{N} m^\kappa a_m$$

and

$$\sum_{m=N_0-[N_0/M]}^{N_0-1} m^\kappa a_m = O(1).$$

Comparing the above inequalities, we can bound

$$\underbrace{\left(1 - \beta(1-c)\frac{(1+1/M)}{(1-1/M)^\kappa} \right)}_{>0} \sum_{n=N_0}^{N} n^\kappa a_n \leq \beta c M^{\kappa+1} \sum_{n=N_0}^{N} n^{\kappa+1} \mathrm{var}_n(\psi_1) + O(1).$$

Letting $N \to +\infty$, we see that $\sum_{n=N_0}^{\infty} n^\kappa a_n < \infty$, which completes the proof. $\square$

When $\kappa = 0$, we have the following corollary, which was stated in the Introduction.

Corollary 5.2. *If $\sum_{n=0}^{\infty} n\, \mathrm{var}_n(\psi_1) < \infty$, then $\sum_{n=0}^{\infty} \mathrm{var}_n(\psi_2) < \infty$.*

Another application of (3) is the following.

Theorem 5.3. *Assume that there exist $c_1 > 0$ and $0 < \theta_1 < 1$ such that $\mathrm{var}_n(\psi_1) \leq c_1 \theta_1^{\sqrt{n}}$ for all $n \geq 0$. Then there exist $c_2 > 0$ and $0 < \theta_2 < 1$ such that $\mathrm{var}_n(\psi_2) \leq c_2 \theta_2^{\sqrt{n}}$ for all $n \geq 0$.*

Proof. By (3), we can write

$$\lambda_n \leq c \exp\left(c_1 \sum_{k=n-[\sqrt{n}]}^{n} \theta_1^{\sqrt{k}}\right) + (1-c)\lambda_{n-[\sqrt{n}]}$$

$$\leq c \exp\left(C\theta^{[\sqrt{n}]}\right) + (1-c)\lambda_{n-[\sqrt{n}]}$$

for any $\theta_1 < \theta < 1$ and some $C > 0$. Using this inequality inductively $[\sqrt{n}]$ times, we can write

$$\lambda_n \leq c \exp\left(C\theta^{[\sqrt{n}]}\right) + (1-c)\left(c \exp\left(C\theta^{[\sqrt{n}]}\right) + (1-c)\lambda_{n-2[\sqrt{n}]}\right)$$

$$\cdots$$

$$\leq c \exp\left(C\theta^{[\sqrt{n}]}\right) \sum_{k=0}^{[\sqrt{n}]} (1-c)^k + (1-c)^{[\sqrt{n}]}\lambda_{n-[\sqrt{n}]^2}$$

$$\leq \exp\left(C\theta^{[\sqrt{n}]}\right) + (1-c)^{[\sqrt{n}]}\lambda_0.$$

In particular, we see that $|\lambda_n - 1| = O\left(\theta_2^{\sqrt{n}}\right)$, where $\theta_2 = \max\{\theta, (1-c)\}$, from which the result follows. □

The following is an easy consequence of the theorem and its proof.

Corollary 5.4. *Assume that there exist $c_1 > 0$ and $0 < \theta < 1$ such that $\mathrm{var}_n(\psi_1) \leq c_1\theta^n$ for all $n \geq 0$ (i.e., ψ_1 is Hölder continuous). Then there exists $c_2 > 0$ such that $\mathrm{var}_n(\psi_2) \leq c_2\theta^{\sqrt{n}}$ for all $n \geq 0$.*

The same conclusion, using a different proof, appears in [3].

References

[1] R. Bowen. *Ergodic Theory of Axiom A Diffeomorphisms.* Lecture Notes in Mathematics. Springer, Berlin, 1975

[2] J.-R. Chazottes and E. Ugalde. *Projection of Markov measures may be Gibbsian.* J. Statist. Phys. **111**(5–6) (2003) 1245–1272

[3] J.-R. Chazottes and E. Ugalde. *On the preservation of Gibbsianness under symbol amalgamation.* This volume, Chapter 2, 2011

[4] D. Ruelle. *Thermodynamic Formalism.* Addison-Wesley, New York, 1978

[5] P. Walters. *Ergodic Theory.* Graduate Texts in Mathematics **79**, Springer, Berlin, 1982

9

Thermodynamics of hidden Markov processes

EVGENY VERBITSKIY

Mathematics Institute, Leiden University, Postbus 9512,
2300 RA Leiden, The Netherlands
and
Johann Bernoulli Institute for Mathematics and Computer Science,
University of Groningen, PO Box 407, 9700 AK Groningen,
The Netherlands
E-mail address: evgeny@math.leidenuniv.nl

Abstract. Hidden Markov processes are extremely popular and have a wide range of applications. It turns out that under mild conditions the distributions of hidden Markov processes fall in the class of the so-called Gibbs measures. Moreover, a number of questions about hidden Markov processes investigated by information theory have a strong *thermodynamic* flavor. In the present article we review some of the implications of thermodynamic formalism to the study of hidden Markov processes.

1 Hidden Markov models and functions of Markov chains

We start by considering the following simple yet illustrative example of a hidden Markov chain.

Example 1.1. Consider a stationary Markov chain $\{X_n\}$, $X_n \in \{-1, 1\}$, with the transition probability matrix

$$\mathbf{P} = \begin{pmatrix} 1-p & p \\ p & 1-p \end{pmatrix}.$$

Let also $\{Z_n\}$, $Z_n = \{-1, 1\}$, be a sequence of independent random variables with

$$\mathbb{P}(Z_n = -1) = \varepsilon, \quad \mathbb{P}(Z_n = 1) = 1 - \varepsilon.$$

Define a process $\{Y_n\}$ by

$$Y_n = X_n \cdot Z_n \quad \forall\, n \in \mathbb{Z}.$$

Entropy of Hidden Markov Processes and Connections to Dynamical Systems: Papers from the Banff International Research Station Workshop, ed. B. Marcus, K. Petersen, and T. Weissman. Published by Cambridge University Press. © Cambridge University Press 2011.

The process $\{Y_n\}$ is a hidden Markov process, because $Y_n \in \{-1, 1\}$ is chosen independently for any n from an *emission* distribution π_{X_n} depending on X_n alone. Some authors [1, p. 185] referred to such processes as *Markov sources*, but this terminology did not become standard. In statistical physics, $\{Y_n\}$ is sometimes referred to as a *copy-with-noise* of $\{X_n\}$ [15]. The process $\{Y_n\}$ is also a realization of the so-called *Boltzmann chain (machine)*, see e.g., [42]. Equivalently, if we consider a Markov chain $\{X_n^{\text{ext}}\}$ with values in $\mathcal{A} = \{(1, 1), (1, -1), (-1, 1), (-1, -1)\}$, and a transition probability matrix

$$\mathbf{P}^{\text{ext}} = \begin{pmatrix} (1-p)(1-\varepsilon) & (1-p)\varepsilon & p(1-\varepsilon) & p\varepsilon \\ (1-p)(1-\varepsilon) & (1-p)\varepsilon & p(1-\varepsilon) & p\varepsilon \\ p(1-\varepsilon) & p\varepsilon & (1-p)(1-\varepsilon) & (1-p)\varepsilon \\ p(1-\varepsilon) & p\varepsilon & (1-p)(1-\varepsilon) & (1-p)\varepsilon \end{pmatrix},$$

then $Y_n = \phi(X_n^{\text{ext}})$ for an obvious deterministic function $\phi : \mathcal{A} \to \{-1, 1\}$. Hence, $\{Y_n\}$ is a *function* of a Markov chain $\{X_n^{\text{ext}}\}$ [19], also known as an *aggregated* [32], *lumped* [43], or *fuzzy* version of $\{X_n\}$ [35]. In dynamical systems literature [33], $\{Y_n\}$ would also be referred to as a *one-block factor* of $\{X_n^{\text{ext}}\}$. In general, there is a one-to-one correspondence between hidden Markov chains and functions of Markov chains. To simplify notation, from now on we consider only the functions of Markov chains.

Suppose that $\mathcal{A}$, $\mathcal{B}$ are finite sets, and $\phi : \mathcal{A} \to \mathcal{B}$ is onto. $Y_n = \phi(X_n)$ is a function of a Markov chain $\{X_n\}$ with values in $\mathcal{A}$. If $\mathbb{P}$ is a translation-invariant measure of the chain $\{X_n\}$, then $\mathbb{Q} = \mathbb{P} \circ \phi^{-1}$ is the translation-invariant measure for the process $\{Y_n\}$. With a minor abuse of notation, we will continue to use $\mathbb{P}$ to describe the joint distribution of $\{(X_n, Y_n)\}$.

2 Memory

In general, processes which are functions of Markov chains are not Markov (of any finite order). Conditions for $Y_n = \phi(X_n)$ to remain Markov are known, see e.g., [19] for an overview and extended list of references.

Therefore, the natural probabilistic question is to understand the structure of the conditional probabilities

$$\mathbb{Q}(y_0 | y_{-n}^{-1}) = \mathbb{Q}(Y_0 = y_0 | Y_{-1} = y_{-1}, \ldots, Y_{-n} = y_{-n})$$

and their eventual limit (if it exists)

$$\mathbb{Q}(y_0 | y_{-\infty}^{-1}) := \lim_{n \to \infty} \mathbb{Q}(Y_0 = y_0 | Y_{-1} = y_{-1}, \ldots, Y_{-n} = y_{-n}).$$

Proposition 2.1. *Suppose that* $\{X_n\}$, $X_n \in \mathcal{A}$, *is a stationary Markov chain with a strictly positive transition probability matrix*

$$\mathbf{P} = (p_{i,j})_{i,j \in \mathcal{A}} > 0,$$

Suppose also that $\phi : \mathcal{A} \mapsto \mathcal{B}$ *is a surjective map, and* $Y_n = \phi(X_n)$ *for every n. Then the sequence of conditional probabilities* $\mathbb{Q}(y_0|y_{-n}^{-1})$ *converges uniformly (in* $y_{-\infty}^0$) *as* $n \to \infty$ *to the limit denoted by* $\mathbb{Q}(y_0|y_{-\infty}^{-1})$. *There exist* $C > 0$ *and* $\varkappa, \theta \in (0,1)$ *such that*

$$\varkappa \leq \inf_{y_{-\infty}^0} \mathbb{Q}(y_0|y_{-\infty}^{-1}) \leq \sup_{y_{-\infty}^0} \mathbb{Q}(y_0|y_{-\infty}^{-1}) \leq 1 - \varkappa, \tag{1}$$

$$\beta_n := \sup_{y_{-n}^0} \sup_{\tilde{y}_{-\infty}^{-n-1}, \tilde{\tilde{y}}_{-\infty}^{-n-1}} \left| \mathbb{Q}(y_0|y_{-n}^{-1}\tilde{y}_{-\infty}^{-n-1}) - \mathbb{Q}(y_0|y_{-n}^{-1}\tilde{\tilde{y}}_{-\infty}^{-n-1}) \right| \leq C\theta^n \tag{2}$$

for all $n \in \mathbb{N}$.

The result of Proposition 2.1 shows that the memory decays exponentially. This result has been obtained in the literature a number of times using different methods, see [2, 8, 16, 22, 23, 21, 32]. Depending on the method employed, the values of θ vary greatly. Note that (2) implies that

$$\theta^* := \varlimsup_{n \to \infty} (\beta_n)^{1/n} \leq \theta. \tag{3}$$

Hence, smaller values of θ in the statement of Proposition 2.1 provide sharper estimates. The following table summarizes upper bounds for θ^* in the case of the hidden Markov chain defined in Example 1.1 for $p = 0.4$, $\varepsilon = 0.1$ (values $\theta_B, \theta_H, \theta_{BP}$ are taken from [49]):

Birch [8]	θ_B	≈ 0.99998495
Harris [22]	θ_H	≈ 0.97223
Baum and Petrie [2]	θ_{BP}	$= 0.94$
Han and Marcus [21]	θ_{HM}	$= 0.58$
Hochwald and Jelenković [23]	θ_{HJ}	$= 0.2$
Fernández *et al.* [16]	θ_{FFG}	$= 0.2$

Remarkably,

$$\theta_{HJ} = \theta_{FFG} = |1 - 2p|$$

are independent of ε.

During the Banff workshop, Yuval Peres explained that the results of [6, 30] could be used as an alternative derivation of the bound

$$\theta^* \leq |1 - 2p|. \tag{4}$$

This inequality is a corollary of a rather deep result on Ising models on graphs, which in the case of Example 1.1 gives the inequality

$$\gamma_n \leq \delta_n, \tag{5}$$

where

$$\gamma_n = \sup_{y_0, y_{-n}^{-1}} \left| \mathbb{P}[Y_0 = y_0 | Y_{-n}^{-1} = y_{-n}^{-1}, X_{-n-1} = 1] \right.$$

$$\left. - \mathbb{P}[Y_0 = y_0 | Y_{-n}^{-1} = y_{-n}^{-1}, X_{-n-1} = -1] \right| \tag{6}$$

and

$$\delta_n = \sup_{x_0} \left| \mathbb{P}[X_0 = x_0 | X_{-n-1} = 1] - \mathbb{P}[X_0 = x_0 | X_{-n-1} = -1] \right|. \tag{7}$$

One can easily show that $\beta_n \leq \gamma_n$. The decay rate of δ_n can be computed exactly, since

$$\mathbf{P}^{n+1} = \begin{pmatrix} \frac{1}{2} + \frac{1}{2}(1-2p)^{n+1} & \frac{1}{2} - \frac{1}{2}(1-2p)^{n+1} \\ \frac{1}{2} - \frac{1}{2}(1-2p)^{n+1} & \frac{1}{2} + \frac{1}{2}(1-2p)^{n+1} \end{pmatrix}$$

and thus

$$\delta_n = |1-2p|^{n+1}, \quad \lim_{n \to \infty} (\delta_n)^{1/n} = |1-2p|,$$

establishing the same result as in [23, 16]. Nevertheless, we conjecture that the inequality (4) between θ^* and $|1-2p|$ is *strict*, the reason being that the process $Y_n = X_n \cdot Z_n$ is more "random" than the original chain X_n, and that should result in faster decay. Hence, estimating γ_n independently should result in a better bound on θ^*. Based on the intuition developed in [6, 30], it seems plausible that the sequence $y_{-n}^{-1} = (y_{-n}, \dots, y_{-1}) \in \{-1, 1\}^n$ maximizing the difference γ_n in (6) is possibly an alternating sequence ± 1.

We complete this section with a short discussion of functions of Markov chains with transition probability matrices $\mathbf{P}$ which are not strictly positive. In this case the result of Proposition 2.1 is no longer valid. One can construct transition probability matrices $\mathbf{P} \geq 0$ such that the process $Y_n = \phi(X_n)$ has discontinuous conditional probabilities. The simplest example is due to Walters [47] and independently van den Berg [34]: letting $Y_n = X_n \cdot X_{n+1}$, where $\{X_n\}$ is a Bernoulli (independent) sequence with values $\{1, -1\}$ and probabilities

$$\mathbb{P}(X_n = 1) = 1 - p, \quad \mathbb{P}(X_n = -1) - p, \quad p \neq \frac{1}{2},$$

results in the process $\{Y_n\}$ whose infinite conditional probabilities are *discontinuous everywhere* [36]. Namely, there exists $c > 0$ such that, for any y_{-n}^{-1}, one can find two infinite continuations $\bar{y}_{-\infty}^{-n-1}, \tilde{y}_{-\infty}^{-n-1}$, such that

$$\varlimsup_{k \to \infty} \left| \mathbb{Q}[Y_0 = 1 | Y_{-n}^{-1} = y_{-n}^{-1}, Y_{-n-k}^{-n-1} = \bar{y}_{-n-k}^{-n-1}] \right.$$

$$\left. - \mathbb{Q}[Y_0 = 1 | Y_{-n}^{-1} = y_{-n}^{-1}, Y_{-n-k}^{-n-1} = \tilde{y}_{-n-k}^{-n-1}] \right| \geq c. \tag{8}$$

Note the difference in behavior of the conditional probabilities in (8) from the one given by (2). The result of (8) means that by changing the configuration arbitrarily far in the past one can significantly alter the conditional distribution of the spin at the origin. In statistical mechanics this phenomenon is referred to as nonquasilocality (discontinuity) of conditional probabilities. Note also that $Y_n = \phi(X_n^*)$, where X_n^* is a Markov chain with states $\{(-1, -1), (-1, 1), (1, -1), (1, 1)\}$, transition probability matrix

$$\mathbf{P} = \begin{pmatrix} p & 1-p & 0 & 0 \\ 0 & 0 & p & 1-p \\ p & 1-p & 0 & 0 \\ 0 & 0 & p & 1-p \end{pmatrix}, \quad \text{and} \quad \begin{aligned} \phi((-1, -1)) &= \phi((1, 1)) = 1, \\ \phi((-1, 1)) &= \phi((-1, 1)) = -1. \end{aligned}$$

The example by Walters and van den Berg shows that discontinuity may arise when $\mathbf{P}$ is only nonnegative. Finally, Chazottes and Ugalde [10] provided some sufficient conditions for functions of Markov chains with $\mathbf{P} \geq 0$ to have continuous conditional probabilities. Recently, their results have been strengthened by Kempton [29].

3 Thermodynamic formalism

The result of Proposition 2.1 immediately places the measure $\mathbb{Q}$ in the class of Gibbs measures [9, 40], as well as in the classes of g-measures [28, 46], chains of infinite order [13], chains with complete connections [25, 31], and countable mixtures of Markov chains [16]. For example, given the exponential decay established by Proposition 2.1, $\mathbb{Q}$ is a Gibbs measure for some Hölder continuous potential $\phi : \mathcal{B}^{\mathbb{Z}} \to \mathbb{R} : \mathbf{y} = (\ldots, y_{-1}, y_0, y_1, \ldots) \to \mathbb{R}$, i.e., there exist some constants $P, C \in \mathbb{R}$, $C > 1$, such that for all $\mathbf{y} \in \mathcal{B}^{\mathbb{Z}}$ one has

$$\frac{1}{C} \leq \frac{\mathbb{Q}(Y_1 = y_1, \ldots, Y_n = y_n)}{\exp\left(\sum_{k=1}^n \phi(\sigma^k \mathbf{y}) + nP\right)} \leq C,$$

where $\sigma : \mathcal{B}^{\mathbb{Z}} \to \mathcal{B}^{\mathbb{Z}}$ is the left shift. In fact, ϕ can be chosen to depend only on $(y_0, y_1, \ldots)$ (the "future") or $(\ldots, y_{-1}, y_0)$ (the "past").

For comparison, the results of [16] on countable mixtures of Markov chains guarantee that there exist a sequence of nonnegative numbers $\{\rho_n\}_{n \geq 0}$ and a sequence of n-step Markov measures $\{\mathbb{Q}_n\}_{n \geq 0}$ such that

$$\sum_{n \geq 0} \rho_n = 1$$

and

$$\mathbb{Q}(Y_0 = y_0 | Y_{-1} = y_{-1}, Y_{-2} = y_{-2}, \ldots) = \rho_0 \mathbb{Q}_0(y_0) + \sum_{n=1}^{\infty} \rho_n \mathbb{Q}(y_0 | y_{-n}^{-1}).$$

Moreover, the ρ_n decay exponentially fast.

Naturally, a large number of useful probabilistic results such as the *Variational Principle*, *Large Deviation Principle*, and *Central Limit Theorem* established for Gibbs measures in dynamical systems or statistical mechanics become readily available.

For example, the variational principle for Gibbs measures immediately guarantees the existence of *asymptotic divergence*: for an arbitrary translation-invariant measure $\bar{\mathbb{Q}}$ on $\mathcal{B}^{\mathbb{Z}}$, the following limit exists:

$$\mathbf{D}(\bar{\mathbb{Q}} || \mathbb{Q}) := \lim_{n \to \infty} \frac{1}{n} \sum_{y_1^n} \bar{\mathbb{Q}}(y_1^n) \log \frac{\bar{\mathbb{Q}}(y_1^n)}{\mathbb{Q}(y_1^n)}. \tag{9}$$

Moreover, $\mathbf{D}(\bar{\mathbb{Q}} || \mathbb{Q}) = 0$ if and only if $\bar{\mathbb{Q}} = \mathbb{Q}$.

Recent work [48] on the *Discrete Universal DEnoiser* (DUDE) stimulated interest in bidirectional modeling of stochastic processes [50], i.e., understanding the properties of two-sided conditional probabilities

$$q(y_0 | y_{-n}^{-1}, y_1^n) := \mathbb{Q}(Y_0 = y_0 | Y_{-n}^{-1} = y_{-n}^{-1}, Y_1^n = y_1^n)$$

and their eventual limits, as well as related quantities such as the *erasure entropies* [44, 45]. The Gibbs measures of statistical mechanics (typically described via two-sided conditional probabilities) seem to be natural candidates to be used in denoising problems (cf. [37]).

As we have argued above, the one-block image of a fully supported Markov measure on $\mathcal{A}^{\mathbb{Z}}$ is a Gibbs measure. What can be said about images (functions) of Gibbs states? It follows immediately from the result of Denker and Gordin [12] that an image of a Gibbs measure with a Hölder continuous potential is again a Gibbs measure for some Hölder continuous potential. The method developed in [12] in combination with a general approach to establishing the preservation of Gibbsianity under renormalization transformations proposed in [15] allows us to obtain the preservation of Gibbsianity under renormalization for potentials in an even larger class: namely, the potentials with *summable variation* [46].

3.1 Entropy

In the past few years great interest has arisen in computation of the entropy of functions of Markov chains [14, 23, 24, 26, 27, 38] and, more specifically, determining the dependence of the entropy on the parameters of the original Markov chain [26, 39, 51]. In cases such as Example 1.1 where $Y_n = X_n \cdot Z_n$, for small $\varepsilon > 0$ one has $\mathbb{P}(Y_n \neq X_n) = \varepsilon$ and, even stronger, by the result of [11],

$$\sup_{k} \sup_{a_0, a_{-k}^{-1}} |\mathbb{P}(X_0 = a_0 | X_{-k}^{-1} = a_{-k}^{-1}) - \mathbb{Q}(Y_0 = a_0 | Y_{-k}^{-1} = a_{-k}^{-1})| \leq C\varepsilon,$$

so it is natural to expect that the entropy-like characteristics of the process $\{Y_n\}$ do not deviate strongly from those of the original process $\{X_n\}$.

We mention two important recent results. First, Han and Marcus [21] established that if the entries of the transition probability matrix $\mathbf{P} = \mathbf{P}(\varepsilon)$ depend analytically on the parameter ε and $\mathbf{P}_0 = \mathbf{P}(0)$ satisfies a certain positivity condition (cf. $\mathbf{P} = \mathbf{P}_0 + \varepsilon \mathbf{T}$ with $\mathbf{P}_0 > 0$ as in Example 1.1), then the entropy

$$h(\mathcal{Y}) = \lim_{n \to \infty} \frac{1}{n} H(Y_1, \ldots, Y_n)$$

is a real analytic function of ε at 0. Han and Marcus [21] give two proofs of this result: the first proof is based on the detailed analysis of the *Blackwell process* $P_n = \mathbb{P}(X_n = \cdot | Y_{n-1}, \ldots, Y_0)$ and the second proof is based on the real analyticity of the topological pressure established in the framework of dynamical systems.

Zuk *et al.* [51] established a rather remarkable *stabilization* property of power series expansions. Consider again a Markov chain with the transition matrix $\mathbf{P} = \mathbf{P}_0 + \varepsilon \mathbf{T}$, $\mathbf{P}_0 > 0$, and $\varepsilon \geq 0$. Since

$$h(\mathcal{Y}) = \lim_{n \to \infty} \left[H(Y_1, \ldots, Y_n) - H(Y_1, \ldots, Y_{n-1}) \right],$$

and both $H(Y_n|Y_1^{n-1}) = H(Y_1,\ldots,Y_n) - H(Y_1,\ldots,Y_{n-1})$ and $h(\mathcal{Y})$ depend smoothly on ε, one can consider Taylor series expansions at $\varepsilon = 0$:

$$H(Y_n|Y_1^{n-1}) = \sum_{k=0}^{\infty} C_k^{(n)} \varepsilon^k,$$

$$h(\mathcal{Y}) = \sum_{k=0}^{\infty} C_k \varepsilon^k.$$

Naturally, $C_k^{(n)} \to C_k$ as $n \to \infty$. However, it turns out that

$$C_k^{(n)} = C_k \quad \text{for all } n \geq \frac{k+3}{2}.$$

This remarkable property allows one to determine C_k for any $k \geq 0$ by a *finite computation procedure*. For Example 1.1, one gets

$$C_0 = -p\log p - (1-p)\log p,$$

$$C_1 = 2(1-2p)\log\frac{1-p}{p},$$

$$C_2 = 2(1-2p)\log\frac{1-p}{p} - \frac{(1-2p)^2}{2p^2(1-p)^2},$$

$$\vdots$$

$$C_6 = 128\frac{(125\lambda^{14} - 321\lambda^{12} + 9525\lambda^{10} + 16511\lambda^8 - 7825\lambda^6 - 17995\lambda^4 - 4001\lambda^2 - 115)\lambda^4}{15(1-\lambda^2)^{10}},$$

where $\lambda = 1 - 2p$, with the general form conjectured to be

$$C_k = \frac{2^{4(k-1)}\sum_{j=0}^{d_k} a_{j,k}\lambda^{2j}}{k(k-1)(1-\lambda^2)^{2(k-1)}}, \quad k \geq 3.$$

Unfortunately, it does not seem possible to derive recurrent formulae for the coefficients $a_{j,k}$.

Zuk [52] extended the results of [51] even further. Suppose that $\mathbb{Q}, \overline{\mathbb{Q}}$ are hidden Markov measures, obtained as functions of Markov chains with transition probability matrices $\mathbf{P} = \mathbf{P}_0 + \varepsilon\mathbf{T}$, $\overline{\mathbf{P}} = \overline{\mathbf{P}}_0 + \varepsilon\overline{\mathbf{T}}$, respectively. Then for ε small enough there exists a power series expansion of the asymptotic divergence rate $\mathbf{D}(\overline{\mathbb{Q}}||\mathbb{Q})$ (cf. (9)).

3.2 Identification of the potential

It is interesting to determine the analytic form of the Gibbs potential for the measure $\mathbb{Q}$. From the general theory, we know that

$$\mathbb{Q}(Y_0 = y_0 | Y_{-1} = y_{-1}, Y_{-2} = y_{-2}, \ldots) = \frac{1}{Z(y_{-\infty}^{-1})} \exp\left(-\sum_{\Lambda \ni 0} U_\Lambda(y_\Lambda)\right),$$

where the sum is taken over all finite subsets Λ of $\mathbb{Z}_- = \{0, -1, -2, \ldots\}$ containing zero, U_Λ depends only on $y_\Lambda = \{y_i\}_{i \in \Lambda}$, and $Z(y_{-\infty}^{-1})$ is a normalizing factor such that

$$\sum_{y_0 \in \mathcal{B}} \frac{1}{Z(y_{-\infty}^{-1})} \exp\left(-\sum_{\Lambda \ni 0} U_\Lambda(y_\Lambda)\right) = 1$$

for all $y_{-\infty}^{-1} = (y_{-1}, y_{-2}, \ldots) \in \mathcal{B}^{-\mathbb{N}}$. Motivated by the result of Han and Marcus [21], one can ask whether, for example, in the case of chains with $\mathbf{P} = \mathbf{P}_0 + \varepsilon \mathbf{T}$ (such as Example 1.1), one can find power series expansions of the conditional probabilities and the potential as well:

$$\mathbb{Q}(Y_0 = y_0 | Y_{-1} = y_{-1}, Y_{-2} = y_{-2}, \ldots) = \sum_{k=0}^{\infty} F_k(\mathbf{y}) \varepsilon^k,$$

$$\mathbb{Q}(Y_0 = y_0 | Y_{-1} = y_{-1}, Y_{-2} = y_{-2}, \ldots) = \exp\left(\sum_{k=0}^{\infty} G_k(\mathbf{y}) \varepsilon^k\right);$$

here $\mathbf{y} = (y_0, y_{-1}, \ldots) \in \mathcal{B}^{\mathbb{Z}_-}$. Using the method developed in [51], it was shown in [49] that indeed such expansions exist. Moreover, $F_k(\mathbf{y})$, $G_k(\mathbf{y})$ depend only on $(y_0, \ldots, y_{-(k+1)})$ for all $k \geq 0$, and if one considers similar expansions of finite conditional probabilities

$$\mathbb{Q}(Y_0 = y_0 | Y_{-1} = y_{-1}, \ldots, Y_{-n} = y_{-n}) = \sum_{k=0}^{\infty} F_k^{(n)}(\mathbf{y}) \varepsilon^k = \exp\left(\sum_{k=0}^{\infty} G_k^{(n)}(\mathbf{y}) \varepsilon^k\right),$$

then the coefficients $F_k^{(n)}$, $G_k^{(n)}$ show similar stabilization: $F_k^{(n)}(\mathbf{y}) = F_k(\mathbf{y})$ and $G_k^{(n)}(\mathbf{y}) = G_k(\mathbf{y})$, but now for all $n \geq k+1$. Using the notation $\lambda = (1-2p)$,

$\sigma_i = y_0 y_{-i}$, $i \neq 0$, one has

$$F_0 = \frac{1 + \lambda \sigma_1}{2},$$

$$F_1 = \frac{-\lambda \sigma_1}{1 + \lambda \sigma_1 \sigma_2},$$

$$F_2 = \frac{2\lambda \sigma_1 [\lambda \sigma_2 \sigma_3 - \lambda \sigma_1 \sigma_2 \{3 - \lambda \sigma_2 \sigma_3\} + 1]}{(1 + \lambda \sigma_1 \sigma_2)^2 (1 + \lambda \sigma_2 \sigma_3)},$$

$$\vdots$$

$$G_0 = \log \frac{1 + \lambda \sigma_1}{2},$$

$$G_1 = \frac{-4\lambda \sigma_1}{(1 + \lambda \sigma_1)(1 + \lambda \sigma_1 \sigma_2)},$$

$$G_2 = \frac{4\lambda \sigma_1 (\sigma_3 \lambda^3 + (\sigma_1 \sigma_3 - 3\sigma_2 - 3\sigma_1 \sigma_2 \sigma_3)\lambda^2 + (\sigma_2 \sigma_3 - 3\sigma_1 \sigma_2 - \sigma_1)\lambda + 1)}{(1 + \lambda \sigma_1)^2 (1 + \lambda \sigma_1 \sigma_2)^2 (1 + \lambda \sigma_2 \sigma_3)}.$$

Again, it seems impossible to guess the general expression.

Statistical mechanics offers a number of other approaches suitable for determining the dependence structure of the process $\{Y_n\}$. Zuk *et al.* [51] observed a close relation between Example 1.1 and the random field Ising model (RFIM): for any $(y_1, \ldots, y_n) \in \{-1, 1\}^n$,

$$\mathbb{Q}(y_1, \ldots, y_n) = \frac{1}{2} \frac{1}{(e^J + e^{-J})^{n-1}} \frac{1}{(e^K + e^{-K})^n}$$

$$\times \sum_{x_1^n \in \{-1,1\}^n} \exp\left(J \sum_{i=1}^{n-1} x_i x_{i+1} + K \sum_{i=1}^{n} x_i y_i\right),$$

where

$$J = \frac{1}{2} \log \frac{1-p}{p}, \quad K = \frac{1}{2} \log \frac{1-\varepsilon}{\varepsilon}.$$

In this way, the cylinder probability is proportional to the partition function of the RFIM with the disorder given by the y's. The random field Ising model in dimension $d = 1$ has been studied rather extensively in statistical physics. Besides the publications already cited in [51], quite interesting results have been obtained in [3–5,18,41]. In particular, rather explicit expressions for the partition function (and hence for the cylinder probability) in terms of an *auxiliary local random field* deserve further analysis.

The cluster expansion method of statistical physics is sometimes employed to determine the properties of renormalized Gibbs states, see e.g., [7]. In dimension $d = 1$, one is in fact able to perform computations rather explicitly and rederive the stabilization results of [49] cited above. In general, cluster expansion deals with partition functions of the form

$$\Xi = 1 + \sum_{k \geq 1} \frac{1}{k!} \sum_{(\gamma_1, \ldots, \gamma_k) \in \mathcal{P}^k} z_{\gamma_1} \ldots z_{\gamma_k} \prod_{i < j} \mathbb{I}[\gamma_i \sim \gamma_j], \tag{10}$$

where $\mathcal{P}$ is a set of the so-called *polymers* with a *compatibility relation* $\sim$ and $\mathbb{I}$ is the indicator function – so the sum runs over all pairwise-compatible k-tuples of polymers, and either complex or real *fugacities* $\{z_\gamma | \gamma \in \mathcal{P}\}$. The cluster expansion method provides a recipe for finding functions $\phi^{\mathrm{T}}(\gamma_1, \ldots, \gamma_k)$ and constants $\{\rho_\gamma\}_{\gamma \in \mathcal{P}}$ such that

$$\Xi = \exp\left(\sum_{k \geq 1} \frac{1}{k!} \sum_{(\gamma_1, \ldots, \gamma_k) \in \mathcal{P}_n^k} \phi^{\mathrm{T}}(\gamma_1, \ldots, \gamma_k) z_{\gamma_1} \cdots z_{\gamma_k}\right) \tag{11}$$

for all $\{z_\gamma\}$ satisfying polydisk conditions of the form $|z_\gamma| \leq \rho_\gamma$.

Consider again the hidden Markov process of Example 1.1. In [17] it is shown that for all $\{y_0, \ldots, y_n\} \in \{-1, 1\}^{n+1}$, the $\mathbb{Q}$-measure of the cylinder $[y_0^n]$ can be represented as

$$\mathbb{Q}(y_0^n) = (1 - 2\varepsilon)^{n+1} \cdot \mathbb{P}(y_0^n) \cdot \Xi_{y_0^n},$$

where $\Xi_{y_0^n}$ has the form of (10): namely,

$$\Xi_{y_0^n} = 1 + \sum_{k \geq 1} \frac{1}{k!} \sum_{(\gamma_1, \ldots, \gamma_k) \in \mathcal{P}_{[0,n]}^k} z_{\gamma_1}^{[0,n]} \ldots z_{\gamma_k}^{[0,n]} \prod_{i < j} \mathbb{I}[\gamma_i \sim \gamma_j] \tag{12}$$

for the set of polymers $\mathcal{P}_{0,n}$ consisting of all *integer intervals* $[a, b] = \{a, a+1, \ldots, b\}$, $a, b \in \{0, 1, \ldots, n\}$, the two intervals γ_1, γ_2 being compatible if and only if $\gamma_1 \cup \gamma_2$ is not an interval, and some appropriately defined fugacities $z_\gamma^{[0,n]} = z_\gamma^{[0,n]}(y_0^n)$. Note that due to the compatibility condition the sum is finite. Combining (11) and (12), one obtains a cluster expansion for the conditional probability:

$$\log \mathbb{Q}(y_0|y_1^{n-1}) = \log p_{y_0, y_1} + \log(1 - 2\varepsilon)$$

$$+ \sum_{k=1}^{\infty} \frac{1}{k!} \left[\sum_{(\gamma_1, \ldots, \gamma_k) \in \mathcal{P}_{[0,n]}^k} \phi^{\mathrm{T}}(\gamma_1, \ldots, \gamma_k) z_{\gamma_1}^{[0,n]} \cdots z_{\gamma_k}^{[0,n]} \right.$$

$$\left. - \sum_{(\gamma_1, \ldots, \gamma_k) \in \mathcal{P}_{[1,n]}^k} \phi^{\mathrm{T}}(\gamma_1, \ldots, \gamma_k) z_{\gamma_1}^{[1,n]} \cdots z_{\gamma_k}^{[1,n]} \right].$$

All the terms can be computed directly. Moreover, the cluster expansion method provides explicit bounds on the domain of convergence of the corresponding expansions. The bounds obtained via the so-called *Gruber–Kunz conditions* seem to be somewhat inferior to the bounds on the domain of analyticity found in [21] for Example 1.1. Nevertheless, the cluster expansion approach is of definite interest, since it provides a generic method for establishing domains of analyticity for the power series expansions discussed in this article.

4 Random fields

Markov random fields on lattices $\mathbb{Z}^d$, $d \geq 2$, are natural generalizations of Markov chains: $\mathbb{P}$ is a (nearest-neighbor) Markov random field if for any finite set $\Lambda \subset \mathbb{Z}^d$

$$\mathbb{P}(X_\Lambda = x_\Lambda | X_{\mathbb{Z}^d \setminus \Lambda} = x_{\mathbb{Z}^d \setminus \Lambda}) = \mathbb{P}(X_\Lambda = x_\Lambda | X_{\partial^+ \Lambda} = x_{\partial^+ \Lambda}),$$

where $\partial^+ \Lambda = \{ \mathbf{n} \in \mathbb{Z}^d \setminus \Lambda : \text{dist}(\mathbf{n}, \Lambda) = 1 \}$, $x_\Lambda = \{x_{\mathbf{m}}\}_{\mathbf{m} \in \Lambda}$, and $x_{\mathbb{Z}^d \setminus \Lambda} = \{x_{\mathbf{n}}\}_{\mathbf{n} \in \mathbb{Z}^d \setminus \Lambda}$. The most famous examples of Markov random fields (MRFs) are the Ising and Potts models.

Markov random fields are of interest not only to the statistical physics community but are routinely used in fields such as image processing as well. The theory of functions of Markov random fields or, more generally, of renormalized Gibbs random fields is substantially more difficult in dimension $d \geq 2$ than it is for the processes in dimension $d = 1$. The main reason is that preservation of Gibbsianity is no longer guaranteed in higher dimensions even for a fully supported Markov random field, see e.g., [20, 35] for concrete examples and [15] for a general introduction to renormalization pathologies and an extensive list of references. Algorithms for denoising based on the ideas of DUDE have been proposed for random fields as well. It might be interesting to understand whether there is a relation between the performance of these algorithms and preservation/loss of Gibbsianity.

References

[1] R. B. Ash. *Information Theory*. Dover, New York, 1965

[2] L. E. Baum, and T. Petrie. *Statistical inference for probabilistic functions of finite state Markov chains*. Ann. Math. Statist. **37** (1966) 1554–1563. MR0202264 (34 #2137)

[3] U. Behn and V. A. Zagrebnov. *One-dimensional Markovian-field Ising model: physical properties and characteristics of the discrete stochastic mapping*. J. Phys. A **21**(9) (1988) 2151–2165. MR952930 (89j:82024)

[4] U. Behn, V. B. Priezzhev, and V. A. Zagrebnov. *One-dimensional random field Ising model: residual entropy, magnetization, and the "perestroyka" of the ground state*. Physica A: Statist. Mech. Appl. **167**(2) (1990) 481–493

[5] J. Bene and P. Szépfalusy. *Multifractal properties in the one-dimensional random-field Ising model*. Phys. Rev. A **37**(5) (1988) 1703–1707

[6] N. Berger, C. Kenyon, E. Mossel and Y. Peres. *Glauber dynamics on trees and hyperbolic graphs*, Prob. Theory Related Fields **131**(3) (2005) 311–340. MR2123248 (2005k:82066)

[7] L. Bertini, E. N. M. Cirillo, and E. Olivieri. *Renormalization group in the uniqueness region: weak Gibbsianity and convergence*. Commum. Math. Phys. **261**(2) (2006) 323–378. MR2191884 (2006k:82063)

[8] J. J. Birch. *Approximation for the entropy for functions of Markov chains*. Ann. Math. Statist. **33** (1962) 930–938. MR0141162 (25 #4573)

[9] R. Bowen. *Equilibrium States and the Ergodic Theory of Anosov Diffeomorphisms*. Lecture Notes in Mathematics, **470**. Springer, Berlin, 1975. MR0442989 (56 #1364)

[10] J.-R. Chazottes and E. Ugalde. *Projection of Markov measures may be Gibbsian*. J. Statist. Phys. **111**(5–6) (2003) 1245–1272, 0022-4715. MR1975928 (2004d:37008)

[11] P. Collet, A. Galves, and F. Leonardi. *Random perturbations of stochastic processes with unbounded variable length memory*. Electron. J. Prob. **13**(48) (2008) 1345–1361. MR2438809

[12] M. Denker and M. Gordin. *Gibbs measures for fibred systems*. Adv. Math. **148**(2) (1999) 161–192. MR1736956 (2001j:37061)

[13] W. Doeblin and R. Fortet. *Sur des chaînes à liaisons complètes*. Bull. Soc. Math. Fr. **65** (1937) 132–148 (in French). MR1505076

[14] S. Egner, V. Balakirsky, L. Tolhuizen, S. Baggen, and H. Hollmann. *On the entropy rate of a hidden Markov model*. In Int. Symp. Information Theory, 2004, p. 12

[15] A. C. D. van Enter, R. Fernández, and A. D. Sokal. *Regularity properties and pathologies of position-space renormalization-group transformations: scope and limitations of Gibbsian theory*. J. Statist. Phys. **72**(5–6) (1993) 879–1167. MR1241537 (94m:82012)

[16] R. Fernández, P. A. Ferrari, and A. Galves. *Coupling, renewal and perfect simulation of chains of infinite order*. Lecture Notes for the Vth Brazilian School of Probability, Ubatuba, 2001

[17] R. Fernández and E. Verbitskiy. *Hidden Markov models and cluster expansion*. Work in progress (2009)

[18] G. Gyorgyi and P. Rujan. *Strange attractors in disordered systems.* J. Phys. C: Solid State Phys. **17**(24) (1984) 4207–4212

[19] L. Gurvits and J. Ledoux. *Markov property for a function of a Markov chain: a linear algebra approach.* Linear Algebra Appl. **404** (2005) 85–117. MR2149655 (2006g:60108)

[20] O. Häggström. *Is the fuzzy Potts model Gibbsian?* Ann. Inst. H. Poincaré Prob. Statist. (in English, with English and French summaries) **39**(5) (2003) 891–917. MR1997217 (2005f:82049)

[21] G. Han and B. Marcus. *Analyticity of entropy rate of hidden Markov chains.* IEEE Trans. Inform. Theory **52**(12) (2006) 5251–5266. MR2300690 (2007m:62008)

[22] T. E. Harris. *On chains of infinite order.* Pacific J. Math. **5** (1955) 707–724. MR0075482 (17,755b)

[23] B. M. Hochwald and P. R. Jelenković. *State learning and mixing in entropy of hidden Markov processes and the Gilbert–Elliott channel.* IEEE Trans. Inf. Theory **45**(1) (1999) 128–138. MR1677853 (99k:94028)

[24] T. Holliday, A. Goldsmith, and P. Glynn. *Capacity of finite state channels based on Lyapunov exponents of random matrices.* IEEE Trans. Inf. Theory **52**(8) (2006) 3509–3532. MR2242362 (2007c:94037)

[25] M. Iosifescu and S. Grigorescu. *Dependence with Complete Connections and Its Applications.* Cambridge Tracts in Mathematics, **96**. Cambridge University Press, Cambridge, 1990. MR1070097 (91j:60098)

[26] P. Jacquet, G. Seroussi, and W. Szpankowski. *On the entropy of a hidden Markov process.* In Proc. Data Compression Conf., Snowbird, UT, 2004, pp. 362–371

[27] P. Jacquet, G. Seroussi, and W. Szpankowski. *On the entropy of a hidden Markov process.* Theor. Comput. Sci. **395**(2–3) (2008) 203–219. MR2424508

[28] M. Keane. *Strongly mixing g-measures.* Invent. Math. **16** (1972) 309–324. MR0310193 (46 #9295)

[29] T. Kempton. *Factors of Gibbs measures for subshifts of finite type.* Preprint (2009)

[30] C. Kenyon, E. Mossel, and Y. Peres. *Glauber dynamics on trees and hyperbolic graphs.* In 42nd IEEE Symp. Foundations of Computer Science, Las Vegas, NV, 2001. IEEE Computer Society, Los Alamitos, CA, 2001, pp. 568–578. MR1948746

[31] S. P. Lalley. *Regeneration in one-dimensional Gibbs states and chains with complete connections.* Resenhas **4**(3) (2000) 249–281. MR1797366 (2001i:60176)

[32] N. Larget. *A canonical representation for aggregated Markov processes.* J. Appl. Prob. **35**(2) (1998) 313–324. MR1641793 (99k:60190)

[33] D. Lind and B. Marcus. *An Introduction to Symbolic Dynamics and Coding.* Cambridge University Press, Cambridge, 1995. MR1369092 (97a:58050)

[34] J. Lörinczi, C. Maes, and K. Vande Velde. *Transformations of Gibbs measures.* Prob. Theory Related Fields **112**(1) (1998) 121–147

[35] C. Maes and K. Vande Velde. *The fuzzy Potts model.* J. Phys. A **28**(15) (1995) 4261–4270. MR1351929 (96i:82022)

[36] A. v. Moffaert. *Non-locality and Irreversibility: Applications of an Extended Gibbs Formalism.* PhD thesis, Katholieke Universiteit Leuven, 1999

[37] T. Moon and T. Weissman. *Universal filtering via hidden Markov modeling.* IEEE Trans. Inf. Theory **54**(2) (2008) 692–708. MR2447158

[38] M. Pollicott. *Computing entropy rates for hidden Markov processes*. This volume, Chapter 7, 2011

[39] T. W. E. Ordentlich. *New bounds on the entropy rate of hidden Markov process*. In Proc. Information Theory Workshop, San Antonio, TX, 2004, pp. 117–122

[40] D. Ruelle. *Thermodynamic Formalism. The Mathematical Structures of Classical Equilibrium Statistical Mechanics; With a Foreword by Giovanni Gallavotti and Gian-Carlo Rota*. Encyclopedia of Mathematics and its Applications, **5**. Addison-Wesley, Reading, MA, 1978. MR511655 (80g:82017)

[41] P. Ruján. *Calculation of the free energy of Ising systems by a recursion method*. Physica A: Statist. Theor. Phys. **91**(3–4) (1978) 549–562

[42] L. K. Saul and M. I. Jordan. *Boltzmann Chains and Hidden Markov Models* (G. Tesauro, D.S. Touretzky, and T. K. Leen, eds.). Advances in Neural Information Processing Systems (NIPS), **7**. MIT Press, Cambridge, MA, 1995

[43] A. Shaikh. *Statistical inference for Markov chains with lumped states*. Biometrika **63**(1) (1976) 211–213. MR0403121 (53 #6934)

[44] S. Verdu and T. Weissman. *Erasure entropy*. In IEEE Int. Symp. Information Theory, 2006, pp. 98–102

[45] S. Verdu and T. Weissman. *The information lost in erasures*. IEEE Trans. Inf. Theory **54**(11) (2008) 5030–5058

[46] P. Walters. *Ruelle's operator theorem and g-measures*. Trans. Amer. Math. Soc. **214** (1975) 375–387. MR0412389 (54 #515)

[47] P. Walters. *Relative pressure, relative equilibrium states, compensation functions and many-to-one codes between subshifts*. Trans. Amer. Math. Soc. **296**(1) (1986) 1–31. MR837796 (87j:28028)

[48] T. Weissman, E. Ordentlich, G. Seroussi, S. Verdú, and M. J. Weinberger. *Universal discrete denoising: known channel*. IEEE Trans. Inf. Theory **51**(1) (2005) 5–28. MR2234569 (2008h:94036)

[49] A.C.C. v. Wijk. *Entropy of Hidden Markov Models*. MSc thesis, Eindhoven University of Technology, 2007

[50] J. Yu and S. Verdú. *Schemes for bidirectional modeling of discrete stationary sources*. IEEE Trans. Inf. Theory **52**(11) (2006) 4789–4807. MR2300356 (2007m:94144)

[51] O. Zuk, E. Domany, I. Kanter, and M. Aizenman. *From finite-system entropy to entropy rate for a hidden Markov process*. IEEE Signal Proc. Lett. **13** (2006) 517–520

[52] O. Zuk. *The relative entropy rate for two hidden Markov processes*. In Proc. ITG Conf. on Source and Channel Coding, Munich, 2006

Printed in the United States
By Bookmasters